Marine Pollution

Marine Pollution

Christopher L. J. Frid

School of Environment, Griffith University, Queensland, Australia

Bryony A. Caswell

Environmental Futures Research Institute, Griffith University, Queensland, Australia

Hermit crab calls a plastic cap its home, Maldives, 2010

OXFORD
UNIVERSITY PRESS

OXFORD
UNIVERSITY PRESS

Great Clarendon Street, Oxford, ox2 6dp,
United Kingdom

Oxford University Press is a department of the University of Oxford.
It furthers the University's objective of excellence in research, scholarship,
and education by publishing worldwide. Oxford is a registered trade mark of
Oxford University Press in the UK and in certain other countries

First Edition published in 2017

Published in the United States of America by Oxford University Press
198 Madison Avenue, New York, NY 10016, United States of America

British Library Cataloguing in Publication Data

Data available

Library of Congress Control Number: 2017943728

ISBN 978–0–19–872628–9 (hbk.)
ISBN 978–0–19–872629–6 (pbk.)

DOI 10.1093/oso/9780198726289.001.0001

Printed and bound by
CPI Litho (UK) Ltd, Croydon, CR0 4YY

For
Professor R.B. (Bob) Clark
1923–2013

Preface

As we finalise the text of this new book the fiftieth anniversary of the wreck of the oil tanker *Torrey Canyon* is fast approaching (18 March 1967). While marine pollution had been occurring for many hundreds, even thousands, of years, it was the spilling of 120 000 tonnes of crude oil on the seas and beaches of south-west Britain, northern France and Ireland that caught the imagination of the developing environmental movement of the late 1960s. In the immediate aftermath of the spill, a professor of marine zoology, a specialist in polychaete worms, was surprised by the lack of scientific knowledge on which to base the environmental response. Professor R.B. (Bob) Clark established a seabird cleaning programme at the Dove Marine Laboratory at the University of Newcastle upon Tyne, where he was the Director. This facility cleaned and rehabilitated hundreds of birds and as the immediate impacts wound down it was transformed into a research programme investigating the effects of oil on seabirds and techniques for clean-up.

Bob, however, had recognised that the lack of scientific knowledge extended beyond seabird welfare. He began to communicate with colleagues around the world on issues about pollution of the seas, establishing first a newsletter and then an academic journal, the *Marine Pollution Bulletin*, to promote better science and sharing of knowledge. These efforts resulted in Bob often being credited with establishing the study of marine pollution as a discipline. As a natural communicator and a passionate educator, Bob went on to produce what became the classic advanced student textbook, *Marine Pollution*, in 1978, a work that was reprinted numerous times and eventually ran to five editions. Reviews frequently cited the measured, clear and authoritative style of the book. It was grounded in science,

illustrated the challenges and trade-offs that faced pollution regulators and was free from the rhetoric associated with the environmental lobby groups.

So what has that to do with this work? As undergraduate students of marine biology, we both took courses in marine pollution. For me it was the first edition of Bob's book that was the guiding text, while a few years later BC was enlightened by the fourth edition.

In 1989 I met the recently retired Bob Clark at the University of Newcastle and subsequently Bob and I worked together for twenty years on a number of projects. Bob was always keen to step out of retirement to give a seminar on marine pollution to my students and to share his experiences. Bob invited Martin Attrill (University of Plymouth) and I to join him in writing the fourth (and subsequently the fifth) editions of *Marine Pollution*. With publication of the fifth edition, Bob declared that he had now been out of active science for too long and would not be doing a sixth edition. The absence of a sixth edition from Bob has created a vacuum for an up-to-date, concise but comprehensive overview of marine pollution that focuses on the impacts of pollution on marine ecosystems.

Eleven years after the publication of the fifth edition, we found ourselves redesigning a course on marine pollution. BC suggested that the students (and wider readership) needed a new, up-to-date book of comparable nature and quality to *Marine Pollution* that incorporated the emerging marine pollution challenges. It seemed fitting that in the year that Bob died we should develop a new book on marine pollution, one that we hope might add to Bob's intellectual legacy that began in 1978.

A lot has changed since the *Torrey Canyon* accident and the first edition of *Marine Pollution*. The scientific knowledge on the effects of pollution, from the cell to the ecosystem, is now vast. This knowledge is used to guide human actions and to regulate potentially harmful activities, but the complexities of the science, the competing demands on the environment and the international nature of the seas continues to provide challenges to effective management. This new volume seeks to review our current knowledge of marine pollution, to understand how that knowledge is gained and used and to examine how in the twenty-first century the calls for holistic, integrated environmental management and the sustainability agenda interact with this knowledge.

CF
2017

Acknowledgements

The final act of committing to paper the text that forms this work builds on years of experience and exposure to marine science and management challenges. Many colleagues, authors, data gatherers and what would now be called 'stakeholders' have been involved and have contributed to the thinking and understanding that underpin the work. We would like to thank them all. It is the joy of being an academic that one can explore issues deeply and range widely.

Certain colleagues, however, have been more closely involved and do deserve individual mention: Bob Clark, Stewart Evans, Margaret Gill, Susan Clark, Tom Mercer, Roger Bamber, John Hall, Odette Paramor, Catherine Scott, Silvana Birchenough, Julie Bremner, Christina Lye, Salvadore Herrando-Perez, Stuart Rogers, Steve Hawkins, Jon Green, Leonie Robinson, Joe Lee, Fred Leusch, Matthew Spencer, Paul Scott, Angela Coe, Rebecca Holt, Rob Marrs, Sarah Clarke, Rachel Hurle, Emily Lee, Nicholas Bury, Heather Fowle, Anja Ramsay and Martin Wilkinson. We would also like to thank the students, many in number, for their enthusiasm and regard for marine ecosystems and who inspire us to continue teaching courses on marine pollution.

The authors would like to thank all the scientists whose efforts over the years have led to our increased understanding of marine pollution and especially those who have given us permission to reproduce copyright material here. Every effort has been made to contact holders of copyright and in materials reproduced in this book. Any omissions will be rectified in future printings if notice is given to the publisher.

We would also like to thank our editors Ian Sherman, Lucy Nash and Bethany Kershaw, the OUP production team and art department; without you this would just be some extensive ramblings on a word processor! Finally we would like to thank Bob Clark for teaching us about marine pollution.

Contents

CHAPTER 1

Introduction to marine pollution

1.1 What is pollution?

Most people know what pollution is. It is a bad thing, it is caused by humans, it is linked to human waste in the environment, it should be controlled, it is preventable, it makes for prominent TV news stories and it ultimately might make the planet uninhabitable, a vision seen in a number of science-fiction books, films and TV shows. You might think that the academic study of pollution would be at odds with these interpretations but these kinds of popular perceptions are actually close to the mark. The widely used academic and political definition of *marine pollution* was produced for the United Nations (UN; see Box 1.1) and is:

Pollution means the introduction by man, directly or indirectly, of substances or energy into the marine environment (including estuaries) resulting in such deleterious effects as harm to living resources, hazards to human health, hindrance to marine activities including fishing, impairment of quality for use of sea water and reduction of amenities.

(GESAMP 1990)

Unpacking this definition we see that pollution arises from substances or energy introduced by man, so is caused by humans and therefore is preventable as it should be possible to control inputs. Pollution causes deleterious effects, which means that it is a bad thing by definition and, of course, 'bad things' happening is the stock of TV news. So while the jury might be still be out on whether pollution will lead to an uninhabitable future Earth, the remaining popular notions of pollution are supported by the UN definition. However, the definition goes further than just labelling harmful human waste as pollution. It is worth reminding ourselves

that energy—for example, heat, noise and vibrations—can also be a pollutant as well and that the deleterious effects do not relate the environment *per se* but to human uses of the environment. The definition, for example, refers to 'living resources' and not the wider concept of biodiversity. It goes on to highlight the effects on human health, the quality of seawater for use (not *per se*) and amenity value. It therefore follows that not everything that humans introduce into the marine environment becomes a pollutant; if it does not affect living resources, human health, the quality of seawater for human use, or the amenity value associated with the sea or coastal environments, it does not pose a problem. The term used to describe these additions is *contamination* and is defined as:

substances (i.e. chemical elements and compounds) or groups of substances that are toxic, persistent and liable to bioaccumulate, and other substances or groups of substances which give rise to an equivalent level of concern.

(Marine Strategy Framework Directive, EU 2008)

Contamination therefore includes naturally occurring substances when they are toxic and of societal concern, including even those that are known to not have an adverse effect. As the definition is restricted to substances; noise, light and heat are not considered as contaminants within this definition.

Humans have been utilising the marine environment for as long as the planet has been populated. Initially foraging on the shore for shellfish and other food, human waste was doubtless left on the shore which the sea removed and 'treated' the waste. As technology advanced, boats and ships were used to transport humans, to reach fishing grounds, populate new habitats and eventually to move goods. With the development of the economy,

Marine Pollution. Christopher L. J. Frid & Bryony A. Caswell.
© Christopher L. J. Frid & Bryony A. Caswell 2017. Published 2017 by Oxford University Press.
DOI 10.1093/oso/9780198726289.001.0001

Box 1.1 What is GESAMP?

GESAMP is the UN-sponsored joint Group of Experts on the Scientific Aspects of Marine Environmental Protection (GESAMP). It is an advisory body, established in 1969, that advises the United Nations (UN) and other international bodies on the science of marine environmental protection. At present (2016) there are 16 experts from a range of countries and with a diverse range of expertise who develop reviews and advice. Working groups that focus on particular topics/issues support the main panel and a wide pool of international experts populate these working groups. All the experts act independently of any government or organisation and provide independent technical expertise. In 2001, GESAMP was subject to a major review and a series of measures have been introduced, broadening the focus from pollution to sustainable ocean management, and increasing the engagement of experts from less economically developed nations.

Box 1.2 The ecosystem services approach

Our earliest ancestors were hunter–gatherers: they gained their food from foraging in the environment. They also left their waste products (bodily, old clothes (hides, woven plant fibres), tools and food-processing residues) in the environment for scavengers and microbes to breakdown and remineralise. Therefore, from earliest times human society has been dependent on and used materials and processes that occur as part of the natural ecosystem. While modern societies and human needs appear very different to those of our distant ancestors we are still highly dependent on our ecosystems. Natural processes still underpin our capture fisheries (which provide around 6% of the global protein demand), over 50% of our terrestrial crops are pollinated by 'wild' pollinators such as bees, large volumes of organic wastes and nutrients from agriculture flow into the seas where they are assimilated and 'treated' in the environment. Referring to the benefits that we derive from ecological processes that naturally occur in ecosystems as 'ecosystem services' has led to new attempts to link ecology and economics via valuations of the benefits biodiversity and *ecological functioning* provided to us.

The ecosystem services terminology has developed since the introduction of the concept in the UN's Millennium Ecosystem Assessment (generally known as the MA). The MA was a UN initiative to assess the state of the natural environment and how humans used it at the turn of the millennium. Launched in 2001, the project produced a series of wide-ranging and highly informative reports in 2005. However, the subsequent global economic crisis meant that the lessons from the MA were largely ignored in favour of traditional economic measures to stabilise a 'business-as-usual' global economic model.

Ecosystem services are usually described in terms of four groups; Provisioning Services (food supply, fresh water, fibre for fuel, clothing, building materials), Regulating Services (carbon sequestration and climate regulation, waste degradation, waste sequestration, disease control), Supporting Services (the carbon cycle, nitrogen cycle, food web, crop pollination) and Cultural Services (amenity use, recreation, education, spiritual value).

more waste was discharged into the sea from ships, from the markets that developed at ports and from the manufacture of goods (and, of course, the waste from a growing human population). As manufacturing developed so the range of waste products also expanded.

Up until the eighteenth century, almost all waste discharged to the sea was probably either organic (e.g. human or animal manure, food waste) or naturally occurring minerals (e.g. the residues from metal smelting). The former would be degraded in the environment but could cause pollution when the level of waste exceeded the local environments capacity to assimilate it. The latter would often cause local impacts, due to metal toxicity for example, but dispersion and dilution in the sea would ultimately render this waste 'harmless'.

With the expansion of human populations and the Industrial Revolution, the scale of waste disposal and the range of substances utilised increased dramatically and began to exceed the environment's capacity to assimilate the waste. This in turn compromised other uses of the sea such as food production and, more recently, its amenity value and increased disease prevalence.

The UN Conference on the Environment and Development (UNCED), held in Rio de Janiero in 1992, recognised the links between the natural environment and the development of human civilisations. For example, it established the framework that recognises that by placing waste in the sea and allowing microbes and the biota to degrade, assimilate and recycle it we do not always need to build and operate expensive treatment works. Cleary other business sectors, such as tourism, also benefit from and depend on a clean and healthy environment; nobody wants to holiday on a beach covered in litter and swim in water unsafe due to the presence of chemicals and human pathogens. It has, for example, been estimated that the global natural environment provided around US$125 trillion of economic benefits to the world economy in 2011, and of this almost US$50 trillion is provided by the marine environment (values are in 2007 US$). This ecosystem services approach has developed a new way of thinking about how we as humans interact with our environment, with the environment providing a series of ecosystem services that support human well-being (see Box 1.2). Given the topic of this book it should be borne in mind that marine

environment provides an estimated US$ 37 million of waste treatment/assimilation services each year (value in 2007 US$). This is money that society does not need to spend building waste treatment plants and operating disposal operations because the marine environment provides this service.

1.2 How do pollutants reach the sea?

Pollution reaches the sea by a wide variety of routes (Figure 1.1); not all pollution enters from discharge pipes. Key pathways carrying pollutants into the sea include direct discharges from industry and municipal works, riverine inputs, atmospheric sources and ocean disposal of pollutants. For many substances, only one of these routes will be important, but other substances will enter via a number of routes and predicting their impacts in the environment and regulating the effects will be made more complex as a result.

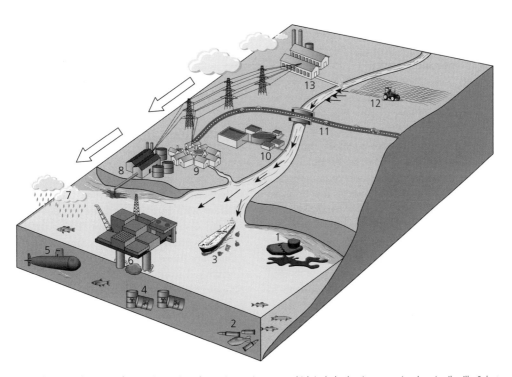

Figure 1.1 Pathways and sources of contaminants into the marine environment which includes but is not restricted to: 1. oil spills; 2. lost or dumped munitions; 3. waste from ships; 4. dumped nuclear and industrial waste; 5. lost or dumped vessels, their cargoes, and power plants; 6. contaminated drill cuttings; 7. washout of atmospheric pollutants, including heavy metals and hydrocarbons; 8. industrial waste discharges; 9. urban wastewater and street drainage; 10. sewage effluent; 11. vehicle exhausts (via the atmosphere); 12. agricultural fertilisers and pesticides; 13. cooling water (waste heat).

The most obvious sources of marine pollution, and those easiest to regulate, are direct discharges, also known as 'point' sources. A factory, a sewage treatment works, a power station or a ship might all place its waste directly into the sea, often via a pipeline. Waste that enters streams, rivers or estuaries eventually ends up in the sea as the natural hydrological flow cycle carries it down stream. Estuaries are, however, complex hydrological and chemical environments and many compounds undergo chemical changes; for example, fine particulate material may, due to changes in the surface chemistry of saline waters, flocculate and precipitate. So what had been a waste in suspension, which would be dispersed, now becomes a solid waste that is deposited and accumulates within estuaries. Pollutants do not just enter the seas and other watercourses from pipelines. There are also indirect or diffuse sources that include the general washing of substances such as artificial fertilisers or herbicides that have been applied over large areas of agricultural land into a watercourse by precipitation wherever land drainage occurs. These indirect or 'diffuse' sources are much harder to regulate and control than direct point-source discharges.

The role of 'feeder' watercourses in carrying pollution into the sea is obvious; far less obvious is the role the atmosphere plays in delivering pollutants to the seas and oceans. Many waste products are emitted into the atmosphere from factory chimneys, car exhausts and waste incinerators. While much of the emitted material is gaseous and remains in the atmosphere (or undergoes chemical reactions in the gaseous phase), many of the substances emitted are either small particulates or they react with water in the atmosphere, dissolving in rain droplets. These particles/water droplets are carried in the atmosphere and, because the oceans cover 70% of the Earth's surface, when the particles drop to Earth or rain falls a large proportion is deposited in the oceans. The atmosphere is the main source through which many heavy metals (e.g. lead) and halogenated hydrocarbons such as polychlorinated biphenyls (PCBs) get into the sea. They enter the atmosphere via combustion and then fall or are 'washed' out into the sea. This route, whereby pollutants enter the oceans indirectly via the atmosphere, is also considered a 'non-point' source.

Once in the sea pollutants can change state. The seas and oceans are dynamic and materials within the sea are moved around by currents and waves. Water movements will tend to disperse the wastes away from the source of input and dilute them as 'clean' and 'contaminated' water become mixed. For many substances these physical processes will be accompanied by usually biologically mediated chemical changes. For example, faecal material will be physically broken down and biologically degraded by bacteria and other organisms, eventually becoming carbon dioxide, water and nutrient/mineral salts. Wastes that are not chemically or physically degraded accumulate in the oceans. Contaminants that remain in solution will eventually become mixed throughout the world ocean. An example of this is the artificial radioactive isotope $_{137}$Cs. This isotope does not occur naturally but was emitted during the 1960s to the 1980s by some nuclear fuel reprocessing facilities. It remains in solution and is used as a tracer for the current flow around the discharge points (see Section 4.7). By contrast, some materials will precipitate, for example by being scavenged on clay mineral particles and will become incorporated into marine sediments at deposition sites and will essentially be removed from the biosphere by geological processes.

1.3 Pollutant behaviour

Once in the environment, not all pollutants behave in the same way and their behaviour greatly influences the extent to which they subsequently interact with humans and human activities. First, we can distinguish pollutants based on how they change over time in the environment. *Persistent pollutants* are those substances that remain essentially unchanged from the form in which they were discharged. Examples include heavy metals, many man-made organic compounds such as herbicides and pesticides and radio-nucleotides. Over time the material is moved away from the discharge point and mixed with 'clean' water so that the concentration declines over time and distance from the discharge point; however, the substance remains chemically the same as when it was discharged. This is in contrast with *biodegradable pollutants* that are subject to biological breakdown and over time are converted into water, carbon dioxide and some inorganic salts. Examples of biodegradable materials include food waste, sewage, paper pulp-mill effluent and oil. The third category is *dissipative pollutants*, which refers to pollutants such as heat, sound and light. Over time, these pollutants are essentially distributed into the wider environment and so are no longer detectable against the background levels.

An old adage of sanitary engineers (the profession that designed urban wastewater (sewage) treatment schemes) was that 'the solution to pollution is dilution'. It is clear that diluting dissipative wastes will speed up their natural dissipation, that diluting biodegradable wastes increases access to oxygen and hence reduces hypoxic effects and dilutes any regenerated nutrients. Even toxic persistent wastes are less likely to have a biological effect when diluted. It is therefore easy to see why the large volume and dynamic nature of the sea, with its mixing by waves and tides, made it appear a logical place to deposit human wastes of all forms.

The second approach to investigating pollutant behaviour is based on how pollutants move around the environment. Some substances remain dissolved in water, some concentrate at the surface, others react with minerals and particles in the water column and so become bound to them. As these particles settle on the seafloor, so the pollutant is deposited with them. Pollutants in the water and at the sea surface are biologically available, as are pollutants in surface sediments, but as more sediments are deposited on top of those bearing pollutants they become less biologically available. This is because most of the biological activity in marine sediments occurs in the surface layers, usually the top few centimetres. However, human activities such as dredging (Section 3.5) can bring these buried contaminated sediments back to the surface. These are often referred to as *legacy pollutants*. In some areas it is possible to take a core through the sediments and then reconstruct the history of contamination from the chemical profile of compounds within the sediment core.

1.4 The history of marine pollution

Archaeological evidence shows that prehistoric human populations used the seashore for food gathering and built habitations nearby. These archaeological investigations have also shown that these builders deposited their refuse in pits or piles, and these middens provide a wealth of evidence about how these people lived (Figure 1.2). Early humans will almost certainly have deposited some of their waste into the sea. During that period, many

Figure 1.2 Newburgh Mesolithic shell middens on the banks of the Ythan estuary in north-east Scotland showing the extensive deposits of shells from consumed shellfish. Source: http://www.geograph.org.uk/ photo/1023519. Image Copyright Martyn Gorman. This work is licensed under the Creative Commons Attribution-Share Alike 2.0 Generic Licence. To view a copy of this licence, see http://creativecommons.org/ licenses/by-sa/2.0.

inland communities were located close to streams and rivers for ready access to water and their waste, which entered flowing water, would eventually end up in estuaries and the sea. While this waste certainly resulted in contamination, the scale of communities and the organic and hence biodegradable nature of the wastes limited any impacts to being very local. The human population was small and the seas vast—the old adage that *the solution to pollution is dilution* would have applied in most cases. The exceptions were where sewage contaminated drinking-water supplies or food resources and so presented a risk to human health (and hence 'pollution') (see Section 3.2).

Humankind next entered the sea to get to offshore fishing grounds and reach distant trading partners. Urban centres developed on estuaries and coasts because from there it was easy to reach the market. Subsequent rapid urban growth occurred following first the agricultural and then industrial revolutions. Thus, waste inputs and their spatial concentration increased. As early as the Middle Ages (1100–1200) in Europe, there are reports of human health issues in ports, and national legislation introduced to control the deposition of human waste in the streets can be traced back to the Parliamentary statute of 1388, the English Sanitary Act (Table 1.1).

Legislative measures to control pollution are often driven or prompted by high-profile events.

In spite of the piecemeal attempts to keep the city clean, the rapid growth and urbanisation of London in the seventeenth and eighteenth centuries led to major problems regarding public health and waste disposal. These led to periodic epidemics of diseases, such as cholera and typhus, but it was the Great Stink of 1858 that prompted major action by the UK Parliament. By the middle of the 1800s the use of the Thames and its tributaries as a means of waste disposal by the huge city of London resulted in an essentially abiotic estuary that was acting as a vast open sewer.

The eminent scientist, Michael Faraday, urged the government to take action and in 1855 he carried out an experiment, dropping pieces of paper into the Thames from various piers as well as a boat on the estuary. At every location the paper sank and disappeared from view within 2 centimetres of the surface. Faraday wrote an open letter, published in the *Times*, to Parliament in which he warned of the dangers of the continuing deterioration of the Thames and the impact of a 'hot season'. Three years, later during the hot summer of 1858, his predictions proved correct. In the warm weather the gases and noxious odours emitted drove most of London's inhabitants indoors. At the then newly built Houses of Parliament, business was initially switched to rooms on the side of the building furthest from the Thames, but as conditions deteriorated Members

Table 1.1 Examples of medieval English legislative instruments to control waste disposal and hence pollution. The emphasis was on creating more pleasant living conditions; that is, the ability to walk the streets without becoming covered in human and animal waste. The link between waste and disease was not established until many centuries later. (Source: http://www.thepotteries.org/dates/health.htm).

Date	Legislation
c. 1225	Acts passed dealing with the repair of sewers and control of nuisances.
1281	City of London Regulation prohibited pigs wandering in the streets, and a further Regulation in 1297 required the removal of pigsties from the streets.
1283	City of London Regulation prohibited tallow-melting in the streets and further Regulations the scouring of furs (1310), flaying of dead horses (1311) and solder melting (1371).
1309	City of London Regulation prohibited the casting of filth from houses into the streets and lanes of the City. People 'ought to have it carried to the Thames or elsewhere out of town'.
1357	Royal Order required that no rubbish or filth should be thrown or put into the rivers of Thames and Flete; all such rubbish must be taken out of the City of London by cart.
1371	Royal Order forbade the slaughtering of oxen, sheep and swine in the City of London.
1388	The first English Sanitary Act dealt with offal and slaughterhouses; prohibited the casting of animal filth and refuse into rivers or ditches and 'corrupting of the Air'.

considered relocating outside London. This idea was dropped in favour of a plan to clean-up London's drains and the Thames. Within only 18 days, legislation was drafted and passed into law requiring the building of new sewerage systems, treatment works and embankments on the estuary.

In many parts of the world, the Industrial Revolution was powered by energy generated from the burning of fossil fuels. In colder parts of the world, fossil-fuel combustion also provided domestic heating, heat for cooking and often light. In the economically depressed conditions after the Second World War, the burning of coal to supply industry, drive transport (steam trains) and heat homes led to high emissions of smoke (a mix of particles, carbon dioxide and a variety of acidic gases including sulphur dioxide and heavy metals). These emissions often led to poor air quality in urban areas and industrial centres, causing respiratory diseases, asthma and long-term chronic health concerns. The combination of low-lying cold, damp air (fog) and this mix of atmospheric pollution became known as 'smog'. In the winter of 1952, a heavy fog descended on London and combined with the atmospheric pollution to form a dense smog. The fog and pollution became trapped close to the ground by an atmospheric inversion and in parts of London visibility was reduced to less than 50 cm. Persisting for five days from the 5 December 1952, the *Great Smog*, or Great Smoke as it became known, is believed to have been directly responsible for 12 000 deaths. The high death toll shocked Londoners and prompted the introduction of the 1956 Clean Air Act that began the process of controlling emissions from factories and industrial premises and restricting the burning of coal in homes.

Following the drive to burn less coal, industry, transport and domestic users switched to burning oil and oil derivatives because these gave a cleaner burn and were less 'dirty' to use. Thus, demands for oil grew and the transport of oil from the oilfields in the Middle East to the markets of Europe increased. Advances in technology allowed ships to get bigger and the first generation of supertankers took to the seas in the 1960s. On 18 March 1967 a problem with the steering on the oil tanker the *Torrey Canyon* resulted in her crashing against rocks in south-west Cornwall. Built in 1959 and with a capacity of 120 000 tonnes, she was at the time the largest ship to be wrecked. Fully loaded, her entire cargo of crude oil was released into the sea damaging large areas of the English, French, Channel Island and Spanish coasts. Approximately 15 000 seabirds died. The scale of the ecological effects and the international nature of the *Torrey Canyon* wreck prompted major changes in international legislation (e.g. the 1972 International Convention on Pollution from Ships), the development of oil-spill-response technology and advances in the science of oil pollution (see Section 4.5).

It is not just dramatic pollution events that prompted a step change in environmental legislation. In 1962 Rachel Carson's book, *Silent Spring*, was published, which not only highlighted the impact of pesticides on the environment, particularly on birds of prey, but also the chemical industry's use of misinformation and attempts to conceal the truth surrounding their polluting activities. The book is widely acknowledged for its role in galvanising US public opinion and prompting changes in US pesticide legislation, the worldwide ban on many uses of dichlorodiphenyltrichloroethane (DDT), and even the creation of the US Environmental Protection Agency.

Ten years after the publication of *Silent Spring* another influential book was published: *The Limits to Growth* by Meadows and colleagues. The authors used early computer models to look at the interactions between an exponentially increasing human population and its demands on resources, including waste assimilation, provided to us by the natural world. The conclusions were stark: the human population was rapidly approaching a tipping point where its demands on the planet would exceed the capacity of the planet. This prompted another call for action, led largely by the growing environmental movement. However, mainstream economists were widely critical of the methods used, the conclusions reached, and lacking a dramatic news event, the book was sidelined as a foundation text of the 'green' movement. More recent analyses show that many of the predictions made by Meadows' models match those that have emerged in the last 30 years.

1.5 How do we manage marine pollution?

The title of this section is rather a misnomer as it is impossible to manage pollution. We can manage human activities that give rise to pollution, and we can manage the impacts of pollution—for example, through mitigation—and in both cases we are seeking to limit the effects that pollution has on human activities. As Section 1.4 showed, attempts to restrict activities that damage the environment through regulation/legislation have a long history and until the latter part of the twentieth century these measures were often local regulations or national legislation. In many instances, marine pollution is international in its effect because the seas often form the boundaries between nations: waste discharged by country X gets washed on to the beaches and fishing grounds of country Y. The global reach of DDT and the number of countries impacted by oil from the *Torrey Canyon* all contributed to the development of international agreements under which countries worked together to apply common standards of regulation and acted together in the face of environmental pressures.

One of the central tenants of pollution regulation has been the adoption of the *polluter-pays principle*. At its simplest, this principle requires that the organisation responsible for placing pollution in the environment pays for the cost of cleaning it up and mitigating the damage. Thus, if an oil spill occurs and affects the coasts of several countries, separate national governments will initiate a response to protect and clean-up its own part of the coastline. Then, once the emergency is over, it will recover the costs from the business/organisation (or their insurers) that caused the spill. The same principle can be applied to damage caused by effluents from industry. However, if a business is regulated and licenced to make a discharge, then provided they conform to the licence conditions (e.g. the nature and quantity of the waste) they would normally be immune from prosecution.

It is tempting to assume that such international agreements would result in the entire ocean being protected to the same extent and in the same way. The UN Convention on the Law of the Sea (UNCLOS, which began life in 1956 and reached its current form in 1994) sets out a series of measures to provide a uniform minimum standard for pollution regulation that covers both nation states and activities on the high seas. However, the protracted process by which UNCLOS was developed means that in the interim many regional or single-issue agreements were drawn up to deal with urgent issues; for example, the 1972 London Convention on the disposal of radioactive waste in the deep sea or the Oslo–Paris Convention on pollution in the north-east Atlantic region.

While the Russian Federation, Canada, the United States and Australia may be sufficiently large countries such that national legislation also provides a degree of regional pollution regulation, in many parts of the world multiple countries share sea areas. In Europe, the European Union acts as a superstate and regulations (in the form of directives) that are passed by the European Commission require the member states (the legal entities) to introduce legislation to enact the regulations. Examples include the Bathing Waters Directive (1976, revised 2006) that requires the waters at all traditional recreational bathing sites to meet certain standards, the Dangerous Substances Directives (1976 and subsequent daughter directives) that set limits for environmental concentrations and discharge limits (zero for the most toxic) for a range of carcinogenic and bioaccumulating substances. More recent directives (the Water Framework Directive and the Marine Strategy Framework Directive) have included both required management actions but they have also set out the frameworks to be used in managing human activities that impinge on the aquatic environment.

The UN Convention on Biodiversity (1992) goes beyond the GESAMP definition of pollution in requiring all signatory states to make provisions to protect biological diversity and the functioning (eco)systems on which they depend or of which they are a part. This requirement has therefore seen a widening of the regulations controlling damaging human activities and a series of new initiatives which seek to protect the wider ecosystem and its functions with the adoption of a more 'holistic' and ecosystem-based approach to management (see Chapter 8). This approach is driven by the recognition that different human activities can interact when they occur together and that these interactions can produce effects that are much more significant than the effects of each activity on its own. Traditional approaches to regulating human actions tend to be based around types of activity and so do not provide the means of assessing and controlling these ecosystem effects.

In parallel with these initiatives aimed at environmental protection there has been increasing pressure on manufacturers to understand and share information on the toxicity and biological activity of chemicals. For example, in both Europe (the REACH scheme: Registration, Evaluation, Authorisation and restriction of CHemicals) and the United States (TSCA Inventory: Toxic Substances Control Act), chemical manufacturers have to provide safety data, including environmental safety, about all new substances developed and marketed. These data can then form the basis of assessments by regulators in advance of issuing discharge/disposal licences or dealing with chemical spills.

The actual process of regulation and enforcement varies between jurisdictions and change periodically in response to political initiatives. In the United States, Canada and Australia, for example, there are both state and federal agencies with environmental portfolios and each has specific areas of responsibility. In the EU there is a European Environment Agency, but it is policy focused and European legislation requires that member states implement the requirements. In England and Wales the Environment Agency licences discharges to the atmosphere and controlled waters (including rivers, estuaries and the coastal sea), while the Marine Management Organisation regulates offshore waste disposal, dredging, fishing and offshore activities including wind farms, oil exploration and extraction. Local governments regulate some discharges, control litter and it has some sea defence responsibilities. Following devolution, the situation in Scotland is more complex.

1.6 Marine pollution: Is the problem 'solved'?

Can marine pollution be considered a problem that has been solved? After all, it is now 150 years since the Great Stink forced the UK Parliament to suspended sitting and half a century since the *Torrey*

Canyon tanker polluted tourist beaches and fisheries in the English Channel. Of some things we can now be sure. We know substantially more about how human activities affect marine organisms. The science that underpins the management of marine pollution is well developed and a very large body of knowledge exists on the impacts of individual pollutants on isolated individual organisms, particularly those species that are favoured for laboratory testing. However, we have very little appreciation of the impacts of contaminants at the population, community and ecosystem levels; we do not yet understand the impacts of long-term exposure to pollutants that cause chronic effects, nor do we understand the effects of contaminant mixtures in the environment that may interact to increase or reduce the effects of each other. We explore these challenges in greater depth in Chapter 2. It is also clear that many areas that were clearly suffering from the effects of pollution are now healthier. The Thames estuary (see Box 1.3), the New York Bight, the Mediterranean, the Baltic are notable examples (see also Chapter 6).

Direct discharges of some of the most toxic substances are now banned in many jurisdictions or have had their usage severely restricted (e.g. DDT, pesticides such as the 'drins', chlorofluorocarbons (CFCs), PCBs, heavy metals; see Chapter 3). However, it has proved impossible to completely stop discharging all of these substances into the marine environment. In some cases there is no practical (or economic) alternative and the human needs have been deemed to outweigh the environmental costs. For some substances, point-source discharges are well controlled in many parts of the world but non-point sources continue to be problematic because it is difficult to trace the material back to a source that can be the subject of enforcement. For many persistent pollutants ceasing their use has limited beneficial affects because they are near permanent additions that persist in the environment, marine organisms and even our own bodies (so-called legacy contamination). The current status of these historic challenges for which the science is well known are explored in Chapter 4.

In addition to the challenges of 'solving' marine pollution by well-established pollutants, we are continually developing new compounds; it is estimated

Box 1.3 The clean-up of the Thames estuary

As early as the thirteenth century local government was introducing measures to try to clean-up pollution in London. Continued population growth and the development of industry in the city undermined such attempts and regulations tended to play catch-up with poor environmental conditions. Sewers were laid as the city developed to take rainwater from the streets and roofs into nearby watercourses (and hence ultimately the Thames estuary). With the introduction (from about 1810) of the water closet, the volume of water entering the cesspools that served to provide rudimentary treatment of the waste increased dramatically and in 1847 an Act of Parliament was passed requiring that all household liquid waste be drained into the sewers. So, over a six-year period, approximately 30 000 cesspools were replaced by flowing systems to carry human, animal and other wastes into the sewers and ultimately the Thames.

The Thames was still used as a source of drinking water so this initiative probably contributed to the death of 14 000 people in the 1849 London cholera outbreak. The deterioration of Thames water quality led to the Great Stink of 1856 when Parliament was unable to sit due to the smell from the estuary. These two events prompted the 1856 Act of Parliament that, among other things, required the interception of all the old sewers and their redirection to a site downstream of London. The subsequent civil engineering scheme that renewed and redirected London's sewers is one of the great engineering achievements of the Victorian age and it saw London's sewage collected and transported to two large

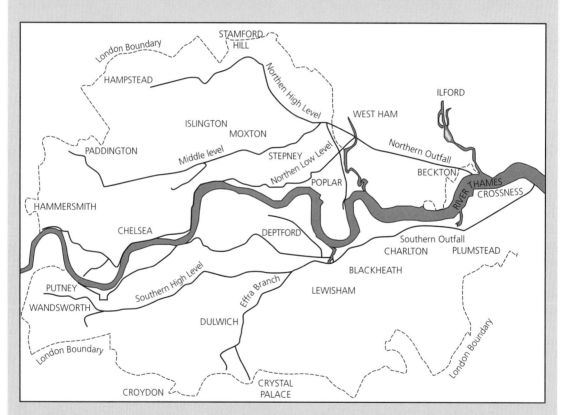

Box 1.3, Figure 1 London and the River Thames showing the routes of main interceptor sewers, designed and installed by Bazellgette, who directed sewage to the treatment works at Beckton and Crossness. Source: https://en.wikipedia.org/wiki/Great_Stink By Philg88; Attribution: Wikimedia Foundation http://www.wikimedia.org (Own work) [CC BY 4.0 http://creativecommons.org/licenses/by/4.0], via Wikimedia Commons

(continued)

Box 1.3   *(Continued)*

sewerage works, at Beckton on the north bank and at Cross-ness on the south shore of the Thames (Figure 1, Box 1.3). At each site, the sewage was stored in large reservoirs and then discharged into the Thames on the ebb tide. Unfortunately, the understanding of tidal circulation in estuaries was rather limited and the effluent from these discharges was carried back into London on the subsequent flood tide and deposited in the lower estuary as banks of sludge. The works were therefore extended in 1891 to introduce settling of the sludge. The liquid effluent was still discharged into the estuary while the sludge was transferred to ships for dumping in the outer estuary (a practice that continued until 1998) (see Section 3.2.2). However, the renewal of the sewers (which prevented contamination of wells in the city) and the subsequent treatment of the effluent led to major improvements in the water quality of the Thames estuary and reduced the incidence of disease.

Over the next few years the city continued to grow, new factories and power stations opened, and the number of people adding their sewage to Beckton and Crossness increased. By the 1950s the Thames was in extremely poor health again with very sparse fish fauna, a seabed dominated by a single

tolerant species of worm for much of its length and periods of anoxia (no oxygen) in warm summers. This prompted another phase of intervention with first Crossness (1964) and then Beckton (1976) treatment works being upgraded, controls being introduced on industrial discharges (including their redirection to the treatment works), and pressure on power stations to reduce the amount of heated water discharged. These measures resulted in increased levels of oxygen in the water and the recovery of initially the benthos (seafloor) followed by the fish communities. The number of species of fish recorded from the London reaches of the Thames estuary increased from 3 in 1964 to 98 in 1980, and from 1980 onwards the number of salmon returning to the Thames has also increased dramatically (Figure 2, Box 1.3).

The story of the Thames estuary shows dramatically how human activities and the changing nature of society place pressures on the environment. The history of pollution regulation also shows that while reactive measures are often quickly outstripped, more holistic and integrated approaches used in recent decades can result in reversals in the decline of even highly degraded marine environments.

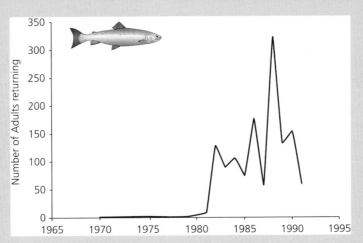

Box 1.3, Figure 2 The number of adult salmon returning to the Thames estuary following major water quality improvements in the 1970s (inset: female Atlantic salmon). Data from: Attrill 1998.

that 15 000 new chemicals are developed throughout the world each day by chemical and pharmaceutical manufacturers (Section 5.1). Only a small proportion of these compounds undergo direct toxicity testing before entering into use and an even smaller

number are effectively screened for their possible effects outside a laboratory environment: interactions with other substances and the whole range of biotic and abiotic components of an ecosystem the affects can be very different beyond the lab. It is only now that some of these compounds are being recognised for having novel and profound effects on the biota, the 'gender-bending' endocrine disruptors being classic examples (see Chapter 5).

We have therefore structured our account of the major categories of marine pollution into those that might be considered 'solved' (Chapter 3) and those that are 'on-going' as issues (Chapter 4). In placing a given issue into the 'solved' chapter we do not wish to imply that this problem has disappeared completely; rather, we suggest that the science around the issue is well developed and that management actions/options are well defined, but it is only if these have been effectively implemented, and any legacy issues dealt with, can the problem really be regarded as having been solved. Of course, new potential pollutants are emerging everyday (Chapter 5) and these pose new challenges to the science underpinning regulation and management (Chapter 7). History shows us that most initiatives to protect the environment, to regulate and manage human impacts, have been prompted by the recognition of negative consequences. What these events were and how they have been responded to vary between different seas and oceans (Chapter 6), but common themes are emerging. A recognition of the challenges of trans-boundary effects, the impact of micro-pollutants (micro-particles to micro-concentrations of potent biologically active substances), and the interaction of pollutants with other pollutants and other forms of human induced changes mean that holistic and coordinated management is becoming an increasingly common means of addressing pollution problems (Chapter 8).

Resources

UN Millennium Ecosystem Assessment A report on the state of the global environment and the ability to support human well-being: http://www.millenniumassessment.org/en/index.html.
UK National Ecosystem Assessment The UK national component of the UN Millennium Ecosystem Assessment:

http://uknea.unep-wcmc.org/About/Conceptual Framework/MillenniumEcosystemAssessment/tabid/112/Default.aspx.
GESAMP The UN group of experts that provide the science to support sustainable ocean management: http://www.gesamp.org.
UNCLOS UN Convention on the Law of the Sea: The gateway site, providing access to the convention text and annexes and supporting documentation: http://www.un.org/depts/los/convention_agreements/convention_overview_convention.htm.
OSPAR The home of the Oslo–Paris Commissions which cover the NE Atlantic, with access to treaties, reports, policy documents, Marne Protected Areas, health of the seas reports, etc.: http://www.ospar.org.
Chemical Registration Europe http://www.hse.gov.uk/reach.
USA http://www.epa.gov/oppt/existingchemicals/pubs/tscainventory/index.html.

Bibliography

Attrill, M.J. (ed) (1998). *A Rehabilitated Estuarine Ecosystem: The environment and ecology of the Thames Estuary*. London: Springer.
Carson, R. (1962). *Silent Spring*. New York, NY: Penguin Classics.
Costanza, R., de Groot, R., Sutton, P., van der Ploeg, S., Anderson, S.J., Kubiszewski, I., Farber, S. and Turner, R.K. (2014). Changes in the global value of ecosystem services. *Global Environmental Change*, **26**, 152–8.
European Parliament and Council (2008). Directive 2008/56/EC of the European Parliament and of the Council of 17 June 2008 establishing a framework for community action in the field of marine environmental policy (Marine Strategy Framework Directive). *Official Journal of the European Union*, **L164**, 19–40.
GESAMP (1990). *The State of the Marine Environment*. Oxford: Blackwell Scientific.
Meadows, D.H., Meadows, D.L., Randers, J. and Behrens III, W.W. (1972). *The Limits to Growth*. Milford, CT: Universe Books.
Turner, G. (2008). A Comparison of the Limits to Growth with Thirty Years of Reality. Canberra: CSIRO.
United Nations, 1992. Convention on Biological Diversity. UN, New York.

CHAPTER 2

Toxicology

2.1 Toxicity, toxicology and ecotoxicology

The *Oxford English Dictionary* defines *toxicity* as a 'toxic or poisonous quality', specifically the relative magnitude of the toxic quality. The word is derived from the Latin *toxicus* meaning poison. The concept of a toxic or poisonous compound has become increasingly prevalent as industrially produced, potentially hazardous chemicals have become commonplace in our daily lives. These compounds may be present in the surrounding natural environment, our homes and workplaces, in the food and water we consume, and in the pharmaceutical compounds we intentionally ingest. This is not to say that there are no naturally toxic compounds in our environment there are many produced by the plants and animals with which we share our world. In fact, humans have used toxic compounds for centuries as weapons, medicines and poisons and as antidotes for poisoning. The first poisons seem to have been used by the Kung San tribe from the Kalahari desert, who used an extract of the chrysomelid beetle on poison arrows since the Neolithic (about 10 000 years BC). The ancient Egyptians had an advanced understanding of poisons that is documented in manuscripts dating back to at least 2000 BC. However, in the present-day the artificially manufactured compounds far exceed natural toxicants in number, complexity and quantity in the natural environment.

The scientific discipline that considers the mechanisms, effects and measurements of the amounts of toxins in the environment is *toxicology*. The science of toxicology addresses the effects of chemicals on human health, known as *medical toxicology*, or the impacts of pollutants on the environment and its inhabitants. The latter is called *environmental toxicology* and is the main focus of this book. This is not to say that many of the toxicants present in aquatic environments do not represent a threat to human health: they represent a very real threat through the food and water we consume and through our physical interactions with the environment. Most of the best-known and regulated toxicants are in fact those that represent a threat to human health. *Ecotoxicology* is a branch of toxicology that considers the impacts of pollutants on ecological systems above the level of the individual. In order to assess the effects of toxicants on the environment and its residents, including ourselves, the species, community and ecosystem effects must be determined because ecological complexity means that toxicity at the individual level cannot be simply scaled up and used to predict the effects on ecosystems. Modern management of natural resources and human activities within natural environments has led to the adoption of a holistic or ecosystem approach which follows on from the recognition that marine ecosystems are highly interconnected (Sections 1.5 and 7.1).

One way to determine whether or not a contaminant has the potential to become a pollutant—*that it has demonstrable detrimental effects*— is to establish its toxicity through experimentation. By definition, only those contaminants that have an adverse effect are actually considered pollutants (Section 1.1). Determining the toxicity of a compound is not a simple task because different species and indeed different individuals within a species respond differently to the same toxicant. Their sensitivity varies. These inherent factors are in addition to the complex nature of toxicant mixtures actually found in the environment, the influence of the physical and chemical features of

Marine Pollution. Christopher L. J. Frid & Bryony A. Caswell.
© Christopher L. J. Frid & Bryony A. Caswell 2017. Published 2017 by Oxford University Press.
DOI 10.1093/oso/9780198726289.001.0001

the environment upon the contaminants, their interactions and the affected organisms (Section 2.2). It is therefore very important to understand their chemistry, their behaviour within the environment and in organisms (e.g. the biochemistry and physiology) and the factors that control their distribution (e.g. the sediment chemistry, geomorphology and hydrodynamics). In order to predict a pollutant's ecotoxicology we need a robust understanding of species–species interactions and species–environment interactions (i.e. their ecology).

Contaminants enter the sea through several different routes and these influence which contaminants and in what quantity they occur (Section 1.2). Once contaminants enter the sea a variety of chemical, physical and biological factors control their fate and the concentrations that marine organisms, and ultimately humans, are exposed to. Other factors that influence the amount of contaminants organisms are exposed to include their behaviour and longevity, or residence time, within the environment (Section 1.3). For example, pollutants that are *persistent* do not degrade but rather remain present in the environment for considerable periods, and so marine organisms are very likely to become exposed to, and negatively impacted by, them. *Biodegradable* pollutants, on the other hand, are more likely to be broken down and so the exposure times are usually much shorter (Section 3.1).

Organisms are affected by toxicants in many different ways. The impacts may be lethal or they may have other effects, referred to as *sub-lethal* effects. The response of organisms may be fast, or acute, being expressed within a few hours or days, or they may be slow, or chronic, only manifesting after protracted periods of weeks or months. In cases where they build up inside of organisms, it could take years (e.g. those stored in fat reserves; Section 3.4). Sub-lethal effects that may manifest over the long term include cancers and tumours (or their precursors), and developmental or growth defects. Lethal effects are definitive and often result from exposure to high concentrations of toxicants over short periods of time. Sub-lethal toxicity may be apparent at multiple different scales from the molecular level up to the whole organism and tend to manifest over longer timescales. Pollutants may affect the behaviour of organisms, their reproduction,

metabolism and physiological processes. Thus, there are many different ways that we can try to measure their effects.

In this chapter we consider how contaminants behave in marine environments and how this affects their interactions with organisms (Section 2.2). We then describe the various different ways in which marine organisms respond to contamination at the individual (Section 2.3.1) and the population level (Section 2.3.2). The approaches used to measure toxicity are equally diverse and reflect the range of known detoxification mechanisms found in marine environments (Section 2.4). These approaches cover many different organisational scales and have differing strengths and weaknesses (Section 2.4) for achieving the ultimate goal of anticipating the effects of contaminant discharges on ecosystems, ourselves and our use of ecosystem services.

2.2 Contaminants in the marine environment

2.2.1 Contaminant interactions with the natural environment

Features of the natural environment affect the toxicity of contaminants by influencing their chemical bioavailability and impacting the uptake processes of aquatic organisms. The temperature and pH of a water body can affect the bioavailability of contaminants because it affects their ionisation state: non-ionised forms are more bioavailable and so will be taken up and bioaccumulated. The redox potential of the ambient environment, whether it is the water column or the sediment, also affects the form and activity of contaminants. The presence of dissolved organic matter changes water-column chemistry and pollutant fate when it binds to *hydrophobic* organic compounds. Particulate organic matter also binds contaminants and can make them more likely to be ingested or transported to the seafloor.

Water temperature directly impacts the bioavailability of contaminants by influencing the balance of ionised and non-ionised forms. Alternatively, the effects of water temperature may be indirect: through its influence upon the permeability of body barriers, for example the epidermis, epithelium or cuticle, or through interference with detoxification mechanisms such as the production of mucus (Section 3.1). It is thought that for poikilotherms, like many marine invertebrates and vertebrates, as temperature increases metabolic rate increases and so does the rate of uptake, meaning that less time is required for the contaminant to reach toxic concentrations in the body. However, for taxa utilising biotransformation (Section 2.3.1), these processes may also speed up and so elevated toxicity may not occur. However, there are, of course, upper thresholds to detoxification beyond which an organisms tolerance is no longer effective.

The toxicity of some metals (e.g. cadmium, arsenic, zinc, mercury and lead) has been shown to increase with temperature although the nature of this relationship varies with exposure conditions and species. For example, mercury accumulates two-times faster with each 10°C temperature increase. This pattern also applies more generally with every 10°C temperature increase corresponding to a two- to ten-fold increase in toxicity, suggesting it is linked with metabolism. As organisms approach their thermal tolerance limits they experience more adverse effects from contaminants; this is thought to be due to increased energy requirements (e.g. for respiration and detoxification) at higher temperatures. Of the organic compounds: polychlorinated biphenyl (PCB) accumulation increases with temperature, and the organochlorines (dichlorodiphenyltrichloroethane (DDT) and methoxychlor) and pyrethroids are more toxic at low temperatures. For example, exposure of the marine unicellular alga *Tetraselmis chuii* to a suite of polycyclic aromatic hydrocarbons (PAHs) for 96 hours showed that the concentrations producing toxic effects were 1.6–5 times higher at 20°C compared with 25°C (Table 2.1); that is, they were more toxic at higher temperatures. Although the concentrations of PAHs tested were not at environmentally relevant concentrations, the temperature increase is comparable with that found in some coastal areas, where PAHs are elevated within sediments, during the summer at temperate latitudes. The relationship between the toxicity of organic compounds and temperature is thought to be related to biotransformation (Section 2.3.1).

Heightened sensitivity to pollutants is observed when an organism is exposed to salinities outside of those for which it is adapted. For fish, crustacea and other invertebrates, tolerance is decreased at lower salinities and this is interpreted to reflect the increasing costs of osmoregulation. At low salinities, metals are more toxic because they are more bioavailable (existing as free ions). The only organic pollutants that are influenced by salinity are the organophosphates which are more toxic at higher salinities. Water hardness may also affect metal toxicity being less toxic in soft water (at constant pH), and the effects are most pronounced for cadmium and moderate effects occur with copper, zinc and nickel.

The pH of the substrate also influences the balance of ionised and non-ionised contaminants and thus their uptake. Ammonia is an extreme example of this (Section 4.3): a rise of 1 pH unit increases the amount of ammonia (NH_3) in solution, which exists in equilibrium with its ionised form, ammonium (NH_4^+), and the toxicity of the medium increases six-fold. For metals, the relationships with pH are complex and non-linear due to interactions with dissolved organic carbon and calcium in the environment, but generally zinc, cadmium and nickel are less toxic at low pH (Section 3.3). For organophosphate pesticides such as trichlorfon toxicity increases between pH 7.5–9.

Benthic taxa in estuaries and coastal lagoons or those living in areas receiving organic wastes may be regularly exposed to low oxygen conditions. Reduced oxygen causes a number of serious impairments to aerobes, especially those with high oxygen demands (Section 3.2). Synergy between the effects of low oxygen and acute toxicity of zinc, copper, lead and phenols has been shown for a range of marine taxa. Increases in toxicity of up to two orders of magnitude have been found. Changes in redox cause metals to become more bioavailable.

Table 2.1 The median inhibition concentration (IC50) for *Tetraselmis chuii* after 96 hours exposure to three polycyclic aromatic hydrocarbons (PAHs) at two different water temperatures. Data from: Vieira and Guilhermino (2012).

Toxicant	IC50 (mg L^{-1})	
	20°C	25°C
Anthracene	3.326	2.145
Naphthalene	1.813	0.992
Phenanthrene	1.316	0.262

Also, enhanced gill ventilation by vertebrates or invertebrates when oxygen is depleted may cause increased water flow over the gills and thus also increased potential to absorb contaminants.

Most organic material breaks down within or upon marine sediments, thus they function as a natural source of organic and inorganic compounds, and they have an important role in the biogeochemical cycling of carbon, nitrogen, phosphorous, sulphur and trace metals. However, the sediments also function as a sink for contaminants. Most pesticides, aromatic and chlorinated hydrocarbons and metals bind to organic matter, fall to the seafloor, and are buried over time as further sediments cover them. Compounds that are slow to degrade—the *persistent* pollutants, for example PCBs, tributyltin (TBT), DDT and metals—may remain within the sediments even after their sources have been removed/controlled. They are then referred to as *legacy contaminants* (Section 1.3). These contaminants may become resuspended and be a cause of future toxicity if sediments are disturbed at a later date (Section 3.5).

The impacts of such natural environmental variables on the biota depend upon the toxicant and species, whether exposure is acute or chronic, and whether the changes in these variables are sudden or gradual. Sudden changes that limit the time available for acclimatisation have the greatest effect. In addition to affecting the bioavailability of contaminants in the environment directly, by interfering with the mechanisms for, and barriers to, their uptake, these extrinsic factors may also influence the organism's energetics. When conditions are suboptimal the environmental and toxicological stress have a synergistic effect. Taxa living at their environmental limits are thus more vulnerable and contaminants that were previously non-toxic may become toxic. Organisms exposed to multiple stressors have elevated metabolic rates and concurrent decreases in contaminant tolerance occur as the energy required for detoxification increases. We have an urgent need to know more about these synergistic relationships because our use of marine ecosystems is expanding and with it the stresses on ecosystems are pushing species to the limits of their tolerance.

2.2.2 Contaminant mixtures

One hundred thousand different contaminants are estimated to be present in the environment. The addition of all of these different compounds means that organisms are rarely exposed to them individually; rather, they are almost always exposed to mixtures. Interactions between the different contaminants can result in *additive* toxicity (a summed effect), or they may have an *antagonistic* effect where their combined toxicity is less than expected, or their toxicity may be *synergistic*. In the latter case, the combination is more toxic than the sum of each individually. These interactions are important because the toxicity may be higher or lower than anticipated based on what is known of their individual toxicity. A synergistic compound can decrease the metabolism of another and so increase its toxicity. Known synergists include a number of different pesticides in terrestrial systems and several metals. For example, the effects of mercury, lead and zinc on the growth of populations of the aquatic protozoan *Cristigera* vary when exposed individually and in combination (Table 2.2). Based on their individual toxicity the predicted additive impact was a 38% reduction in growth. When combined, mercury and zinc showed a synergistic effect: their combined toxicity exceeded their summed individual toxicity by 10%. Conversely, mercury and lead combined had lower toxicity than expected and as the mercury concentration increased this antagonism became more pronounced. The three metals in combination caused a 68% reduction in *Cristigera* growth and so was almost twice that of their summed toxicity (Table 2.2).

Other known antagonistic combinations include selenium decreasing the toxicity of mercury, similarly selenium and cadmium; zinc and cadmium, manganese and lead, and copper and lead, in each case the latter reduces the toxicity of the former. Other synergistic combinations include zinc and copper, zinc and nickel, and manganese and copper. A good understanding of the behaviour of mixtures of toxicants is important because regulations usually assume additive effects, which is an important misconception.

Table 2.2 Mean percent reduction in growth of the ciliated protozoan *Cristigera* after exposure to HgCl, PbNO$_3$ and ZnSO$_4$ individually and in combination. * Indicates combined toxicity that exceeds individual toxicity (synergism); # indicates lower combined than individual toxicity (antagonism); † difference compared with additive toxicity: negative values indicate lower toxicity, positive higher toxicity and 0 indicates no difference. Data from: Gray and Ventilla (1973).

Toxicant (concentration)	Growth decrease (%)	Difference from additive (%)[†]
HgCl (0.005 ppm)	12.1	0.0
PbNO$_3$ (0.300 ppm)	11.8	0.0
ZnSO$_4$ (0.250 ppm)	14.2	0.0
HgCl (0.005 ppm) + PbNO$_3$ (0.300 ppm)	21.8#	−2.1
HgCl (0.005 ppm) + ZnSO$_4$ (0.250 ppm)	35.5*	9.2
PbNO$_3$ (0.300 ppm) + ZnSO$_4$ (0.250 ppm)	25.9	0.0
ZnSO$_4$ (0.250 ppm) + HgCl (0.005 ppm) + PbNO$_3$ (0.300 ppm)	67.8*	29.7
HgCl (0.0025 ppm) + PbNO$_3$ (0.300 ppm)	14.5#	−9.4

2.2.3 Biological variability in toxicity

The impacts of toxicants on organisms are very variable which presents serious challenges to the design and application of toxicity tests and their results. Between species, toxicity can vary by several orders of magnitude. This includes taxa that are closely related, that is, in the same family or even the same genus. Differences in species' sensitivity to pollutants are attributable to their innate biology and the ecological interactions that govern their exposure to pollutants. They can also vary depending on the type of test used, especially its duration, and the endpoints measured (Table 2.3).

While consistent 'environmental' conditions are maintained for laboratory toxicity testing and the biological variations between the test subjects are kept to a minimum (Section 2.4), the variations in biological and environmental history are unknown for wild caught test subjects. An individual's biological history can influence its health and thus its response to a pollutant; for example, an animal that has recently reproduced or moulted, or one that has previously been exposed to environmental stresses such as extremes of temperature, salinity or oxygen. Prior exposure to contaminants, either those being tested or others, can also affect an organisms sensitivity to the toxicants being tested because it may have been stressed by prior pollutant exposure or it may have become acclimatised to it (Section 2.3.2). Similarly, if test subjects originate from a population with inherited tolerance (Section 2.3.2), the toxicity would be underestimated. Biological parameters such as age, gender, nutritional state (well-fed individuals are usually less impacted; Section 2.2.3), reproductive state and other ontogenetic factors can also produce variability within a species response.

Table 2.3 Biological factors that contribute to variability in the toxicity of pollutants between and within marine species.

Causes of variability within species
Age
Gender
Reproductive state
Nutritional state
Duration of exposure
Environmental stress: salinity, water temperature, oxygen supply
Genotypic adaptation
History of pollution exposure (acclimatisation)
Environmental history (and geographic origin)

Causes of variability between species
Biology
Ecology: life habit, feeding behaviour, behaviour in presence of toxicant
Detoxification mechanisms
Ecotoxicological endpoint measured
Test duration and exposure type

2.3 How do organisms cope with toxicants?

As noted in Section 2.1, organisms of different species, and even those within a species, do not respond in the same way to toxicants. This may be a product of their ecology or their behaviour. For example, their living habit or feeding behaviour may limit their exposure to a particular toxicant. Many species exhibit mechanisms for detoxification (Section 2.3.1). These may include physiological acclimatisation or they may involve inherited genetic resistance (Section 2.3.2). Broadly, these mechanisms involve rendering the contaminant inert by binding to proteins and similar molecules, or they may metabolise it into less toxic products. Sometimes these products are excreted but some are usually retained within the body. Antioxidants and stress proteins may also protected against toxicant damage (Sections 2.3.1 and 3.2.1). The behavioural attributes that may lessen or increase the exposure of an organism to a pollutant might include its living habit, which may raise or lower its exposure depending on its chemical affinities and distribution in the environment (Section 2.2.1). For example, sediment-dwelling organisms will be exposed to more contaminants bound to sediment particles (such as metals or halogenated hydrocarbons) compared with taxa that inhabit the water column. Demersal fish, such as flounder, plaice or sole, which spend considerable periods of time lying on or buried within the sediment, have a high incidence of ulcers, hyperpigmentation, lymphocystis, liver nodules and papillomas that may be caused by chronic exposure to pollutants such as heavy metals and PAHs. This is not to say that all sediment dwellers will be affected by toxicants. Exposure will also be governed by their absorption of the contaminants; for example, the nature and density of biochemical receptors and the presence of any competitive substrates.

The mechanism by which a species feeds may also affect its exposure to a contaminant; for example, suspension feeders which have very high filtration efficiency, such as bivalve molluscs, can quickly become exposed to all of the compounds within a water body (those in aqueous form or bound to suspended particles). These filtration rates can be very high; for example, when the oyster population in Chesapeake Bay, Virginia, was at its historic maximum the oysters could filter all water in the estuary in 3 to 4 days (Section 6.4), and the filter feeders in the Bay of Brest can completely filter the water in 15 to 30 days. Filter feeders are therefore exposed to most contaminants and accumulate many of them. A deposit feeder on the other hand is mostly exposed to compounds that are locked up in the sediments including the persistent legacy contaminants (Sections 3.3–3.5). Species at higher trophic levels can receive an elevated contaminant load through their diet; for example, if they consume worms with a high body burden.

2.3.1 Detoxification mechanisms

There are a number of strategies used by aquatic organisms to prevent the adverse impacts from toxicants. The first defence includes the use of barriers to reduce their biological uptake. Marine taxa may avoid taking up heavy metals by stimulating mucus production: the mucus contains anionic mucopolysaccharides that bind metals. This mechanism is also used in the gut to bind and excrete contaminants. Nereid and tubificid worms, bivalves, gastropods and some fish have been observed to produce mucus when exposed to metals including cadmium, copper, zinc, mercury and silver. In flounder, rainbow trout and African catfish the mucus can bind between 40% and 70% of the metals to which they are exposed. In fish, mucus is produced on the gills and in the gut. Impervious barriers—for example, cuticles, carapaces, tests, shells or scales—may also help to limit the contaminant uptake.

Once pollutants enter the body they may become bound by specialised proteins or biominerals and be stored as non-toxic forms (Section 2.3.1). However, once pollutant storage exceeds capacity, toxic effects may occur. So, in order to prevent toxicity, mechanisms are needed to expel any accumulated contaminants or limit their uptake from the environment in some way. Alternatively, pollutants may be biotransformed and most of their by-products will be excreted from the body. Biotransformation

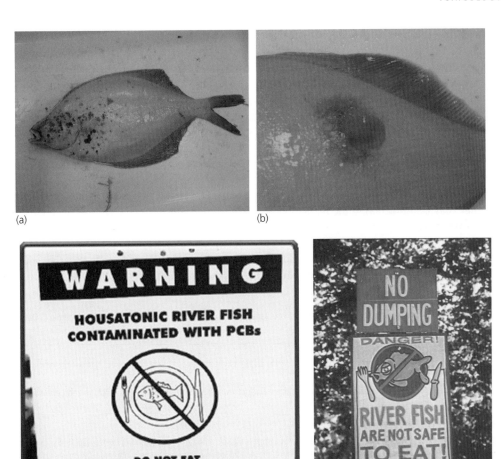

Figure 2.1 (a) Severe hyperpigmentation on the ventral surface of dab (*Limanda limanda* from Liverpool Bay). Hyperpigmentation is a chronic fish disease which reduces fitness, and is caused by both environmental conditions and exposure to contaminants. © Jonathan Molesworth 2014. (b) Close up of ulceration on the ventral surface of dab from Liverpool Bay. © Jonathan Molesworth 2014. (c) Warning sign for fishermen and hunters on the Housatonic River, Connecticut. Credit: U.S. Geological Survey Department of the Interior/photo by C.M. Custer 2007. (d) Signs on Woonasquatucket River, Rhode Island warning of possible dioxin contamination of river fish, © C.M. Custer.

occurs in three phases: phase I involves the oxidation, reduction or hydrolysis of the pollutant; in phase II, biotransformation enzymes link the metabolites from phase I with substances in the body that increase their solubility. The products of phase II biotransformations can be toxic themselves causing gene damage and/or the formation of cancers. In phase III, the metabolites of phase II transformations are excreted by the multixenobiotic transport system (Section 2.3.1.2).

2.3.1.1 Binding and storage of contaminants

A variety of detoxification mechanisms exist within animals, one of which is the production of the intracellular protein metallothionein (MT). These are cysteine-rich low molecular weight proteins that occur within organelles known as the Golgi apparatus (that process proteins), and sometimes in the lysosomes (organelles that breakdown waste proteins and other compounds). Metallothioneins occur in a number of different variants and all contain cysteine residues with thiol groups (–SH) that can bind metals. In this way, MTs prevent metals causing damage through their interference with enzyme mediated reactions. In mammals, MTs are produced in various tissues after exposure to metals, oxidative stress and glucocorticoids (steroid hormones which mediate MT induction). Metallothioneins are thought to have three main functions: (1) to regulate the supply of the essential trace metals zinc and copper, (2) to protect against toxic metals such as cadmium and mercury, and (3) to protect against oxidative damage by scavenging free radicals. They may also function in the regulation of nerve cell growth.

Although not all species have been sampled, most animals are believed to contain MTs and the amount varies between species, organs and individuals depending on their age, gender and exposure history. Three main classes of MTs have been described based on the structural arrangement of cysteine residues. Class I are those found in mammals, some crustaceans and molluscs; class II occurs only in invertebrates and plants, and class III are the phytochelatins only found in plants. In mammals, MT class I, four main variants have been identified. Variant 1 and 2 occur in various tissues but are abundant in the liver and kidney; variant 3 also predominates in vertebrate brains; and variant 4 occurs in the squamous epithelial cells (that line organs and blood vessels). They have different functions in different parts of the body (e.g. detoxification in the kidney). A selection of marine taxa that contain MTs are shown in Table 2.4.

Table 2.4 A selection of marine taxa shown to contain at least one variant of metallothionein or metallothionein-like compound. Data from: Isani and Carpene 2014; Braune and Scheuhammer 2008.

Phyla/Class	Species (common name)
Chordates	
Mammals	*Zalophus californianus* (Californian sea lion); *Tursiops truncates* (common bottlenose dolphin); and *Cephalorhynchus commersonii* (Commerson's dolphin).
Sauropsids	*Caretta caretta* (loggerhead turtle); and *Chelonia mydas* (green turtle)
Actinopterygians (ray finned fishes)	*Gobius niger* (black goby); *Alepocephalus rostratus* (Risso's smooth-head); *Channa punctata* (spotted snakehead); *Acanthopagrus schlegeli* (blackhead seabream); *Acanthopagrus latus* (yellowfin seabream); *Cynogloassus arel* (tonguesole); *Saprus aurata* (gilt-head bream); *Dicentratus labrax* European seabass); *Scophthalmus maximus* (turbot); and *Pleuronectes platessa* (plaice).
Aves	*Phalacrocorax carbo* (cormorant); *Phalacrocorax auritus* (double-crested cormorant); *Oceanodroma leucorrhea* (Leaches storm petrels); *Fratercula arctica* (Atlantic puffin); *Larus argentatus* (herring gull); *Rissa tridactyla* (black legged kittiwakes); *Cepphus grylle* (black guillemots).
Echinoderms	
Echinoids	*Strongylocentrotus purpuratus* (purple sea urchin)
Molluscs	
Cephalopods	*Loligo forbesi* (veined squid); *Nototodarus gouldi* (red-arrow squid); *Sepia officianalis* (cuttlefish); *Todarodes pacificus* (Japanese flying squid); and *Octopus vulgaris* (common octopus)
Bivalves	*Mytilus edulis* (common mussel); *Mytilus galloprovincialis* (Mediterranean mussel); *Scapharca inaequivalvis* (arcid clam); *Crassostrea virginica* (eastern oyster); *Macoma balthica* (Baltic tellin); *Crassostrea gigas* (Pacific oyster; *Placopecten magellanus* (Atlantic deep sea scallop); *Mercenaria mercenaria* (hard clam); and *Pecten maximus* (great scallop).
Gastropods	*Patella vulgata* (common limpet)
Polyplacophorans	*Cryptochiton stelleri* (chiton)
Arthropods	
Malacostracans	*Cancer pagarus* (edible crab); *Callinectes sapidus* (blue crab); *Homarus americanus* (American lobster); *Panulirus argus* (Caribbean spiny lobster); *Scylla serrate* (giant mud crab); *Tigriopus brevicornis* (copepod); and *Orchestia gammarellus* (amphipod).
Annelids	
Polychaetes	*Hediste diversicolor* (ragworm); and *Perinereis nuntia* (nereid worm)

In plants, the primary metal binding compounds are the phytochelatins also referred to as class III MTs. These are peptides, not proteins, formed from the transformation of glutathione via phytochelatin synthase. Phytochelatins are found widely throughout the plant kingdom but are rare in animals (although phytochelatin synthase has been found in some nematodes). Phytochelatins function similarly to MTs they are cysteine rich and detoxification is via binding to the thiol groups. Phytochelatin production is induced by the presence of cadmium, zinc, mercury, copper, silver, nickel, gold, lead and arsenic. Some plants, fungi and cyanobacteria also contain class II MTs.

Metallothionein-like proteins have also been isolated from marine invertebrates and vertebrates, and these proteins have many properties in common with MTs except that they are usually of lower molecular weight. The cephalopod molluscs, the veined squid *Loligo forbesi*, the red arrow squid *Nototodarus gouldi*, the Japanese flying squid *Todarodes pacificus*, the cuttlefish *Sepia officianalis* and the common octopus *Octopus vulgaris* contain MT-like compounds that are of lower molecular weight than MT. Other non-MT metal binding compounds have also been found. For example, *Hediste diversicolor* produces a myohaemoerythrin-like protein (similar to the respiratory pigment of invertebrates) in the intestinal and coelomic cells when exposed to acute and chronic concentrations of cadmium, and *Nassarius* snails from contaminated environments contain a non-MT cadmium-binding protein within their cell cytosol.

Metallothionein synthesis does not occur immediately upon exposure to metals; the first intracellular line of defence against metal toxicity is provided by glutathione. This is a low molecular weight tripeptide produced immediately after exposure to metals. It contains numerous thiol groups that bind heavy metals and so is structurally similar to MT. Within invertebrates, glutathione production is induced by copper, iron, manganese, zinc, silver, cadmium, mercury and lead. Glutathione is synthesised in the cells of animals, plants, fungi and some bacteria, and it has a number of other functional roles including the transport of metals and amino acids, the metabolism of toxins, the synthesis of DNA and proteins.

Although invertebrates contain MT and MT-like compounds, the biomineralisation of metals is more common. This process occurs within intracellular lysosomes. Metals bind to lipofuscin which is an insoluble granular matrix formed by the peroxidation of lipid membranes and the degradation of lysosomal MT. Metals react with the acidic groups on the lipofuscin granules and as they grow the metals become trapped and accumulate within the granules (e.g. Box 2.1, Figures 2 and 3). The metals are then ejected from the cell by exocytosis of the lysosomes and ultimately are excreted from the body. However, it is not clear to what extent this excretion occurs as taxa have been found with very high body burdens. Lysosomal granules have been found bound to copper, cadmium, zinc, calcium, aluminium, manganese and iron. In many marine invertebrate species, metals are also accumulated as mineralised granules, intra- (Box 2.1, Figures 1a, 3) and extracellularly (Box 2.1, Figure 2b), of four types: iron-rich (related to iron metabolism), copper sulphate granules (Box 2.1, Figure 2b; thought to function in haemocyanin metabolism in molluscs and crustacea), and calcium/manganese- carbonate concretions (involved in shell deposition), and phosphate concretions (Box 2.1, Figure 3; possibly involved in intracellular calcium ion detoxification). Similar to lysosomal granules, these may be eliminated by exocytosis. The production of metalbound granules is common among bivalve and gastropod molluscs. These granules are localised in the digestive glands of gastropods, the gut of nereid worms and in the gills, kidneys and mantles of bivalves. Phosphate granules in gastropods contain manganese, iron, cobalt, nickel and zinc, and copper sulphate granules have been found bound to copper, cadmium and silver.

Many taxa utilise more than one mechanism for binding and removing metals from the body. For example, littorinid snails use copper sulphate granules, iron granules and MTs. Metallothioneins and lysosomes are both involved in the production of lipofuscin granules, showing that the different processes of metal homeostasis and detoxification are not completely independent. These are the main metal-binding mechanisms found in marine organisms: a few other less common methods exist (see bibliography).

2.3.1.2 *The biotransformation of contaminants*

Phase I biotransformation processes include the actions of enzymes such as the cytochrome P450 enzymes that metabolise a range of compounds formed within the body—for example, hormones, steroids, vitamins, fatty acids, and compounds within bile. They also detoxify contaminants in the heart, liver and kidneys. The cytochrome P450 protein family comprises approximately 21 000 discrete proteins (coded for by the cytochrome 450 gene) containing a haeme group. Different forms of cytochrome P450 with a detoxification role exist in the different organelles of plants and animals, and many genes coding for cytochrome P450 have been identified in marine invertebrates. The cytochrome P450 enzymes are also known as the *mixed function oxidases or oxygenases* (MFOs). They occur within the microsomes of the endoplasmic reticulum. Mixed function oxidases (MFO) catalyse oxidation-reduction reactions whereby one oxygen atom is introduced into the product of the reaction and the other is reduced to water. When exposed to pollutants the MFOs are produced in large quantities and are thought to detoxify organic compounds. In a series of oxidation reactions, the insoluble organic compounds are converted into soluble products that can be excreted from the cell. Mixed function oxidases are produced in fish, mammals and invertebrates exposed to hydrocarbons. Their production is induced when exposed to PAHs, heterocyclic amines, PCBs, DDT, polybrominated biphenyls (PBBs), dioxins and a range of other pesticides.

Many types of fish produce the MFO ethoxy resorufin *O*-deethylase (EROD) in their livers in response to PAHs, PCBs and polychloroazobenzenes. Arylhydrocarbon hydroxylase, another enzyme in the cytochrome P450 family, has been identified in marine crustaceans and fish, but it has not been found in bivalves, algae or echinoderms. Mussels produce other MFOs, particularly benzo[*a*]pyrene oxydase and benzo[*a*]pyrene mono-oxygenase, when exposed to PAHs. Phase I biotransformation is also provided by the fluorescent bile compounds in fish which are induced by hydrocarbons. The bile compounds also function to increase the solubility of insoluble compounds so that they can be excreted (a process which requires several weeks).

Phase II biotransformation occurs in the cell cytosol and membranes and in this phase enzymes function to catalyse conjugation reactions which attach water-soluble polar molecules to the pollutants. Increasing their solubility in this way means they can be excreted from the cell. In animals the phase II enzymes include glutathione-*S*-transferase (GST), UDP-glucuronosyl-transferase and sulfotransferase. Glutathione-*S*-transferase is induced when exposed to organic pollutants, for example, PAHs, organochlorine pesticides and PCBs in fish, molluscs and corals. Glutathione-*S*-transferase combines glutathione with electrophilic compounds (the contaminant or endogenous compounds); and, the products are metabolised in a series of reactions with cysteines to produce mercapturates (*S*-(*N*-acetyl)-*L*-cysteine conjugates) that are excreted in the bile or urine. However, both phase I and II biotransformation processes can produce metabolites that can be more toxic than their reactants (being mutagenic and carcinogenic) and which can accumulate in the body.

Phase III proteins termed *multidrug-resistance proteins* (MDRs) are involved in the multixenobiotic response to xenobiotics or contaminants. The term multidrug-resistance proteins was coined in medicine to describe the detoxification of drugs although is actually a general metabolic process non-exclusive to drugs. The MDRs transport the by-products of phase I and II biotransformations across cell membranes and out of cells via a family of proteins called adenosine triphosphate (ATP) binding cassette transporters that occur within the membranes of eukaryotic cells. Multixenobiotic resistance and genes coding for the ATP-binding cassette proteins have been identified in echinoderms, gastropods, worms, sponges, bivalves and fish. Studies with different species of *Mytilus* have shown that the production of these proteins correlates with the degree of contamination, and they have been induced in response to organochlorine pesticides, PCBs and PAHs. Some environmental contaminants—for example, synthetic musk—have the potential to inhibit the expression of MDRs, and these are referred to as chemosensitisers. Multixenobiotic resistance seems to be particularly important in juvenile life history stages that do not employ phase I and phase II biotransformation.

2.3.1.3 Antioxidants

Within aerobes, exposure to pollutants can result in the production of reactive oxygen species. These reactive oxygen species are also produced as the by-products of regular enzyme processes and can cause substantial damage to DNA, proteins and lipids, and they can affect the availability of reduced and oxidised compounds through their influence on the intracellular redox state. In this way reactive oxygen species interfere with normal cell functions such as enzyme behaviour. The damage they cause is protected against through (1) enzyme-mediated oxidation, reduction, hydrolysis, and/or (2) through the non-enzymatic oxidation of pollutants. Enzymatic protection is provided by superoxidases, peroxidase and catalases. Non-enzymatic protection is provided by MTs and glutathione (Sections 2.3.1.1–2.3.1.2) that, in addition to metal binding, also function to protect against some of the damage from reactive oxygen species.

2.3.1.4 Stress proteins

Stress proteins or heat shock proteins (HSPs) are a ubiquitous group of proteins that are induced by substantial temperature increases. They are also referred to as *stress proteins* because they are now known, since their discovery, to be synthesised in response to other environmental stresses such as low oxygen, salinity changes, contaminants such as metals and oxidative stress. They function in protein homeostasis and perform roles in protecting and repairing proteins when damaged by environmental stress. They have been sequenced from many vertebrates and invertebrates. Heat shock proteins are also produced in plants exposed to heavy metals and have been found in the intertidal algae *Fucus vesiculosis* and *Chrondus crispus* that are regularly exposed to environmental extremes, the green algae *Chlamydomonas*, and the cyanobacteria *Synecocystis*. Stress proteins can also provide cross tolerance: those living in stressful environments where stress proteins are induced regularly by, say, temperature produce them more quickly when exposed to other stresses such as pollutants; however this response does vary between species and even populations of the same species.

2.3.2 The inheritance and cost of tolerance

Taxa that utilise the mechanisms outlined in Section 2.3 for detoxifying or rendering environmental contaminants inert are said to be tolerant of, or resistant to, these compounds. Variations in tolerance can be genetic or may be the result of previous exposure. Physiological acclimatisation (the phenotypic response) from prior exposure may produce variations in tolerance between populations, whereas heritable genetic adaptation (the genotypic response) produces variations between taxa. Acclimatisation can occur after chronic exposure to contaminants (e.g. populations from the Fal estuary, Box 2.1). Genetic adaptation has been observed in some fish after just one or two generations which would obviously occur quickly in taxa with very short generation times (e.g. microalgae or zooplankton). Heritable tolerance can be demonstrated in the lab and tolerant populations are found in areas experiencing long-term pollution. Also, tolerance to a particular contaminant can be conferred if tolerance to another already exists and this is termed *co-tolerance*. This could be because they are structurally similar or due to their use of similar detoxification mechanisms. Co-tolerance has been found for cadmium and copper, lead and copper and organic compounds such as PCBs and DDT. Co-tolerance has the added benefit of reducing the energetic costs of tolerance.

Just because a population exhibits tolerance that does not mean there will be no impact: acute exposures like accidental spillages will have a substantial impact because the resident species are not primed. Acclimatisation or adaptation requires time and exposure to non-lethal concentrations. Acute pollutant exposures can result in mortality and extirpation (local extinction), and so there may not be any survivors to develop tolerance. Most natural populations are exposed to complex mixtures of contaminants, and it is not clear how mixtures affect populations (Section 2.2.2). Complex mixtures may inhibit or slow the development of resistance.

Tolerance to pollutants of course comes at a cost. The processes that protect against damage from toxicants, outlined in Section 2.3, require energy, for example, to produce defensive compounds or enzymes to break down the toxicants.

These additional energetic costs mean there is less energy available for other processes. For example, the secretion of mucus to limit heavy metal uptake has a significant energetic cost, of up to 32% of the energy obtained through an organisms diet. Environmental factors such as changes in temperature or salinity, deoxygenation and exposure to other contaminants may cause additional physiological stress and so raise the energetic requirements even further. The normal processes of growth, moulting and reproduction may be impacted because there is less energy available. Thus the fitness of an organism and a population may be reduced. Fish populations with tolerance to copper have lower survival rates, poorer body condition, slower growth, lowered fecundity and smaller offspring than those from uncontaminated sites. Similar results have been found for insects, and behavioural changes such as reduced prey capture have also been found in fish.

While genetically inherited tolerance can be transferred between generations, so can the costs: multiple generations may be affected by having reduced fitness. Higher tolerance to pollutants may result in reduced genetic diversity within a population. Increased mutation rates and pollutant-induced selection results in the loss of genetic material associated with the sensitive genotypes, and mutation increases in tolerant genotypes. Reduced population sizes within contaminated areas and the resulting changes in physical dispersal patterns, and thus gene flow, through the environment also cause a loss of genetic diversity. This has been shown for freshwater fish and crayfish species that had lower genetic diversity in contaminated sites compared with reference sites. Lost genetic diversity has also been shown for horseshoe crabs, *Carcinoscorpius rotundicauda*, the gastropod *Littorina brevicula*, the fish *Microgadus tomcod*, sun fish *Lampris auritus*, flounder and killifish from contaminated sites.

Box 2.1 The Fal estuary in Cornwall

The Fal estuary in Cornwall has received high levels of heavy metal pollution draining from copper and tin mining activities on the Carnon River (Section 3.3) for many years. Tin is thought to have been mined in this area since the Bronze Age and copper since Roman times. Deep mining in the area has been practised since at least 1740 with peak production occurring in 1850. By 1900, mining for copper ore had declined substantially, but a recommencement of tin mining at the Wheal Jane tin mine during the 1970s showed corresponding increases in zinc inputs. Today the area is a UNESCO (2006) World Heritage Site designated for its mining landscape that encloses the eighteenth-century mining infrastructure. Closure of the Wheal Jane tin mine in 1991 led to an acidic mine-water pollution event in 1992 which contained high levels of iron, zinc and silver, and the discoloured mine water covered an area greater than 6.5×10^6 m^2 extending from the mine to Falmouth Docks. Subsequently, this led to the construction of an active lime treatment plant to deal with the acidic mine wastewater, and overall the treatment costs have exceeded £20 million.

Over time, the creek has seen slow declines in copper pollution but no declines in zinc. Although mining activities have ceased, the drainage of copper and zinc from the disused mines continues (Box 2.1, Figure 1). In 2007, sediment metal concentrations remained high and the concentrations of metals within marine worms reflected those within the sediments (Rainbow et al. 2007).

Local oyster fisheries have existed in the Fal estuary since the sixteenth century and specimens collected from the estuary can have a noticeable green colouration due to the high copper concentrations present in their blood and tissues. The distinct colouration of oysters was recognised more than 150 years ago and the intensity of colour corresponds to the copper concentration. In these oysters, these metals have been found bound within intracellular membrane bound vesicles. Oyster haemocytes contain both copper and zinc and by isolating metals within the blood cells the oysters in the creek are able to tolerate body burdens 10–100 times higher than those from unpolluted sites (Box 2.1, Figure 1). The large amounts of copper in oysters, around 3 mg g^{-1} dry body weight, from the estuary caused an outbreak of human copper poisoning when the oysters were exported to western France in 1862.

Restronguet Creek is the most polluted tributary within the Fal estuary, and the metal levels here are more than an order of magnitude higher for zinc and copper compared with 'clean' reference sites (Table 2.5; Box 2.1, Figure 1). Despite these elevated metal concentrations the fauna are typical of other local estuaries although some species are

continued

Box 2.1   *Continued*

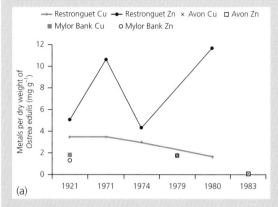

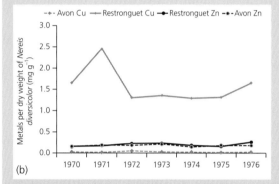

Box 2.1, Figure 1 Copper and zinc within animals from Restronguet Creek, sites near the mouth of the creek at Mylor Bank, and clean reference sites in the Avon estuary in winter–spring. (a) Residues in oyster *Ostrea edulis* tissues between 1921 and 1983; (b) residues in ragworm *Hediste diversicolor* between 1970 and 1976. Sediment samples tested in 2007 had metal concentrations of 3.39 ± 0.29 mg g^{-1} Zn and 3.27 ± 0.26 mg g^{-1} Cu per dry weight of sediment, and 3.39 ± 1.77 mg g^{-1} Zn and 1.93 ± 0.57 mg g^{-1} Cu per dry weight of worm. Data from Bryan et al. (1987) and Rainbow et al. (2009).

Table 2.5 Median lethal concentrations for sexually immature *Hediste diversicolor* from four different populations after exposure to Cu for 96 hours (at 17.5 salinity). *Polluted site at Restronguet Creek, Fal estuary, Cornwall, UK. Data from: Bryan and Gibbs (1983) and Pook et al. (2009).

LC50 96h	Restronguet Creek*	River Avon	Teign estuary	Percuil estuary
Cu (µg L^{-1})	2300	540	1022	1052
Zn (µg L^{-1})	94 000	56 000	–	–

from the creek was around 75% lower and the female animals were half the weight of those from the reference site. This shows that although the tolerant populations could inhabit the polluted areas they had lower fitness as a result.

The tolerance of *H. diversicolor* has been shown to be heritable. A study of the worms from Restronguet Creek found that they produced mucus when exposed to silver and copper, which may have a detoxification role (Section 2.3.1). The worms also had intracellular lysosomes containing copper, the extracellular granules in the epicuticle contained copper and sulphur (Box 2.1, Figure 2a), and spherocrystals that bind zinc were present in the gut wall (Box 2.1, Figures 2b and 3). Worms from the clean site did not produce mucus, lysosomes, granules or spherocrystals containing copper or zinc (Box 2.1, Figures 2–3). However, they did produce MT-like proteins after exposure to the metals. Most of the zinc storage in worms from Restronguet Creek occurred in the spherocrystals. Ninety percent of the copper in worms from the creek was bound as an insoluble form compared with only 30% in worms from the reference site. Biomineral storage of both copper and zinc are clearly important feature of the detoxification process for this species.

Shore crabs, *Carcinus maenas*, from Restronguet Creek also show a high tolerance to copper and zinc compared with the other local populations of the crab. The crabs' adaptations are not thought to be inherited because their wide scale larval dispersal means the larvae could frequently originate from outside the creek. The crabs that survive the high metal concentrations are probably physiologically acclimatised, and the mechanisms of tolerance could therefore differ between individuals. The crabs tolerance to zinc increases with body size, and the accumulation rates are similar between crabs from the creek and those from reference sites. It has been suggested that the zinc tolerance may be derived from improved excretory mechanisms or lower

notably absent (e.g. *Cerastoderma edule, Hydrobia ulvae, Melinna palmata, Mytilus edulis* and *Patella vulgata*). Ragworms *Hediste diversicolor* from sites in the creek contain copper concentrations similar to the sediments, but the worms survive where individuals from other populations of this species cannot (Table 2.5). *Hediste diversicolor* has developed tolerance to copper and zinc, and is co-tolerant to Cd. Measurements of the scope for growth (Section 2.4.4) of tolerant worms was 46–62% lower than for the intolerant worms from reference sites. Also, the fecundity of *ragworms*

continued

Box 2.1   *Continued*

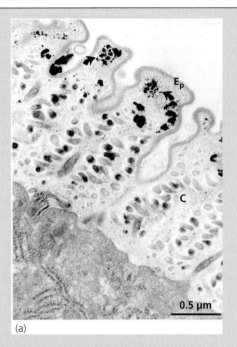

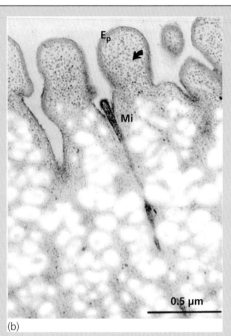

(a)   (b)

Box 2.1, Figure 2   *Hediste diversicolor.* Ultrastructure of the tegument. (a) Restronguet Creek: the epicuticle (Ep) was loaded with numerous dense extracellular granules (arrows) (C collagen; Ep epicuticle). Magnification ×40 000. (b) Blackwater: the epicuticle (Ep) contained only small dense granules (arrows) (Mi microvilli of epidermal cells). Magnification ×60 000. All granules contained mostly Cu and S; lysosomes from Restronguet Creek worms contained copper and sulphur in the apical region, and iron, aluminium and phosphorous elsewhere; lysosomes in worms from Blackwater contained zinc, bromine, calcium and copper. *Marine Biology*, Trace metal detoxification and tolerance of the estuarine worm *Hediste diversicolor* chronically exposed in their environment, 143, 2003, 731–44, C. Mouneyrac, O. Mastain, J.C. Amiard, C. Amiard-Triquet, P. Beaunier, A.-Y. Jeantet, B.D. Smith and P.S. Rainbow, © Springer-Verlag 2003 with permission of Springer.

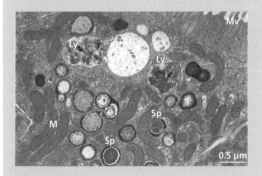

Box 2.1, Figure 3   *Hediste diversicolor.* Intestinal cells of Restronguet Creek worms. Ultrastructural aspect of intracellular concretions, the mineralised lysosomes (Ly) containing clusters of ferritin (arrows) and the spherocrystals (Sp) (M mitochondria; Mv microvilli). Magnification ×30 000. XRF showed lysosomes contained mostly iron, and spherocrystals contained zinc, phosphorous, copper and iron. *Marine Biology*, Trace metal detoxification and tolerance of the estuarine worm *Hediste diversicolor* chronically exposed in their environment, 143, 2003, 731–44, C. Mouneyrac, O. Mastain, J.C. Amiard, C. Amiard-Triquet, P. Beaunier, A.-Y. Jeantet, B.D. Smith and P.S. Rainbow, © Springer-Verlag 2003 with permission of Springer.

body permeability limiting metal uptake. Crabs from the creek are also tolerant of elevated copper. Copper tolerant strains of S*crobicularia plana, Corophium volutator, Nephtys hombergi* and *Fucus vesiculosis* are also present in the creek although the mechanisms of their tolerance have not been established. A variety of other metal tolerant invertebrate and vertebrate populations with differing detoxification mechanisms may be found in the bibliography.

2.3.3 The bioaccumulation of contaminants

Contaminants that are stored, via biomineralisation, as opposed to being biotransformed or excreted can build up within organisms increasing their *body burden*. In such cases older animals will tend to have larger contaminant burdens than younger animals because they have been exposed for longer. This phenomenon is termed *bioaccumulation* and may be defined as the increasing accumulation of a compound in the body of an organism over time. Bioaccumulation is particularly problematic with persistent pollutants such as metals and organochlorine hydrocarbons. Depending on the bioavailability of the bioaccumulated pollutants, they may cause sub-lethal toxicity (Section 2.1). Metal bioaccumulation has been observed in bacteria, plants and invertebrates. In Restronguet Creek (Box 2.1), copper accumulation is thought to have caused mortality of crabs as a consequence of consuming *H. diversicolor* with significant copper body burdens, and sub-lethal effects have also been found for zebrafish fed prey from the creek in laboratory experiments.

Within food webs the growing pollutant body burdens can lead to *biomagnification*, the process whereby pollutants increase or are magnified throughout the food chain. The transfer efficiency of energy through a food chain is approximately 10% between each trophic level, and this means that a herbivorous grazer has to consume ten units of plant biomass to produce one unit of grazer biomass; a first-order predator consumes ten units of grazer biomass per unit of predator. So assuming, for simplicity, that all the material ingested is absorbed, the grazer will be ingesting 10 times as much pollutant as each plant absorbed, the first-order predator will ingest 100 times, and the second order predator 1000 times, and so on. Thus, assuming that once ingested these compounds remain in the body, the concentration of contaminants is magnified up the food chain. Thus, if plants passively take up any *elevated* contaminants from the environment, the load in higher predators will be magnified. This simple model is widely used to highlight the dangers of metals (and other bioaccumulating substances) for human consumers (Figure 2.1). The contaminants that biomagnify tend to have hydrophobic qualities that allow them to accumulate in the body's fat stores. Examples of those that can biomagnify include DDT, PCBs, polybrominated diphenyl ethers (PBDEs) and heavy metals such as mercury (Section 3.3.1).

The fate of contaminants in the food web depends upon the processes of detoxification. Mechanisms that limit uptake such as the secretion of mucus, or those which bind and transport the contaminants out of the body (e.g. biomineralisation followed by excretion) can reduce the potential for bioaccumulation. Taxa that utilise biotransformation mechanisms, to detoxify pollutants, also have a lower likelihood of accumulation. However, the toxic metabolites from phase I and II biotransformation could transfer toxicity up the food chain (Section 2.3.1). Although these processes may not completely eliminate the possibility of bioaccumulation they can reduce it. A review of metal bioavailability (Figure 2.2) found that trophic transfer is highly variable between taxa and metals. Differences in the physicochemical form of the metal within a foodstuff (i.e. where and how it is sequestered; Section 2.3.1, Figure 2.2) and the digestive processes of the consumer affect the amount of contaminant transferred.

2.4 How do we measure toxicity?

So, why do we measure toxicity? Well, because we produce and release artificial and natural contaminants into the environment everyday we need a means of determining which compounds and in what quantities we can safely discharge them without adversely impacting natural ecosystems and human health. For many compounds their toxicity was not fully known before their release into the environment, and the knock-on or long-term impacts on the biota only manifest after a period of continued release. For example, the effects of TBT on marine snails (Section 3.3) and DDT on birds (Section 2.4.4) decimated natural populations causing population collapse and local extinctions. So, we need to regulate contaminant production (by registration) and release (by consent to discharge and by environmental risk assessments). Traditionally,

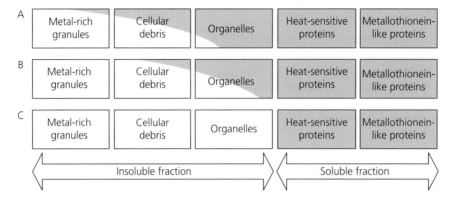

Figure 2.2 The distribution of metal contamination stored in marine prey taxa between five different detoxification compartments. Separated into those that are bioavailable or not to consumers. Grey shaded areas indicate the accumulated metal fraction that is available to the following: A. prey trophically available to neogastropods; B. prey available to predators with weaker digestion than neogastropods; C. prey trophically available to a suspension feeding planktonic copepod. Reprinted from *Environmental Pollution*, 159, P.S. Rainbow, S.N. Luoma, W.-X. Wang, Trophically available metal—A variable feast, 2347–349, Copyright (2011), with permission from Elsevier.

the regulation of contaminants has focused on agrochemicals because these are applied to crops in large quantities that run off into watercourses and the sea, but now there are new challenges for regulation such as the large quantities of pharmaceuticals discharged into the environment which are subject to very little regulation, despite being produced and released for decades (Section 5.2).

Paracelsus, a Swiss German philosopher, botanist and physician in the fifteenth century who is considered to be the founder of toxicology, said that 'poison is in everything, and no thing is without poison. The dosage makes it either a poison or a remedy.' While it is relatively straightforward to establish a toxic effect—that is, through mortality—it is not as simple to determine what concentrations are harmless. Most conventional toxicity testing begins with the toxic dose and proceeds to determine what is known as the *dose–response* relationship. This describes variations in the effects of a toxin on an organism with changing concentration, or dose, within a defined period of time. As addressed earlier, this is complicated by the different features of the toxicant, the organism and the environment (Section 2.2).

The sub-lethal or chronic effects of pollutants on organisms are more difficult to determine than simple mortality. The effects are often not externally visible and they may not manifest for some time. Toxicants may impact organisms at the molecular or the cellular level, they may interfere with metabolic processes directly or indirectly, they may have physiological manifestations, and they may disrupt reproduction at various stages. For example, by causing the disruption of egg formation, the feminisation of male animals, the abnormal growth of sexual organs, or they may interfere with the process of reproduction by masking chemical signalling. They may also affect bodily systems at multiple levels, for example by disrupting genes, the nervous system, the chemosensory system, the endocrine system, detoxification processes or an organism's locomotion.

2.4.1 The measurement and expression of toxicity

Because of the challenges associated with high variability in toxicity results it is very important that test conditions are reported and documented, and a number of standardised testing protocols are provided by the regulatory authorities for this purpose. These include those from the Organisation for Economic Co-operation and Development (OECD), the International Council for the Exploration of the Sea (ICES), the International Organisation for Standardisation, the Oslo–Paris Convention for the Protection of the Marine Environment of the North East

Atlantic, the US Environmental Protection Agency, the American Society for Testing Materials and the American Public Health Association. The protocols specify the organisms to be used and their condition, the exact test conditions, the duration of the test, the method for toxicant application and the chemical standards to be used, and they also supply the methodology for determining toxicity and analysing the data. It is hoped that this level of standardisation enables the production of consistent laboratory results. Most available standard protocols are for single-species tests in the laboratory. Extensive databases are available that summarise the effects of toxicants and these are provided by the US Environmental Protection Agency, the European Commission, and various other bodies (see Resources).

The effect of a toxicant can be quantified in several different ways depending on the nature of the impact, acute or chronic, the toxicant and the test species. These toxicity end-points include the *lethal dose* (LD) or *lethal concentration* (LC) that cause mortality, *effect concentrations* (EC) that cause sub-lethal effects and, for plants, where mortality is harder to establish *inhibition concentrations* (IC) that interfere with regular processes such as growth. These toxicity endpoints are usually expressed together with the duration of the toxicity test and the proportion of organisms impacted. So, you will often see toxicity expressed as the LC50: the lethal concentration for 50% of the individuals tested. Other common

endpoints include the LC10 and LC20 the lethal concentrations for 10% and 20% of the test sample, respectively. Similarly, the effect and inhibition concentrations are often quoted at these percentage levels. An alternative endpoint expressed specifically when toxicity varies with the duration of exposure is the lethal time for the specified percentage of the individuals in which case both the test duration and the toxicant concentration are also provided.

2.4.2 Toxicity testing approaches

Standard toxicity tests to determine the lethal concentrations of the different toxicants expose organisms to a range of concentrations, or doses, and record the number that suffer mortality. The concentration range is precisely defined through a series of tests. The response of organisms, as a percentage, is plotted against the different concentrations to summarise the dose–response relationship (Figure 2.3). These data are used to generate the LC50 concentrations. Test durations, also known as the *exposure time*, for determining the acute effects of toxicants tend to be either 48 or 96 hours. For aquatic organisms, exposure is through the water they inhabit. Toxicity data are collected at several stages throughout the test and toxicity can then be specified for different exposure times. Toxicity tends to increase with the duration as more contaminants are taken up into the body over time.

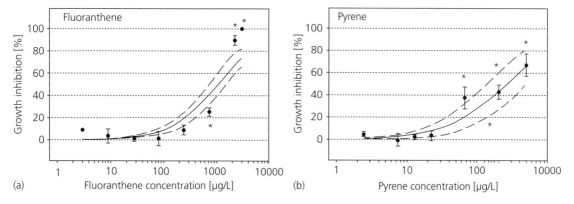

Figure 2.3 Does response curves for growth inhibition of the marine diatom *Thalassiosira pseudonana* exposed to the polyaromatic hydrocarbons (a) fluoranthene and (b) pyrene. Black points indicate mean values and bars show standard error. Solid line = the fitted dose–response curve, broken line = 95% confidence intervals. Reprinted from *Gene*, 396, S.K. Bopp and T. Lettieri, Gene regulation in the marine diatom *Thalassiosira pseudonana* upon exposure to polycyclic aromatic hydrocarbons (PAHs), 293–302, Copyright (2007), with permission from Elsevier.

Toxicant exposures are compared with control treatments in which the test subjects are exposed to identical conditions but without the toxicant. In some cases, negative control treatments may be used to eliminate any impacts from carrier compounds that have been used to increase the solubility of the toxicant.

As the science of ecotoxicology has developed, an appreciation of the extent and diversity of the sub-lethal toxic responses of organisms has grown. Sub-lethal endpoints are more sensitive than lethal endpoints and so can forewarn of possible effects, such as mortality, *before* they occur. Examples of behavioural sub-lethal endpoints include changes in locomotor activity such as fish swimming speed or valve closure in bivalves upon detecting the presence of a toxicant. Sub-lethal endpoints are ecologically more important because they can help us to understand the mechanisms of toxicity and help predict what the impacts might be at the population, community or ecosystem level. For example, reproductive impairment affecting the continued local presence of a species, or behavioural changes leading to changes in the food web. A couple of examples of whole organism behavioural endpoints were described earlier, but the sub-lethal effects can be measured at different organisational levels such as the molecular, cellular, tissue, metabolic and physiological. The use of sub-lethal biomarkers indicative of exposure to toxicants has advanced over the last decade and these methods are described in detail in the next section. For sub-lethal endpoints, the effect concentrations, and the dose–response data are recorded and treated in the same way as for lethal endpoints. Sub-lethal effects require longer to manifest and so tests usually exceed 96 hours.

Additional sub-lethal endpoints used to quantify a non-lethal toxic effect include the *lowest-observable effect concentration* (LOEC) or the *no observable effect concentration* (NOEC). The LOEC is the lowest concentration at which an adverse effect is found, and the NOEC is the highest concentration at which no effect is observed, compared with the controls. The predicted no-effect concentration (PNEC) is used together with the predicted environmental concentration (PEC) for environmental regulation to calculate a risk quotient (PEC/PNEC) for a contaminant. These calculations are described by the European Commission's technical guidance. The PNEC calculation incorporates the number of bioassay results used, the trophic level, the species represented, and the existence of long-term toxicity data (although this is not long-term from an ecological perspective).

Traditionally, single-species ecotoxicity tests have been the norm and so the PNEC is usually based on data from individual not multi-species tests (although it may include data from more than one species). It is intended that these data come from the most sensitive species and therefore assumes that all other species in the community are less sensitive than this species. A number of safety margins are incorporated into the advice provided by regulators. These margins are presumed to account for genetic variability in tolerance, and for this purpose they incorporate a two-fold differential tolerance between populations. However, the results of some studies show variability in species tolerance up to eight-fold and so the safety margins may not account for all such variance. The PNEC calculation also incorporates an assessment factor that ranges from 1 to 1000 for extrapolation between the laboratory results and the impacts anticipated in the natural environment. Different values are used depending on the type (short- or long-term, field- or lab-derived) and amount of data available. Although multi-species or community data are less common, there is increasing appreciation of the difficulty and inappropriateness of applying results from individual species to communities and ecosystems demonstrates the increasing need for multi-species programmes.

Toxicity tests may be static closed systems, with renewals of test media if needed, or they may be flow-through open systems. The choice of whether to use an open or closed system depends on the stability and volatility of the contaminant and whether it will bind to substrates (will it be lost to the atmosphere or to inert substrates?) that may impact the degree of exposure of the test organisms and so could confound the determination of toxicity. Continuous or intermittent flow through systems are typically used for long-term exposures and can ensure better water quality (oxygenation, ammonia removal) and hold a higher density of test subjects. However, they are more expensive.

2.4.2.1 Biomonitoring

Biomonitoring uses a biological response to indicate changes in the environment. *Indicator species* including plants, invertebrates and fish are used to characterise or monitor environmental changes. The species used are also referred to as *bioindicators* or *sentinels*. Biomonitoring programmes are used to assess the risks to the environment and ultimately human health as part of environmental risk assessments. The idea of a sentinel species safeguarding human health is not new: the archetype is the canaries in cages that miners historically took with them into the mines to detect the presence of noxious gases.

The first signs of pollution occur at the individual level and might include the disappearance of a species, the accumulation of a contaminant within a species, evidence of exposure or any adverse biological impacts (Section 2.4.4). The impacts of pollutants on populations, communities and ecosystems take longer to manifest, are more difficult to trace to source, and are difficult and expensive to reverse so early detection is desirable. Changes in the health status or behaviour of animals collected in the field can indicate exposure to toxicants that might affect the composition of communities or whole ecosystems and the processes that occur within them. The presence or absence of a species may also be indicative of pollution or another key change in the environment. For example, in aquatic systems the presence of the opportunist *Capitella capitata* is often indicative of sedimentary organic enrichment. Also, species that bioaccumulate toxicants from the environment, can indicate the amounts of contaminants present. The concentrations within sentinel species can indicate the amounts that are biologically available (e.g. Box 2.2), and combined, these data can indicate the level of risk.

Biomonitoring requires a reference site(s) for comparison with the contaminated sites of interest. This site(s) should have similar natural environmental parameters but it should be free from contamination. Finding reference site can be challenging because of the prevalence of contamination in coastal ecosystems today (Sections 1.4–1.6). Pristine sites are very rare and so instead a relatively clean site must be used. These reference sites must be proximal, of similar temperature and sediment type (e.g. particle size, sediment organic matter and sediment oxygen). For estuaries it is also important that they have a similar salinity regime.

Active biomonitoring involves the transplantation of artificially reared organisms of known biology and history, comparable size, weight and stage of sexual maturity into the natural environment to determine whether or not it is contaminated. These controls on the test organisms help to minimise intraspecific variation. Organisms are transplanted in enclosures (e.g. cages) for a predefined period of time. This approach is used widely in environmental toxicology with bivalve molluscs and fish. It is considered preferable to passive biomonitoring because the organism's history is known and so any prior adaptation to pollutants is minimised, but at the same time organisms are tested under 'real' environmental exposure conditions that incorporate all the natural physicochemical factors that influence contaminant bioavailability. The Biological Integrators Network of the Mediterranean Coast that began in 1998 (run by the *Institut Français de Recherche pour l'Exploitation de la MER* (IFREMER)) monitors contaminant bioaccumulation in mussels. Because mussels occurred in low abundance at some of the desired sample sites, *Mytilus galloprovincialis* were transplanted in cages along 18 000 km of the French coast. The programme monitors metals, persistent organic pollutants and radio-nucleotides in mussel tissues. This approach can also be used to determine the reasons for any natural absences; for example, if mussels were transplanted and they continue to grow then their natural absence is not due to the impacts of water quality on adult animals.

The species used in biomonitoring tend to be those identified as *model species* for laboratory studies (Section 2.4.3). Mussels have been widely used as sentinels because they are prolific, easy to culture and their biology is well known. Also, being suspension feeders mussels filter the water column with high efficiency and therefore are exposed to the full range of dissolved or suspended particulate toxicants within a locality. Being sessile they are good bioindicators of local conditions and they are easy to transplant for active monitoring programmes. Because mussels have low levels of enzymatic biotransformation (Section 2.3.1) they can accumulate many toxicants and so analyses of their tissue residues can provide valuable quantitative data on local contaminant levels.

In order to gain a true appreciation of the impacts of pollutants on marine organisms the various different ecosystem components need to be considered, and while water column toxicity is assessed using filter feeders such as mussels, those in the sediments require different types of tests. This is very important because contaminants are known to bind to sediment particles in particular fine organic particles (Section 2.2.1).

In Europe, sediment testing with the infaunal bivalve *Scrobicularia plana* and the ragworm *H. diversicolor* have been developed. These species represent a considerable proportion of the biomass within the muddy substrates of the intertidal zone, and being deep infaunal burrowers they also play a role in the resuspension of contaminants into the water column and/or their burial. The two species use different detoxification mechanisms, different feeding strategies, and have different burrowing behaviour and so provide different information on toxicity. Tests methodologies using sediments often do not reflect the *in*

Box 2.2 Biomonitoring with mussels

A biomonitoring programme called '*mussel watch*' started in the United States in 1976 (proposed by Goldberg 1975) and has been used to monitor contaminants around the coast of the United States until this day. Similar programmes were used in the United Kingdom, The Netherlands, France, Canada, Australia, Japan, India, South Africa and Russia. These programmes have been used to detect metals, chlorinated hydrocarbons, organic pollutants and radionuclides in efforts to minimise the threats posed to human health. The United States mussel watch programme was replaced by an updated monitoring programme, the National Status and Trends mussel watch programme, in 1986 which is now coordinated by the National Oceanographic and Atmospheric Administration. The United States programme samples 300 sites each year along the coast of the United States, Alaska,

Hawaii and Puerto Rico. The programme uses naturally occurring populations of several different bivalve species which reflects their differing geographic distributions: for example, *M. edulis* (Box 2.1, Figure 1) is used on the north-east coast of the United States, *Crassostrea virginica* on the south-east coast, and *M. edulis* and *M. californianus* are used on the west coast. Determinations of the concentrations of 140 contaminants are made from mussel tissues. Those measured include DDT and its breakdown products, dieldrin, 51 different PCBs, 65 PAHs, butyltin, cadmium, copper, chromium, lead, zinc, mercury, silver and zinc. Any differences in contaminant bioaccumulation between sentinel species need to be quantified if they are to be used in combination for biomonitoring. For example, mussels and oysters accumulate metals at different rates: oysters accumulate 2.5 times more cadmium, 10 times more copper, 15 times more zinc and 50 times more silver than mussels. Intra-specific variation is minimised by always sampling equivalent sized mussels in the winter prior to spawning. In the United States the programme is used for routine monitoring and to identify hotspots of pollution, but it has also been used to monitor the effects of pollution events such as *Deepwater Horizon* oil spill (Section 4.5) and natural disasters. The mussel watch data are used as the basis for regulatory decisions. The Asia–Pacific mussel watch programme (run by government agencies and tertiary institutions) spans ten Asian countries and monitors the levels of tributyltin, PBDEs, PCBs, DDT, chlordane compounds and organochlorines in *Perna viridis* and *Mytilus edulis*. The French mussel watch programme run by IFREMER uses *Crassostrea gigas* and *Mytilus galloprovincialis* to monitor coastal pollution. Other species that have been used are the freshwater bivalve *Dreissena* in the Great Lakes and *Chama sinuata* in the Florida Keys.

Box 2.2, Figure 1 The common mussel, *Mytilus edulis* that is one of the most popular sentinel species for the mussel watch programme in Europe.

situ conditions (salinity, oxygen supply and organic matter) within the sediments. However, these factors are known to affect contaminant bioavailability and so have a major impact upon their toxicity. Further development of sediment testing schemes is needed if the results are to be used for reliably predicting the community or ecosystem effects.

Fish are regularly used in biomonitoring because of their critical role in food web processes, and being part of human our diets they represent a pathway for toxicants to enter the human body. Fish represent some challenges as a test species (Section 2.4.3) because they are highly mobile and therefore have unknown and possibly inconsistent exposure histories. Thus, they may be better suited to active biomonitoring programmes. Fish are often used as indicators of exposure to organic pollutants some of which they bioaccumulate (PCBs, polychlorinated dibenzoparadioxins (PCDDs) and polychlorinated dibenzofurans (PCDFs)).

2.4.3 Test species

Toxicology, in medical and genomic research, utilises a suite of *model organisms* that are well known in terms of their biology, life history, genome and behaviour; for example, for plants, rockcress *Arabidopsis thaliana*, for bacteria, *Escherichia coli*, for yeast, *Saccharomyces cerevisiae*, for insects, the fruit fly *Drosophila*, and for vertebrates, rats and mice. In aquatic studies, the freshwater crustacean *Daphnia pulex* and the green alga *Rhaphidocelis subcapitata* are regularly used. For marine systems the model species include the mussel *Mytilus edulis*, the ragworm *Hediste diversicolor* and the crustacean *Ceriodaphnia*. These so-called model species are just that: they are an example from which we hope to extrapolate. The advantages of using models is that they have established biology and so it is possible to identify any effects accurately; they are of known exposure, nutritional and reproductive history and they are easy to acquire and maintain in the lab. However, being model species they are not always relevant to the organisms inhabiting the 'real' environment. Also, the populations of model species maintained and bred only in the laboratory may undergo selection for particular characteristics and so may be less

genetically diverse and less reflective of the attributes of the natural populations. This is especially the case for taxa with fast generation times such as bacteria or algae. If toxicity tests are mostly going to be completed with surrogates or models then their selection is obviously important. A number of criteria should be considered:

1) *Model suitability* depends upon whether the response can be extrapolated to other taxa. This will be determined by their sensitivity to pollutants and their natural geographic range. Species with broad geographic ranges will be more widely applicable.
2) *Species ecology*: the determination of clear effects will be possible for well-known species, and understanding of an organism's detoxification mechanisms can help to understand their susceptibility to toxicants.
3) *General tolerance* of environmental variation means a species will be more tolerant to manipulation, and for biomonitoring studies a greater range of habitats may be used. Similarly, some tolerance to pollution will mean that organisms can survive long enough for the sub-lethal effects to manifest or, in the case of biomonitoring, long enough to bioaccumulate contaminants. Also, for tolerant taxa the organism's recovery or environmental remediation may also be studied.
4) Species that have a *clear response* after exposure to pollutants that varies with concentration. If a species is sensitive enough to exhibit change shortly after exposure it will have greater utility as an early warning system.
5) Test species should be *abundant and ubiquitous* if they are to be sourced from the field, or if using lab-cultured species they should be easy to rear.
6) *Sessile species* are preferred for biomonitoring as they have continuous exposure to pollutants throughout their lives. In this way they can represent the pollution history of an area. Also, sessile species are easier to manipulate in laboratory tests.

Species sensitivity is a balance between having a general tolerance of environmental variations but being responsive enough to detect contamination quickly (see points 3 and 4 above). Often euryhaline species are used because of their broad tolerance to environmental conditions and anthropogenic factors, and they tend to be widely distributed and so are readily available and broadly applicable. However, it could be argued that using generally tolerant taxa underestimates the response of the more sensitive members of the community (because sensitivity to environmental change and pollutants often go hand-in-hand). The use of laboratory-reared strains is confounding for the same reason. If test subjects originate from different geographic locations then they may be acclimatised to environmental regimes that differ from those being tested, and/or they may be genetically distinct from the populations residing in the target area. Thus, as much as possible, test species should be from or at least representative of the local habitats of interest.

Toxicity testing programmes are often a trade-off between the practicalities, economics and relevance of the test data. The large variability in toxic response between and within taxa makes the application of model species data for anticipating effects at the population, community and ecosystem level challenging. In recent years some progress has been made in the development of multi-species tests (Section 2.4.5) that increase the ecological relevance. However, such tests are not presently required for regulation.

2.4.4 Biomarkers

Several of the biomolecules produced for detoxification purposes (Section 2.3) have been developed as environmental biomarkers of contamination. Depledge (1994) defined a biomarker as 'a biochemical, cellular physiological or behavioural change which can be measured in body tissues or fluids or at the level of the whole organism that reveals the exposure at/or the effects of one or more chemical pollutants'. This differs from bioindicators that consider change at the organism level and above, for example, populations, communities or ecosystems. Different types of biomarkers exist: they may record an effect on an individual, for example, behavioural changes or metabolic changes; it may be a physiological change such as eggshell thinning or feminisation; changes may occur in cells and tissues or it may be a biochemical change. Biomarker determination as an ecotoxicological tool began to develop in the 1980s and it was a contrasting approach to

those used previously; that is, chemically determining the concentrations of contaminants in water, soil or air. Biomarkers provide information on the effects of contaminants that allow us to establish whether they are pollutants (with negative effects). The advantage of biomarker data over a chemical determination alone is that (1) the biological effects can be determined, and (2) antagonistic and synergistic (Section 2.2.2) effects are incorporated because it is the summed impacts of chemical and biochemical interactions that are being measured. Biomarkers can provide advance warning of pollution *before* an irreversible effect has manifested and so can help prevent ecosystem degradation.

Biomarkers can be classified by the organisational level—for example, effects on an individual's behaviour, physiology, tissues, cells or their biochemistry (Figure 2.4)—or they may be classified as those produced after exposure or in defence, and those that record an effect or damage (the latter are also known as mechanistic biomarkers). *Biomarkers of exposure* are those produced as part of the detoxification mechanism. *Biomarkers of effect* are those that provide evidence of damage to the organism (at the molecular, cellular or organism level) that threaten its survival; for example, damage to genetic material, cell membranes or the impairment of reproductive processes. Although molecular biomarkers are very sensitive to pollutants (Figure 2.4) these are the least ecologically relevant and typically only describe short-term change. The main biomarkers from each classification are provided in Table 2.6. The power of detection is increased further if the most sensitive species in the community/ecosystem are used.

Several molecular biomarkers are general indicators of stress—for example, glutathione, weakened lysosomal membranes and stress proteins—and so can be induced by pollutants or environmental stressors. Also, most biomarkers are non-specific: they respond to multiple pollutants and thus mostly cannot be used to identify the exact toxicant of concern. However, a few specific biomarkers of marine pollutants exist including eggshell thinning and DDT; δ-amino levulinic acid dehydratase inhibition in blood indicates the presence of lead; and, bile fluorescent compounds are produced in the presence of hydrocarbons. While biomarkers are valuable tools for identifying environmental degradation, the lack

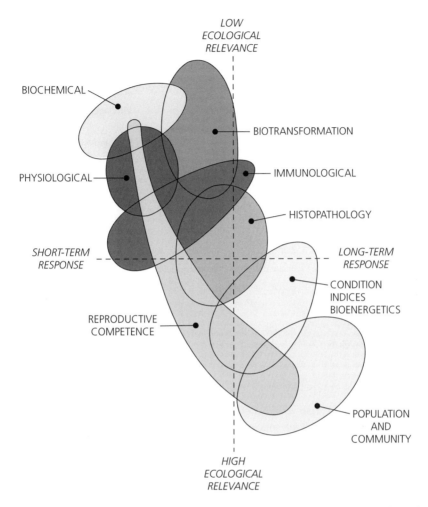

LOW
ECOLOGICAL
RELEVANCE

BIOCHEMICAL

BIOTRANSFORMATION

PHYSIOLOGICAL

IMMUNOLOGICAL

HISTOPATHOLOGY

SHORT-TERM
RESPONSE

LONG-TERM
RESPONSE

CONDITION
INDICES
BIOENERGETICS

REPRODUCTIVE
COMPETENCE

POPULATION
AND
COMMUNITY

HIGH
ECOLOGICAL
RELEVANCE

Figure 2.4 Different biological responses of fish to pollutants along gradients of temporal response and ecological and toxicological relevance (the latter in grey shading–darker greys indicating higher relevance). Reprinted from *Marine Environmental Research*, 28, S.M. Adams, K.L. Shepard, M.S. Greeley, B.D. Jimenez, M.G. Ryon, L.R. Shugart, J.F. McCarthy and D.E. Hinton, The use of bioindicators for assessing the effects of pollutant stress on fish, 459–64, Copyright © 1990 Published by Elsevier Ltd., with permission from Elsevier.

of specific biomarkers makes it difficult to make management decisions that target specific pollutants.

2.4.4.1 Biomarkers of exposure

2.4.4.1.1 Behaviour as a biomarker

The behavioural effects of pollutants include avoidance, impaired locomotion, feeding and changes in burrowing and reproduction. These changes in behaviour can be used as biomarkers of exposure. Pollutants can influence organisms behaviour by interfering with their chemical sensing of the environment or by disrupting physical locomotion through

its effects on the nervous system (neurotoxicity). For example, manganese affects Norwegian lobster foraging by decreasing the permeability of the cuticle and affecting odour diffusion. High concentrations of copper and cadmium can affect fish neuromast cells and so sensing through the lateral line.

Organisms that can sense contaminants can avoid contact and thus can reduce their exposure (seen in fish, crustacea, insects, molluscs and annelids), and those that do not receive a higher dose. For sessile taxa such as bivalve molluscs, the avoidance may be subtler; for example, closing their valves to limit

Table 2.6 Molecular biomarkers of pollutant exposure and effect on marine organisms. * Indicates those considered core biomarkers by the European Commission. Data from: Amiard et al. (2012). See text for an explanation of each biomarker and examples.

Exposure biomarkers	Indicates	Effect biomarkers	Indicates
*Metallothioneins**	Binding of heavy metals	*Acetylcholinesterase activity** (AChE)*	Neurotoxicity
Glutathione	Binding of heavy metals	*Vitellogenin*	Reproductive impairment
Lipofuscin and other granules	Metal biomineralisation	*DNA adducts*	Genotoxicity
Enzymatic antioxidant defences (superoxidases, peroxidases, catalases)	Oxidative stress	*Micronucleus formation*	Genotoxicity
Non-enzymatic antioxidant defences (glutathione)	Oxidative Stress		
Heat shock proteins	Protein damage	*Lysosomal membrane stability**	Cellular damage
Ethoxyresorufin O-deethylase (EROD)	Biotransformation of organic pollutants		
*Fluorescent aromatic compounds in bile**	Biotransformation of organic pollutants		
Phase II enzymes (glutathione S-transferase, UDP-glucuronosyl-transferases, sulfotransferases)	Biotransformation of organic pollutants		
Phase III enzymes (P-glycoproteins and multidrug–resistance proteins)	Transport of the by-products of transformation out of the cell/body		

their exposure to the toxicants. Fish avoid metals and PAHs at environmentally relevant concentrations. However, if exposed to pollutants (e.g. metals, PAHs, PCBs, pesticides and pharmaceuticals), sensory interference can occur causing changes in swimming speed affecting schooling behaviour, feeding and predator avoidance. Changes in locomotion have also been observed for reptiles and crustacea. Feeding is also impaired, from both sensory and locomotor effects, in polychaete worms, fish, insects, crustaceans, snails and bivalves. Infaunal burrowing (in crustaceans, bivalves and annelids) is adversely impacted by hydrocarbons, pesticides and metals. Reproductive and migratory behaviour can change when exposed to metals and organic contaminants (seen in fish). Similarly, they may interfere with chemosignalling during reproduction (e.g. fish and algae).

2.4.4.1.2 Metabolic biomarkers

Measurements of metabolic rates can indicate the health status of an organism and therefore its potential for growth and reproduction. There are a number of biomarkers for measuring the energetic

status of an organism which can be done at the molecular, cellular and organismal level. Assays of the static energy reserves in the form of glycogen or lipids can indicate the chemical energy available within the body. Similarly, the adenylate energy charge indicates the chemical energy available as adenosine triphosphate (ATP). The cellular energy allocation approach uses information on the energy available and its consumption and so reflects the net cellular energy budget. In this section we will consider metabolic impairment at the whole-organism level using an approach termed the *scope for growth*.

The Scope for Growth (SfG) biomarker is a method that quantifies the energy available for the growth of an organism. As identified earlier, the detoxification of pollutants requires energy (Section 2.3.2) therefore it is no longer available for important processes such as growth and reproduction. Increased energy consumption follows exposure to stress, which could be in the form of a pollutant. The SfG approach considers the amount of energy acquired through food minus that required for basal metabolism (i.e. respiration) and the amount excreted. The remaining energy represents the amount available for growth and reproduction.

The SfG is an instantaneous measure of an organisms energy status: positive values indicate optimal conditions and negative values indicate stress as energy reserves are used up in additional activities such as detoxification and repairing the damage caused by toxicants. Animals with a lower SfG are therefore in poorer health and those with higher SfG have greater reserves for growth and reproduction. The approach was originally developed with organisms transplanted into contaminated environments (e.g. active biomonitoring; Section 2.4.2), or those suspected to be contaminated, which were compared with reference sites (Figure 2.5). The SfG approach was developed with bivalves because they are valuable sentinels that are relatively easy to manipulate (Section 2.4.3). Studies show that the SfG of aquatic animals are negatively impacted by PCBs, DDT, PAHs, other organochlorines (e.g. dieldrin, gamma-HCH) and TBT.

A survey of 38 east coast United Kingdom sites found large differences in the SfG of the common mussel *M. edulis*. At these sites mussels were exposed to a range of different pollutants (PAHs, TBT, DDT, dieldrin, gamma-HCH, PCNS, cadmium, lead, selenium and arsenic were found in mussel tissues). The SfG of mussels was approximately two- to four-fold lower on both sides of the Irish Sea compared with clean reference sites. The Irish Sea mussel populations had poorer recruitment, smaller maximum size, and three- to ten-fold slower growth compared with sites to the south and north, respectively. Determinations of contaminant tissue concentrations showed that most of these differences were attributable to PAHs from the combustion of oil and accidental spillages (Figure 2.5). It was also shown that sewage impacted mussel SfG at these sites (Figure 2.5). The link with PAH concentrations was also shown for mussels from the west coast of the United Kingdom in an earlier study.

The SfG method is a fast, sensitive and integrated biomarker and, if used in conjunction with measurements of pollutant tissue residues, it can be used to determine potential stressors. As for other biomarkers, SfG is affected by natural environmental and biological stressors, food and oxygen supply, water temperature, body size/age and reproductive state, as well as pollutants. However, biomarkers such as SfG that indicate the health of an organism, specifically its energetics, provide the advantage of contributing to predictions of population-level effects in a way that biochemical biomarkers cannot at present (Figure 2.5).

In addition to SfG, assessments of the health of an organism can be determined using body condition indices. Such indices are used regularly to assess the health of invertebrates and vertebrates, and the indices include the mass of organs such as the liver or gonads as a proportion of the total body mass. Organisms from polluted environments are usually of poorer condition because of the toxic effects themselves, or because of the diversion of energy towards detoxification. Body condition indices are used for fish, molluscs and mammals. Indices of size/weight have been developed for *H. diversicolor* that can indicate the effects of toxicants in estuarine sediments, which usually contain substantial quantities of legacy contaminants, on the body condition of infaunal taxa.

2.4.4.1.3 Molecular biomarkers

Nine main molecular biomarkers of exposure are used in ecotoxicological assessments, and two of these have been adopted by the European Framework as part of its suite of four core biomarkers for determining the exposure of aquatic organisms to pollutants (Table 2.6). The first of these metallothionein (MT) is a biomarker of metal exposure that has been verified in many field studies. Natural variations in MT induction can occur due to biological features that affect metal uptake, for example, body size, age, gender or other ontogenetic factors such as the moult cycle of crustaceans. Similarly, environmental stress factors such as low temperatures, extreme salinity, anoxia, starvation, handling in the lab, the presence of antibiotics or herbicides can cause low-level MT inductions that obscure the 'signal' from heavy metal exposure. These natural variations, which are a challenge to all toxicity tests regardless of methodology, produce a degree of background 'noise' from which the toxicity of any pollutants need to be distinguished. Consequently the choice of taxa, sample region (i.e. which bodily organ) and the methodology of determination need to be consistent.

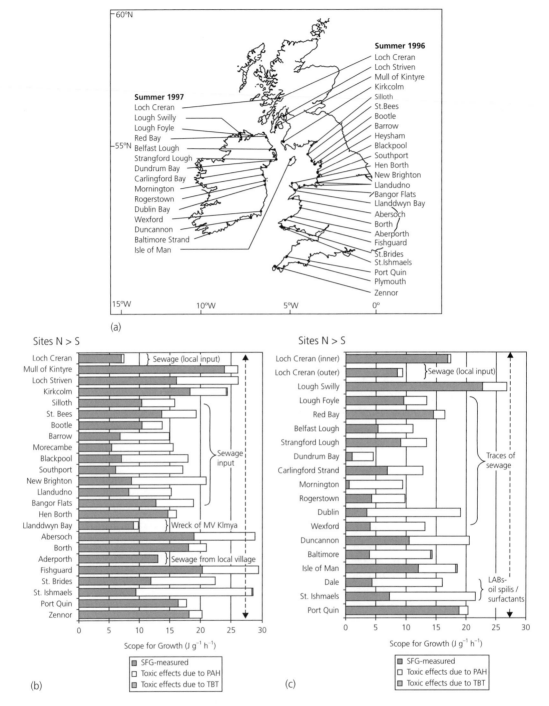

Figure 2.5 The scope for growth (SfG) of *Mytilus edulis*, of 4 cm length, collected across the shore during low water neap tides from 38 sites around the east coast of the UK. (a) Sites sampled. (b) SfG of mussels from the eastern Irish Sea and (c) east and west Irish Sea partitioned between the two main contaminants found in mussel tissue (the partitioning of SfG is based on these tissue residues and the established effects of selected toxicants on *M. edulis* SfG). In (b) to (c) known pollutant inputs are shown on the right. Broken line indicates maximum potential growth under 'normal' conditions using standardised methodology. The 'clean' reference sites are at Loch Creran, Lough Striven, Lough Swilly, Lough Foyle, Port Quinn and Zennor. LABs = linear alkylbenzenes, PAH = polyaromatic hydrocarbons, and TBT = tributyltin. Reprinted from *Marine Environmental Research*, 53, J. Widdows, P. Donkin, F.J. Staff, P. Matthiessen, R.J. Law, Y.T. Allen, J.E. Thain, C.R. Allchin and B.R. Jones, Measurement of stress effects (scope for growth) and contaminant levels in mussels (*Mytilus edulis*) collected from the Irish Sea, 327–56, Copyright (2002), with permission from Elsevier.

The antioxidant enzymes, the superoxidases, peroxidases and catalases, produced in cells under conditions of oxidative stress are employed as non-specific biomarkers using the *total oxyradical scavenging capacity method* (Table 2.6). The production of reactive oxygen species may vary seasonally due to changes in the environment and the availability of food. Non-enzymatic antioxidant defences such as the production of the tripeptide glutathione are also used as a biomarker of oxidative stress (Section 2.3.1).

Heat shock or stress proteins have been sequenced from many vertebrates, invertebrates and plants and their concentration is taken as an indicator of physiological state (Section 2.3.1.4). Stress proteins are indicative of damage to proteins and are produced continuously during the stress making them good biomarkers. Many different types of stress protein are produced but variability in the response between contaminants, species and individuals limits their use as a biomarker.

The enzymes used in biotransformation processes (Section 2.3.1) are utilised as biomarkers for the breakdown of organic pollutants such as pesticides, hydrocarbons and PCBs. EROD activity and the presence of aromatic fluorescent compounds in bile are used as biomarkers of biotransformation in fish. Glutathione-*S*-transferase (GST) measurement is used successfully as a biomarker in molluscs. While GST has been shown to be present in fish, contradictory patterns of induction have been found and so it is not yet considered a robust biomarker for fish. The phase III biotransformation enzymes (multidrug–resistance proteins and P-glycoproteins) are biomarkers of organic contamination in mussels, oysters and gastropods. These phase III enzymes may be particularly valuable biomarkers for the exposure of early life stages.

2.4.4.2 Biomarkers of effect

2.4.4.2.1 Physiological biomarkers

Physiological biomarkers of toxicological effects may include the feminisation of embryos (e.g. intersex; Section 5.3), abnormal growth of reproductive organs (e.g. imposex), or the thinning of bird eggshells. The latter is induced in various species of birds in the presence of DDT, its degradation product dichlorodiphenyldichloroethylene (DDE), and the organochlorine pesticide dicofol that is chemically similar to DDT. This phenomenon was first observed in peregrine falcon eggs after exposure to DDT. When peregrine eggshells are more than 17% thinner than usual they break more easily and thus reproductive, and subsequently population, failure occurs. Eggshell thinning of up to 50% has been observed for brown pelicans exposed to DDE. The mechanism of toxicity is linked with interference in calcium transportation. Peregrine falcons had completely disappeared from the United States and Canada by the 1960s as a result of eggshell thinning after DDT exposure. The birds were reintroduced from Europe 12 years later and populations have since seen an increase. Bald eagle and osprey populations were also impacted in the United States and the sparrowhawk was impacted in Europe. The bald eagle, America's national bird, experienced near extinction due to DDT use. Many other birds of prey have been affected, with raptors and fish eaters being the most sensitive because both DDT and DDE bioaccumulate (Section 2.3.3) in their prey. The effects of DDT application were widespread as it became widely distributed throughout the environment (Section 3.4). The eggshell biomarker is specific to DDT and chemically similar pesticides.

2.4.4.2.2 Biomarkers of tissue/cellular damage

Biomarkers of histocytological changes, in tissues and cells, include disease and damage that is externally visible, internal pathologies and those associated with the gills. The most commonly used histocytological biomarkers are:

- Externally visible signs of disease: fin erosion, vertebrate skeletal malformation, abnormalities of fish operculum (the flap of skin covering the gills) and epidermal hyperplasia (excessive cell growth: a precursor to tumour formation).
- Internal signs of disease: liver, gill and kidney histopathology, degeneration of oocytes, embryo damage and macrophage aggregates (produced in fish as part of an immune and detoxification response).
- Effects at the cellular level: integrity of the lysosome, the concentration of intracellular lipopigments in the liver and peroxisome proliferation.

The internal and cellular histocytological bio-markers are more specific, but the most ecologically relevant are the internal pathologies. Cell pathology has the advantage of providing an early warning of the impacts of pollutants, especially those that are the precursors of disease, and external pathologies are the simplest to measure. These biomarkers are frequently used with fish and hepatic lesions and macrophage aggregates have been used in a number of established monitoring programmes in the United States (by the National Oceanic and Atmospheric Administration, State Fisheries services and US national estuary programme). Pollution monitoring of the North Sea by German scientists has used lysosomal integrity and liver tumours, both recommended biomarkers by ICES.

The histocytology of mollusc digestive glands has also been used as a biomarker, although less frequently than in fish. Histological abnormalities were observed in oysters and soft-shell clams after the *Amoco Cadiz* oil spill in Brittany (Section 4.5). Liver damage, necrosis and tumour formation have been found in benthic but not pelagic fish. In the Potomac River, 50–60% of fish had liver tumours in areas receiving discharges containing PAHs compared with 10% of fish at the reference site. In the same river, 10–37% of fish had skin tumours. Mass mortalities of clams, *Chlamys varia*, in Narragansett Bay were recorded after exposure to sub-lethal levels of silver and these stopped the production of byssal threads, meaning that the mussels could not reattach to their substrate. Studies with *Mytilus edulis* have also found disruption of byssus production by toxicants.

2.4.4.2.3 Molecular biomarkers

Acetylcholinesterase (AChE), the enzyme that breaks down acetylcholine (a nerve transmitter) in animal neurones, is a biomarker for exposure to neurotoxins. For example, the organophosphate nerve gases used in warfare have been seen to induce AChE in human neurones. AChE induction has been shown to occur in animals after exposure to organophosphate pesticides, heavy metals, detergents and some algal toxins and can indicate changes in environmental salinity and physiological condition.

Vitellogenin, the precursor to egg yolk formation, is used as a biomarker of endocrine disruption (Section 5.3). Vitellogenin occurs in the blood and haemolymph of female vertebrates and invertebrates and is produced throughout life. Within marine taxa it is a very sensitive indicator of the exposure to endocrine-disrupting pollutants. Vitellogenin induction has been identified in the blood of male fish in response to a range of endocrine-disrupting compounds that mimic oestrogen (e.g. PCBs, natural phytoestrogens and TBT). These biomarker effects are often observed together with features such as the development of oocytes in the testis of male fish. Presently vitellogenin is used as an indicator in fish and some molluscs, and it is measured directly from the blood plasma using an enzyme-linked immunosorbent assay. This is often done in conjunction with histological examination of the gonads to identify abnormalities.

Animals exposed to pollution have weakened lysosomal membranes in the liver and/or digestive tract. Because lysosomes play a role in accumulating and detoxifying a range of pollutants (Section 2.3.1)—for example, heavy metals, PAHs and the heterocyclic compounds found in many pesticides—they are susceptible to damage associated with this exposure and so are also a biomarker of general stress. Lysosomal membrane instability is indicated by the activation of hydrolytic enzymes and assessment of their concentrations has been developed as a fast and sensitive indicator for stress and the effects of pollutants on lysosomal membrane stability (Table 2.6). Lysosomes occur in all cells of all taxa but are more abundant in specific regions. In mussels, cells in the epithelium of the digestive gland and the granulocytes in the haemolymph are lysosome-rich. Either cell type can be used to assess lysosomal membrane stability in mussels and can detect exposure to micromolar amounts of copper, the pesticide Aroclor 1254 and dimethylbenzoanthracene. Standard methods are provided by the ICES for determining lysosomal enzyme activity (N-acetyl-ß-hexosaminidase and ß-glucuronidase) for *Mytilus edulis, M. galloprovincialis, M. trossulus* and *Perna viridis*. Bivalves have low multi-function oxidase activity (Section 2.3.1) and are particularly sensitive to damage from organic pollutants because they accumulate them within their lysosomes. In flounder, lysosomal membrane stability is affected by exposure to DDT and PCBs. The lysosomal membrane stability approach is very sensitive to low concentrations of organic pollutants and metals.

The failure of antioxidant defence systems to eliminate reactive oxygen species causes lipid peroxidation which damages cell membranes, and the breakdown products of the final stages of lipid peroxidation can be used as biomarkers of damage. The most important of these breakdown products malondial-dehyde can be detected using thiobarbituric acid (Table 2.6). Lipid peroxidation can be affected by environmental factors such changes in temperature and salinity in some bivalves and fish and can also indicate that exposure to PCBs and heavy metals has occurred.

DNA damage is a biomarker of *genotoxicity* (causing mutations of genetic material). Oxidative stress, as mentioned previously, can damage DNA molecules, interfering with regular cell functions, and can cause DNA mutations that lead to cancers. The biomarkers used to identify DNA damage in aquatic taxa include the products of the oxidation of the DNA bases guanine, adenine and thymine.

2.4.5 Population- and community-scale ecotoxicology

As we have seen so far in this chapter, most toxicity data are collected from laboratory tests with one or two species. Of course, in nature it is very rare for one species to exist in isolation. The ultimate aim of ecotoxicology is to understand how the release of contaminants will impact populations, communities and ecosystems. However, the ecosystem effects can be unpredictable and a change in one species may have a relatively small effect, on the other members of the community, or it may be very large and perhaps disproportionate to the pollution impacts. We do not yet understand ecosystem and species dynamics well enough to anticipate these effects with high confidence. So, if we are to get better at understanding the effects of pollutants above the level of the individual we need to study the

impacts on multiple species (sometimes referred to as test batteries).

The types of impacts pollutants have on ecosystems include changing the structure of communities by altering the species composition or population demographics and changing the genetic composition by selecting for tolerance. Also, pollutants may interfere with ecosystem processes or functions, such as bioturbation through the loss of sediment-dwelling species or by causing shifts in species behaviour. The interactions between and within species may change and/or cause their local disappearance. When a species is lost, its ecological role or position (or *niche*, for example its habits) becomes vacant and another species may take its place. This could be a direct substitution of one species for another, a species in the community may fill the vacant niche, or several species may compete to occupy the vacancy. These changes could result in changes in the food web or other ecosystem processes (e.g. bioturbation or the provision of biogenic habitat). For example, the loss of a phytoplankton species may cause the collapse of the zooplankton communities that feed upon it and this effect may cascade up the food web to affect higher consumers as well. The importance of multi-species exposures is recognised in European legislation which requires that species from three different trophic levels are tested, but this remains an oversimplification of community or ecosystem effects. The importance of including multiple trophic levels is also reflected in calculations of the PNEC (Section 2.4.2). Similarly, most biomarker studies use only individual biomarkers but, considering their generality, battery approaches are also desirable for these studies.

In semi-natural settings, artificial mesocosms such as ponds or enclosures within estuaries or coastal waters can be used to investigate the effects of pollutants on populations and communities. Enclosures represent a more natural state than ponds. Smaller-scale studies are referred to as microcosms. These are typically laboratory studies, of limited volume, and so also complexity, and test duration compared with mesocosms. Although there is no standard size delineation, one categorisation for microcosms is greater than 15 m^3 and for mesocosms

Table 2.7 Examples of marine mesocosm and projects, their specifics and the purposes they are used for. Data from: Grice and Reeve (2012).

Project	Details	Purpose
Marine Ecosystem Research Laboratory (MERL), University of Rhode Island	14 land based tanks of 7 m³ volume (1.8 m diam. × 5.4 m depth). Filled with 30 cm sediment and water from Narragansett Bay; flow through or static with mixing.	Exposure to hydrocarbons and metals.
Controlled Ecosystem Populations Experiment (CEPEX), British Columbia	Bags moored in Saanich Inlet. Two sizes: volume 68 m³ (2.4 m diam. × 16.1 m length), and volume 1300 m³ (9.5 m diam. × 23.5 m depth). Nylon reinforced polyethylene.	Toxicity studies with plankton and pollutants (metals, hydrocarbons).
Loch Ewe Enclosures, north-west Scotland	Cylindrical tubes open at the top, tapering to a cone. Volume 100 m³ (3 m diam. × 17 m depth) or 300 m³ (4.7 m diam × 17 m depth). Suspended from flotation collars, moored to seabed. Vinyl reinforced with polyester.	Pollution experiments.
Den Helder enclosures, The Netherlands	Anchored in the harbour. Two sizes: 1.5 m³ volume (0.75 m diam. × 3.5 m depth), and 16 m³ volume (0.75 diam. × 20 m depth). Plastic bags attached to flotation buoys.	Response of plankton to pollutants.

15 m³ to about 1000 m³. A review of some of the different marine mesocosm projects used for ecotoxicological studies on planktonic and sediment dwelling organisms is provided in Table 2.7. Mesocosms can be used to verify the established responses and relationships found in laboratory investigations.

Enclosures represent a more natural state than ponds. The advantages of mesocosms are that they can bridge some of the discrepancies between laboratory and field toxicological studies. These artificial communities of organisms allow for replication in a way that is almost impossible when sampling in the natural environment. However, although the data from mesocosm studies can improve on laboratory toxicity data, mesocosms are still models for changes in the natural environment and cannot incorporate all of the natural variability.

A range of different types of mesocosms have been used for long-term aquatic research and while they may provide more realistic results than laboratory studies, challenges remain. Detecting change at higher levels of organisation (Figure 2.4) requires greater temporal scales than those required to establish the effects at the individual level. Population change simply takes longer to become apparent than individual effects, particularly at higher trophic levels where taxa typically live for longer. Thus, the use of mesocosms must consider the life span and life cycle of the species involved. There should also

be time for species interactions such as changes in the trophic structure to manifest. In order to maximise the ecological relevance, mesocosms must incorporate multiple species, functional groups and habitats. The indirect effects of pollutants at higher organisational levels (e.g. competition or predator–prey interactions) and the recovery from exposure may be determined using mesocosms. Mesocosm experiments can incorporate the physicochemical variables that influence the availability of contaminants through binding to organic matter or sequestration by plants or chemical degradation (Section 2.2.1).

Another approach that helps to improve understanding of the impacts on multiple species is to use biomonitoring (Section 2.4.2). Biomonitoring describe semi-realistic exposures but usually only for a single species; although other species might have interacted with the bioindicators species in the field, the impacts on these other species are not considered. However, a number of community-based biomonitoring approaches have been developed. Examples of some of those used to-date include:

• The Pollution-Induced Community Tolerance (PICT) method that measures the tolerance of communities to pollutants. Tolerant communities are interpreted to have undergone prior exposure in a similar manner to biomarkers of exposure. This approach has mostly been used for microbial communities and algal biofilm communities which are easy to manipulate.

- A number of ratios have been developed as indicators of benthic disturbance that may be derived from benthic community data and these include the relative abundance of: opportunist polychaete/ amphipod and polychaete/copepod ratios. The basis of these ratios is that crustaceans are sensitive to organic enrichment, metal and hydrocarbon pollution whereas polychaetes tend to be tolerant. These ratios have been used to detect organic enrichment associated with oil pollution throughout Europe. Similar ratios, nematode/copepod, have been developed for meiofaunal communities. A number of more complex indices exist which use the proportions of sensitive and tolerant species within a benthic community (e.g. the AMBI).
- Biotic indices such as AMBI (the AZTI Marine Biotic Index) are used to assess the health of communities inhabiting soft substrates. Species are first classified based on their tolerance to disturbance and the index is then applied to field-derived community composition data in an area of suspected contaminated or along a gradient. The approach has been used throughout Europe successfully to detect contamination from submarine outfalls, drill cuttings, mining wastes, the development of coastal structures and sewage discharges. Although the AMBI seems to be highly sensitive to disturbance, the power of AMBI can decrease when assessing very low diversity communities.

2.5 Synthesis

- The majority of aquatic toxicity data have traditionally been derived from the exposure of model species to individual contaminants over short periods in laboratory settings. In other words, ecotoxicology has for the most part taken the path of least resistance. This vast body of work has provided data and a structure for assessing the potential impacts of contaminant discharges (through consents to discharge and environmental impact assessments; Chapter 7), and it has informed the regulation of hazardous chemicals through the registration of new compounds (although see Chapter 5 for discussion of emerging

pollutants which do not have suitable regulations in place). The key challenge for *eco*toxicology is establishing the effects that pollutants will have in the 'real' environment but, as we have seen, many different abiotic and biotic factors influence the exposure of organisms to pollutants (Section 2.2) and their toxicity (Section 2.3).

- Biomarkers provide information on the health status of an organism, and they are sensitive indicators of exposure to pollutants. However, it is difficult to establish the relevance of biomarker results at the molecular level upon populations of organisms (Figure 2.4). To be ecologically useful the individual-based biomarkers must be understood within the context of the whole organism, its reproduction and the transference of inherited traits. All biomarkers provide early warnings but only those which give information on the reproductive processes and its likelihood of success are really useful for understanding population level change.
- Although the majority of biomarkers are non-specific and lack relevance, they can provide the sum of pollutant effects and so are useful tools for indicating exposure. Although the problems associated with unknown exposure history and differential tolerance remain. These challenges can be overcome using test organisms with known exposure history/tolerance. Practicalities mean models have to be used, but these models should be environmentally relevant to the exposure region of interest, and they should incorporate multiple species that occupy different ecosystem roles. Species arrays or batteries can be used to anticipate the effects at higher organisational levels. If species responses are understood then biomonitoring may be used.
- Differences in species biology, behaviour and detoxification mechanism mean that species tolerance is variable (Section 2.3.2). Also, prior exposure to pollutants influences a population's ability to tolerate them because of acclimatisation and genetic adaptation within a species. Tolerance may lead to false positive results in the field and false negative results in the laboratory if using tolerant strains.
- The differences in tolerance between populations mean that in order to determine toxicity we need to know the biology and pollutant exposure

history of the test subjects. All of the different approaches for assessing toxicity addressed in this chapter share the challenge of minimising biotic and abiotic variation while trying to maintain ecological relevance.

- There are many factors governing the toxicity of the contaminants present in our environment to individual organisms, and as we progress up the levels of organisation from individual to species to communities and ecosystems, our predictive power, based on our current knowledge of toxicity, decreases. Because ecotoxicology has traditionally focused on single-species testing under laboratory conditions that do not represent real environments. Our understanding of ecotoxicology is currently limited by: information from too few species, a lack of diversity in the trophic position and the life habits of these species, a focus on individual not multi-species systems, knowledge of the scientific links between molecular changes and organism effects, insufficient consideration of tolerance, exclusion of contaminant interactions, and insufficient testing using sediments. These limitations combine to reduce the environmental relevance of ecotoxicity data. Thus, it becomes necessary to extrapolate from the available data to anticipate the impacts on whole ecosystems and there is much uncertainty in such extrapolations. Multi-species assessments such as those conducted using mesocosms are important because they are more ecologically relevant, provide better replication than can be achieved through biomonitoring, and can help to improve forecasts and, hopefully in turn, the evidence base for regulation.

- Ecological systems are very complex and our understanding of their dynamics is still limited, as is our ability to understand and predict the impacts of anthropogenic disturbances such as pollution. However, pollution can affect biodiversity, biogenic habitat formation, the structuring of trophic systems and natural biogeochemical cycling and primary production through its effects on sediment-dwelling fauna and flora. Ultimately these changes in ecosystem qualities and their functioning affect our use of ecosystem services from the sea. As our climate is rapidly changing,

the potential arises for the synergistic effects of multiple stressors to enhance the effects of pollutants on marine organisms, communities and ecosystems. The potential for synergy between their combined effects (e.g. increased temperature, reduced oxygen supply, changing ocean acidity and patterns of primary production) and those of the cocktail of thousands of contaminants make understanding the effects of pollutants on marine ecosystems more important than ever before. Many species are being pushed towards the furthest extent of their natural geographic ranges and at these suboptimal conditions, contaminants may become pollutants.

Resources

Pesticide Action Network of North America (2014). PAN Pesticides Database—Chemical toxicity studies on aquatic organisms. http://www.pesticideinfo.org.

European Chemicals Agency (2016). Registered Substances Information. http://echa.europa.eu/web/guest/information-on-chemicals/registered-substances.

European Center for Ecotoxicology and Toxicology of Chemicals. 2016. ECETOC Aquatic Toxicity Database. http://www.ecetoc.org.

United States Environmental Protection Agency (2016). Ecotoxicology Database. https://cfpub.epa.gov/ecotox/.

UNESCO (2016). Cornwall and West Devon Mining Landscape. http://whc.unesco.org/en/list/1215.

Bibliography

Adams, S.M., Shepard, K.L., Greeley Jr, M.S., Jimenez, B.D., Ryon, M.G., Shugart, L.R. and McCarthy, J.F. (1989). The use of bioindicators for assessing the effects of pollutant stress on fish. *Marine Environmental Research*, 28, 459–64.

Addison, R.F. (1996). The use of biological effects monitoring in studies of marine pollution. *Environmental Reviews*, 4, 225–37.

Amiard-Triquet, C., Amiard, J.-C. and Rainbow, R.S. (2012). *Ecological Biomarkers: Indicators of Toxic Effects*. New York, NY: CRC Press.

Amiard-Triquet, C., Rainbow, P. S. and Roméo, M. (2011). *Tolerance to Environmental Contaminants*. New York, NY: CRC Press.

AZTI tecnalia (2016). *AMBI: AZTI Marine Biotic Index*. http://www.azti.es/ambi-azti-marine-biotic-index/.

Bopp, S.K. and Lettieri, T. (2007). Gene regulation in the marine diatom *Thalassiosira pseudonana* upon exposure to polycyclic aromatic hydrocarbons (PAHs). *Gene*, 396, 293–302.

Braune, B.M. and Scheuhammer, A.M. (2008). Trace element and metallothionein concentrations in seabirds from the Canadian Arctic. *Environmental Toxicology and Chemistry*, 27, 645–51.

Bryan, G. W. and Gibbs, P. E. (1983). Heavy metals in the Fal estuary, Cornwall: A study of long-term contamination by mining waste and its effects on estuarine organisms. Occassional publication of the Marine Biological Association of the UK, no. 2, 1–112.

Bryan, G.W., Gibbs, P.E., Hummerstone, L.G. and Burt, G.R. (1987). Copper, zinc and organotin as long-term factors governing the distribution of organisms in the Fal estuary in southwest England. *Estuaries*, 10, 208–19.

Depledge, M.H., and Fossi, M.C (1994). The role of biomarkers in environmental assessment (2). Invertebrates. *Ecotoxicology*, 3, 161–72.

Goldberg, E.D. (1975). The Mussel Watch—A first step in global marine monitoring. *Marine Pollution Bulletin*, 6, 111–111.

Grice, G.D. and Reeve, M.R. (2012). *Marine Mesocosms: Biological and Chemical Research in Experimental Ecosystems*. Dordrecht: Springer Science and Business Media.

Gray, J.S. and Ventilla, R.J. (1973). Growth Rates of Sediment-Living Marine Protozoan as a Toxicity Indicator for Heavy Metals. *Ambio*, 2, 118–21.

Isani, G. and Carpene, E. (2014). Metallothioneins, unconventional proteins from unconventional animals: a long journey from nematodes to mammals. *Biomolecules*, 4, 435–57.

Langston, W.J. and Bebiano, M.J. (2013). *Metal Metabolism in Aquatic Environments*. Dordrecht: Springer Science and Business Media.

Millward, R.N. and Grant, A. (2000). Pollution-induced tolerance to copper of nematode communities in the severely contaminated Restronguet Creek and adjacent estuaries, Cornwall, United Kingdom. *Environmental Toxicology and Chemistry*, 19, 454–61.

Mouneyrac, C., Mastain, O., Amiard, J.C., Amiard-Triquet, C., Beaunier, P. Jeantet, A.-Y., Smith, B.D. and Rainbow, P.S. (2012). Trace-metal detoxification and tolerance of the estuarine worm *Hediste diversicolor* chronically exposed in their environment. *Marine Biology*, 143, 731–44.

Pook, C., Lewis, C. and Galloway, T. (2009). The metabolic and fitness costs associated with metal resistance in *Nereis diversicolor*. *Marine Pollution Bulletin*, 58, 1063–71.

Rainbow, P.S., Amiard, J.-C., Amiard-Triquet, C., Cheung, M.-S., Zhang, L., Zhong, H. and Wang, W.-X. (2007). Trophic transfer of trace metals: subcellular compartmentalization in bivalve prey, assimilation by a gastropod predator and *in vitro* digestion simulations. *Marine Ecology Progress Series*, 348, 125e138.

Rainbow, P.S., Luoma, S.N. and Wen-Xiong, W. (2011). Trophically available metal: a veritable feast. *Environmental Pollution*, 159, 2347–349.

Rainbow, P.S., Smith, B.D. and Luoma, S.N. (2009). Differences in trace metal bioaccumulation kinetics among populations of the polychaete *Nereis diversicolor* from metal-contaminated estuaries. *Marine Ecology Progress Series*, 376, 173–84.

Vieira, L.R. and Guilhermino, L. (2012). Multiple stress effects on marine planktonic organisms: Influence of temperature on the toxicity of polycyclic aromatic hydrocarbons to *Tetraselmis chuii*. *Journal of Sea Research*, 72, 94–8.

Walker, C.H. (2014). *Ecotoxicology: Effects of Pollutants on the Natural Environment*. New York, NY: CRC Press.

Wang, X. Yan, Z. and Liu, Z. (2014). Comparison of species sensitivity distribution for species from China and the USA. *Environmental Science and Pollution Research*, 21, 168–76.

Widdows, J., Donkin, P., Staff, F.J., Matthiessen, P., Law, R.J., Allen, Y.T., Thain, J. E., Allchin, C.R. and Jones, B.R. (2002). Measurement of stress effects (scope for growth) and contaminant levels in mussels (*Mytilus edulis*) collected from the Irish Sea. *Marine Environmental Research*, 53, 327–56.

CHAPTER 3

'Solved' problems?

3.1 Introduction

Mummies from ancient Egypt contain almost the same level of inhaled particulates as modern humans, implying significant exposure to poor levels of air quality (i.e. breathing in polluted air). In ancient Egypt this would have included smoke from cooking fires and metalworking and dust from mining and sand storms. Pollution has thus been affecting human health since the dawn of civilisation and while there is no evidence of laws focused on combatting pollution from ancient Egypt, in both the Roman Empire and medieval Europe laws were enacted to control the effects of pollution on the populace. It is therefore not unreasonable to postulate that 2000 years later those particular pollution problems would be solved.

A theory has been proposed that the decline and fall of the Roman Empire was precipitated by widespread lead poisoning. The ancient Mediterranean civilisations, for example, Greek, Roman and Byzantine, certainly used lead at an industrial scale. In fact, ice cores from the Greenland ice sheet show that atmospheric levels of lead (from lead and silver mining and smelting) in the period 500 BC to 300 AD (2500–1700 years before the present) were up to four times higher than the natural background levels. Chemical analyses of lead residues in sediments from the Roman period suggest that lead concentrations in drinking water may have exceeded modern standards of acceptability by as much as 100 times. The theory holds that lead exposure arising from the use of lead water-pipes and lead eating and drinking vessels resulted in widespread mental health problems and rising crime rates and that this undermined effective governance. This idea, while capturing the popular imagination, seems unlikely as the reason for the slow decline and eventual fragmentation of the Empire. Furthermore, documentary evidence from the period shows that Romans were aware of the problem of lead poisoning and took steps to try to avoid it; for example, by using terracotta (earthenware) pipes for drinking water. This is probably the first documented recognition of a water pollution issue and metal poisoning has remained a pollution issue well into the twenty-first century.

The inhabitants of Minimata Bay in Kyushu, Japan, traditionally made their living from fishing and collecting shellfish. In 1952, a new factory opened in the bay that manufactured vinyl chloride and acetylaldehyde. The production processes used large quantities of mercury that were discharged in a liquid effluent to the bay. The effluent contained inorganic mercury and methylmercury (around 5% was as methylmercury). Once in the bay, bacteria in the sediments converted much of the inorganic mercury to methylmercury. As early as 1953, the local fishing families began showing signs of a mystery illness that became named Minimata disease. The symptoms included neurological problems, shaking, fits and large increases in birth defects. In 1957 fishing was banned in Minimata Bay and the disease quickly declined, but by then 43 people were dead and over 700 were left with permanent disabilities. It took until 1959 to demonstrate that mercury was the cause of Minimata disease and until 1960 for the factory to be proven to be the source and to be subject to regulation. By 1959, sediments in the bay contained 200 ppm of mercury, bivalve shellfish 10–39 ppm and fish 10–55 ppm. While the Japanese government officially concluded that the company was

Marine Pollution. Christopher L. J. Frid & Bryony A. Caswell.
© Christopher L. J. Frid & Bryony A. Caswell 2017. Published 2017 by Oxford University Press.
DOI 10.1093/oso/9780198726289.001.0001

liable in 1968 (12 years after the discovery of the disease), it was not until 1973 that the first court settlement resulted in compensation for certified victims. Since then, the battle has continued to certify individuals affected as victims and to secure compensation for uncertified victims. As recently as 29 March 2010, a further 2123 uncertified victims reached an agreement and received compensation from the Chisso Corporation who owned the factory.

Thus, in antiquity the impact of lead poisoning was recognised, in the Middle Ages the health and environmental impacts of sewage were identified and regulations concerning them imposed (see Section 1.4), while in the 1960s the publication of Carson's *Silent Spring* drew attention to the effects of modern agriculture and, in particular, the impacts of organohalide pesticides (see Section 1.4). Growing environmental awareness in the late twentieth century enabled major efforts to control these classes of emissions along with other obvious industrial pollutants; for example, heat, noise and acidity. By the start of the twenty-first century, economically developed nations had legal frameworks in place and mechanisms for controlling metals and other industrial wastes (see Section 3.3), sewage and wastewater (Section 3.2) and highly toxic synthetic compounds (Section 3.4). For the purposes of this book we therefore characterise these as 'solved' problems. Society no longer expects to see widespread poisoning of people from their food (e.g. Minimata), the loss of key fish or shellfish resources due to toxins (e.g. oyster farms wiped out by tributyltin (TBT)) or the widespread death of a species (e.g. sewage-induced eel grass wasting disease). The science of the effects of these pollutants is well known, the control/treatment technology is available and in most places the regulatory framework (Chapter 7) imposes controls designed to limit the impacts to acceptable levels.

3.2 Oxygen-demanding wastes

3.2.1 Sewage: A historical perspective

Humans, like all animals, produce waste in the form of urine and faeces. While some pollution problems can be dealt with by removing the problem at its source—for example, the outlawing of ozone-depleting aerosols in the 1980s—humans will always produce bodily wastes, and the more humans there are the more waste there will be. However, early humans were nomadic hunters and these wastes were simply deposited in the environment over the wide areas they roamed. With the advent of agriculture and settled human populations, wastes were produced and accumulated at settlements and a disposal solution was required. Initially this consisted of depositing the wastes in the surrounding area, and from this developed the practice of using animal and human wastes to boost crop fertility. This practice created new opportunities for pathogens and parasites to be recycled from human wastes back into the human food chain. As small agricultural villages grew into market towns and cities, the populations became more and more remote from unsettled areas. This created practical difficulties for moving wastes from source to field, and the quantities being produced began to exceed the ability of the nearby agricultural systems to assimilate them. This triggered two further changes: surface watercourses outside towns and cities started to become contaminated with human and animal wastes running off the saturated soils. In the towns and cities, cesspits (essentially deep holes in the ground, often dug in the garden or back yard for convenience) were developed as a means of waste disposal without the necessity for long-distance waste transport. However, the cesspits also leached waste into the groundwater and watercourses. Thus, from a fairly early stage of social development, the disposal of human wastes became a challenge, with drinking-water sources (wells and rivers) becoming contaminated, and parasites and pathogens being recycled into populations via crops contaminated by the use of the waste as fertiliser. Unfortunately, the use of cesspits also encouraged the practice of disposing of other wastes like food waste and general refuse, along with the human bodily wastes.

In smaller towns and villages, the waste management systems that developed in Europe in the Middle Ages took the form of the 'night soil' cart. Outhouses (i.e. small buildings separate from the dwelling house) contained 'earth closets', essentially a box mounted beneath a wooden seat where solid waste was deposited in the box and a layer of soil placed over it. Periodically, the boxes were emptied, into a cart, by the 'night-soil man' who transported it out of town and sold it to local farmers as fertiliser. Liquid waste simply leaked from the box into the soil beneath.

As town houses developed from single-storey to multi-storey dwellings, it became less convenient to have to move from an upstairs room to an outside building to access the earth closet. For many people this led to the use of 'chamber pots' which were frequently emptied onto the street below from a convenient window. In high-status buildings 'the long drop' meant a part of the building that overhung a convenient or redirected stream or pond and a seat was fashioned such that the waste simply dropped down into the water below.

The addition of human excrement to the animal wastes in the street was quickly recognised to be undesirable. By as early as 1596 a 'water closet' had been developed that used flowing water to move waste down a pipe from inside the house to an

(a)

(b)

Figure 3.1 (see also Plates 1 and 2) The Roman latrines at Housesteads Fort on Hadrian's Wall in northern England. (a) The excavated latrine building, and (b) an artist's reconstruction. Water from a diverted stream flowed in at the upslope end and waste was carried away to the nearby stream. Cisterns of water were used for hand-washing while sponges on sticks (in the bowl in the middle background) were used in place of modern toilet paper. Source: (a) https://commons. wikimedia.org/wiki/File:Housesteads_ latrines.jpg. By Steven Fruitsmaak (Own work) [Public domain], via Wikimedia Commons; (b) Original painting by Ronal Embleton, copyright Frank Graham.

external watercourse. Using water to wash away waste had been practised for at least 4000 years. It is recorded in images from ancient Egypt and Roman latrines, built in England about 2000 years ago, were designed to use the flow of a diverted stream to carry the waste away (Figures 3.1a and 3.1b, and Plates 1 and 2).

As the use of water closets spread, pipes were laid to carry the waste to nearby streams/rivers that then transported the waste away, ultimately to the sea. These were the first sewerage systems. Of course, this system did not remove the public health risks: foul water leaked into the groundwater and so could contaminate wells and drinking water taken from the river would be directly contaminated. It also impacted the ecology of the receiving waters.

In order to move waste from the source more efficiently, in the nineteenth century the simple water closet was superseded by the flush toilet. This used engineered solutions to deliver a sudden burst or flush of water that efficiently carried away solid wastes. In the United Kingdom, Thomas Crapper, a manufacturer of toilets and plumbing fittings, invented and patented one of the early flush toilet designs but, contrary to the popular myth, he did not invent the flush toilet itself. However, seeing his company name on many European sanitary fittings meant that American servicemen in Europe during the First World War adopted his name as a slang/colloquial term for the toilet.

With the development of indoor toilets, the practice of the co-disposal that had originated with the cesspit continued. Rags, toilet paper, sanitary products, contraceptives and cleaning products were and still are regularly flushed down the toilet. To improve the flow in moving waste down the pipes, baths, sinks and showers also had their wastewater plumbed into the sewers for disposal, and in many places the surface water drains diverted rainwater into the sewers providing further dilution and momentum into the sewage mix.

As urban populations grew the effects on the receiving waters became more apparent. With major ports and cities located on estuaries and with all the accumulated wastes also being carried into estuaries and enclosed coastal bays, these often suffered the most dramatic effects. These included the disappearance of fish and associated collapse of fisheries and the emissions of foul odours, particularly hydrogen sulphide. In England, the stench from the Thames was sufficient to cause Parliament to suspend sitting on a number of occasions in the nineteenth century (see Section 1.4).

Following a number of major outbreaks of cholera in London in the nineteenth century, the link between sewage-contaminated drinking water and disease was established and efforts were made to pipe the sewage waste away and into flowing bodies of water. This reduced the chances of contaminating drinking water wells. However, this exacerbated the effect on the receiving waters. The dominant environmental impacts of bodily wastes derive from the oxygen required for their biological breakdown. The organic components are broken down by microorganisms, liberating nutrients, water and carbon dioxide. The most efficient form of this process is aerobic. However, if oxygen is limited, anaerobic bacteria will utilise the organic matter, liberating methane and hydrogen sulphide gases instead of carbon dioxide.

Thus, the initial drive to 'treat' sewage had been motivated by public health concerns: disease and parasites. This prompted the development of the sewerage network where water and gravity move the waste from the source through pipes to an acceptable discharge point. However, it quickly became apparent that this triggered anoxic effects (lack of oxygen) in the aquatic environment and that this caused further public health problems, particularly the emission of noxious and toxic gases such as hydrogen sulphide and methane. This prompted the development of treatment works where the sewage was given a degree of treatment prior to discharge. The treatment regime focused on the oxygen demand of the waste but in doing so also dealt with most of the pathogen problems as well. The combination of water closets, piped sewerage systems and sewage treatment works provided the controlled aerobic treatment of wastes that became the standard pattern of sewage management in many towns and cities in the late nineteenth century.

3.2.2 Traditional sewage treatment

The organic components of sewage undergo microbial breakdown ultimately becoming bacterial biomass, nutrients, water and carbon dioxide. If oxygen is available this process is carried out by aerobes. The amount of oxygen required to support the complete breakdown of organic wastes is known as its biological oxygen demand (BOD). The standard protocol measures the BOD over five days at 20°C. A typical sewage effluent would have a $BOD_{5\,day}$ of 600 mg O_2 l^{-1} meaning that in breaking down the organic matter present in 1 litre of the effluent, the bacteria would use 600 mg of oxygen. By comparison, a typical beer has a BOD of 70 000 mg O_2 l^{-1} and the effluent from a traditional three-stage sewage treatment plant has a BOD of 20 mg O_2 l^{-1} or less.

Under atmospheric pressure and at 20°C, freshwater in the environment when fully saturated with oxygen contains approximately 9 mg O_2 l^{-1}. This implies that 1 litre of untreated sewage effluent would use all of the oxygen available in 67 litres of water and in doing so would make the water anoxic and unable to support life. Typically, aquatic invertebrates can survive on 2–3 mg O_2 l^{-1}, hardy fish species on 3–5 mg O_2 l^{-1}, while sensitive species such as salmon require 6–7 mg O_2 l^{-1}. So, a simple calculation suggests that if the final concentration of oxygen in the environment is going to be sufficient to support salmon, then only 2 mg O_2 l^{-1} is available for breaking down the waste and so 1 litre of sewage needs to be mixed into 300 litres of water.

In reality this number is much smaller as the bacteria do not use all the oxygen instantaneously and oxygen is continually being mixed back into the water from the overlying atmosphere (and is more rapid in areas of high flow or wave action). For simplicity, consider an effluent released into

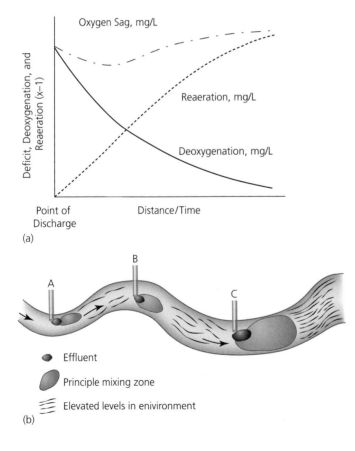

(a)

(b)

Figure 3.2 (a) Organic wastes are degraded in the environment by bacteria, this process uses oxygen and so places a biological oxygen demand on the receiving water. This is seen as a reduction in the level of oxygen in waters downstream of the discharge known as the oxygen sag. The bacterial oxygen demand is not instantaneous as bacterial populations need to increase, and this delay produces a time lag (and a downstream displacement) of the peak oxygen demand. Similarly, as the waste is moved away from the point of discharge, natural processes, such as aeration and mixing, introduce oxygen. At some point downstream the bacterial degradation and the re-aeration will combine to mitigate any impact. The zone in which the effluent has a measureable effect is known as the mixing zone. Inset: Different levels of oxygen-demanding wastes cause different severities in the oxygen sag and, in extreme cases, lead to anoxia. (b) When multiple inputs are located in close proximity, the downstream discharges may be made into the mixing zone from an upstream discharge and so increase the chances of significant impacts.

a river, that is, the flow is always one way downstream. Once released, microbes in the effluent start to break it down, releasing nutrients and carbon dioxide and using up oxygen. As time passes the number of bacteria increase and the rate of effluent breakdown and oxygen use increases. However, the effluent is also being washed further from the discharge point and mixed (diluted) with the surrounding water, while oxygen is continually being absorbed from the atmosphere. These processes occur in what is known as the mixing zone (Figure 3.2a). The degree to which the rate of oxygen use exceeds the rate of oxygen renewal (by dissolving from the atmosphere and mixing with unpolluted waters) defines a zone of decreased oxygen levels known as the *oxygen sag*. In designing effluent treatment schemes the critical parameter is the extent of the oxygen sag (i.e. to what level does oxygen availability decline) and its location: does it prevent migratory fish passing through, does it wash over critical habitats, etc.? Multiple discharges along a

river raise the possibility of mixing zones overlapping (Figure 3.2b).

In estuaries and in the sea the situation is more complex: tidal flows oscillate and the mixing zone continually moves, potentially even upstream of the discharge, and can slow the movement of the waste away from the discharge.

A traditional sewage works, based on the classic nineteenth-century approach, comprises four stages (Figure 3.3). Any particular treatment works may provide only some of the stages depending on the volumes of waste and the nature of the receiving environment, that is, the extent of the mixing zone and the degree of oxygen sag. Treatment works that discharge into rivers generally use all four stages of treatment because a relatively clean effluent is required and the volume of the receiving environment is small. Discharges to estuaries and the coast often only use the early stages of treatment as the dilution capacity and dynamic nature (moving water, continually reoxygenated by breaking

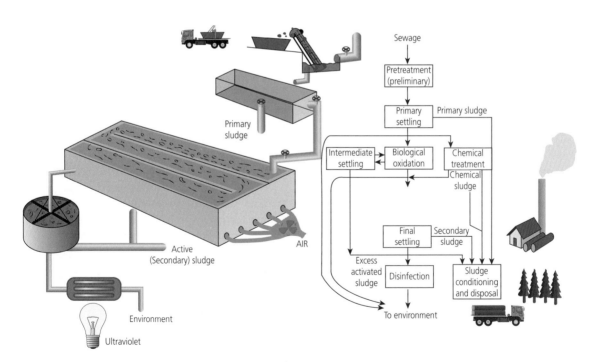

Figure 3.3 A schematic representation of the classic stages in a sewage treatment works. A liquid effluent may be discharged, depending on the nature of the receiving environment and the degree of treatment needed, after preliminary, primary, secondary or tertiary treatment. Sludge generated by the treatment process requires further additional treatment.

waves) means that the final waste breakdown can occur without significant environmental impacts.

3.2.2.1 Preliminary treatment

Preliminary sewage treatment does not really provide any actual treatment of the waste. The incoming wastewater passes through a screen, usually a series of bars 4–5 mm apart, and then through a grit trap (an area in which flow speeds are reduced and heavy particles settle to the bottom).

Large rigid items become caught on the screens and so do not enter the works where they might damage equipment. These include tree branches and other large items washed into the storm drains. Small items and soft material is pushed through, or macerated on the screens, passing into the grit separators.

The heavy particles that settle out in the grit trap along with the material that collects on the screens is periodically removed and generally sent to landfill. The effluent may be discharged at this point or it may progress to primary treatment. Discharges of preliminary treated effluents are usually only done as small discharges into coastal waters. The macerated effluent has a higher surface-area-to-volume ratio than the raw sewage particulates and this increases the rate of breakdown in the environment.

3.2.2.2 Primary treatment

The slurry of the now macerated effluent is passed into settlement tanks where it is left for 24–36 hours. During this time the suspended solids settle out of the supernatant liquid. However, this liquid still maintains a high BOD because it contains colloidal material and bacteria in suspension and soluble wastes.

The supernatant liquid is pumped away for secondary treatment and the settled material is pumped out and this is known as *primary sewage sludge*. Primary sludge has a high BOD but also contains much of the 'litter' and refuse that has entered the sewerage system including sanitary products, barrier contraceptives, 'disposable' nappies, litter washed into drains from the street, needles and hypodermics, paper, dead pets and any of the wide variety of things people place in drains or down toilets. As such it is not suitable for use as a fertiliser.

The organic material in the primary sludge is also chemically contaminated with metals and a variety of organic compounds derived from detergents, cleaning agents, pharmaceuticals, cosmetics and pesticides that were in the effluent which will preferentially adsorb on to the organic particles. While this helps remove the contaminants from the effluent entering secondary treatment, it increases the difficulty of safely disposing of the primary sludge.

3.2.2.3 Secondary treatment

Secondary treatment is designed to stimulate the microbial breakdown of the waste and this is achieved by oxygenating the liquid effluent. A number of configurations can be employed for this including cascades and flow channels or trickling the waste over particles of coke which have a complex and pitted surface giving them a high surface area and this is a good environment for microbial communities to develop. Secondary treatment yields a liquid effluent of significantly lower BOD and a sludge comprising flocs (aggregations or clumps) of microbial biomass known as secondary sludge. The liquid effluent is frequently discharged at this stage if the receiving water provides the high dilution/dynamic conditions, for example, like those for estuarine and marine discharges.

3.2.2.4 Tertiary treatment

Tertiary treatment of the liquid effluent from secondary treatment is often only provided when very high water quality is required—that is, a low-volume receiving water (small river or lake) or an environmentally sensitive location—but in some countries (e.g. Switzerland) it is routine and hence very widespread. In the United Kingdom and Europe, for example, marine outfalls rarely receive tertiary treatment but rivers and lake discharges do.

Tertiary treatment involves trickling the waste through filter beds of coke and sand; these provide further reductions in BOD but also physically filter the effluent to provide a very low BOD and low-turbidity final effluent for discharge. Filtered particles are primarily microbial biomass and so are usually added to the secondary sludge.

3.2.2.5 Additional treatment options

In the latter part of the twentieth century, changes in society led to the emergence of new issues for sewage treatment to deal with. When originally designed, the focus of most traditional schemes had been BOD reduction and the avoidance of hypoxia in the receiving environment, with associated gas production (of hydrogen sulphide and methane) and fish deaths. The process removes most pathogens, but this was a secondary benefit and they accumulated in the primary and secondary sludge.

The increase in contact water sports at offshore locations raised concerns about human contact with the residual pathogen load. Discharges tended to be located away from bathing beaches, but surfers, divers and wind surfers are also now not confining themselves to the traditional bathing beaches. Because most marine discharges only receive primary treatment, the risks are higher. Incorporating a disinfection stage into the treatment process was a response to this. The most usual forms are to disinfect the effluent chemically as a final treatment stage before discharge. Chlorination (a similar process to that used in public swimming pools), ozone and peracetic acid are the most commonly used disinfection agents (Section 5.4.1). Both ozone and peracetic acid are powerful oxidising agents and there is concern that, given the complex mix of chemicals within a sewage effluent, the end products are difficult to predict. This is even more the case for chlorine where the potential for the formation of biologically active organohalide is significant. The fourth widely used disinfection technology is UV light but this is not effective in effluents with high suspended particulate loads—that is, from primary sewage treatments—because the UV light cannot penetrate.

The second issue that prompted additional treatment regimes was nutrients. The recognition that many areas were suffering from eutrophication due to a combination of nutrients entering from agricultural run-off and sewage effluents prompted the imposition of nutrient discharge standards for environmentally sensitive areas. Nutrients are usually removed biologically by passing the effluent through algal ponds or reed beds. The growing plants use the nutrients and so strip them from the effluent. They can also be removed chemically—by ion exchange technology, for example—but this is generally more costly.

3.2.2.6 Sludge disposal

Secondary sewage sludge essentially comprises microbial biomass and as such it makes an excellent fertiliser and soil conditioner. However, it cannot be regarded as pathogen and parasite free and so its use is restricted to municipal parks, roadside landscaping, forestry, grazing land and crops where it

is unlikely to enter the food chain such as cereals (only the seed is eaten) and animal feed crops.

Primary sludge is more problematic to deal with not least because its composition varies. In rural locations where the incoming effluent contains only low levels of toxic materials, the sludge can be used as a soil conditioner/fertiliser on non-food crops; that is, forestry, arable grassland, etc. In the United Kingdom about 30% of the sludge is used in this way. The remainder is unsuitable either because of the levels of contaminants, particularly metals and organohalides, or it is produced in urban centres where the logistics of getting it to suitable crops are prohibitive.

The options for dealing with primary sewage sludge in urban areas are limited. The most widespread practice is incineration, either alone, which requires a high energy cost to initiate the burn, or with other industrial wastes, for example, in oil refinery flares, industrial waste incinerators or in municipal waste incinerators. The local population are rarely enthusiastic about such incinerators. The sludge is burnt, yielding water and carbon dioxide, but there is a risk that in the combustion process halogens or other compounds present might react, forming toxic residues. These and the metals present in the sludge will either be emitted in the hot exhaust gases (and so become airborne contaminants) or be retained in the ash for subsequent disposal (usually as landfill). Some sludge is sent directly to landfill sites from where it may break down, liberating methane than can be collected and used.

Some sewerage works are equipped with anaerobic sludge digesters; these large bioreactors take the sludge and ferment it in a bioreactor to produce methane which is burnt on site to generate energy to run the plant. Any excess electricity generated is sold to the electricity grid. Sludge digesters and incinerators have a high capital cost and also contribute greenhouse gases to the atmosphere.

Until the late twentieth century much of the sludge produced by sewerage treatment works at coastal cities was disposed of at sea, either discharged to deep water by pipe or dumped offshore from a specialised tanker. This was the accepted practice in the United Kingdom, United States, Canada and much of Europe. However, this was increasingly seen as an unacceptable option by society and has now ceased in developed countries (see Box 3.1).

Box 3.1 Marine sludge disposal: A best practical environmental option?

Until 1998 the United Kingdom used an additional method of sewage sludge disposal: that of dumping at sea. In the 1990s, the United Kingdom disposed of 11.5×10^6 tonnes (t) wet weight of sludge per year at 12 marine disposal sites (Box 3.1, Figure 1). This practice is now discontinued but the effects at the dumping sites are expected to persist for some years.

Placing the sludge on the seafloor potentially contributes to marine productivity: as the sludge becomes degraded, the nutrients and carbon dioxide remain on the seafloor where they do not immediately interact with sensitive biotic systems. The slow, continual release of nutrients become dispersed by the currents, preventing any eutrophication effects, and the carbon dioxide dissolves in seawater to form carbonate ions. It was estimated that if all the sewage sludge disposed of into the sea in the mid-1990s had been incinerated, the UK's emissions of carbon dioxide would have increased by 5%.

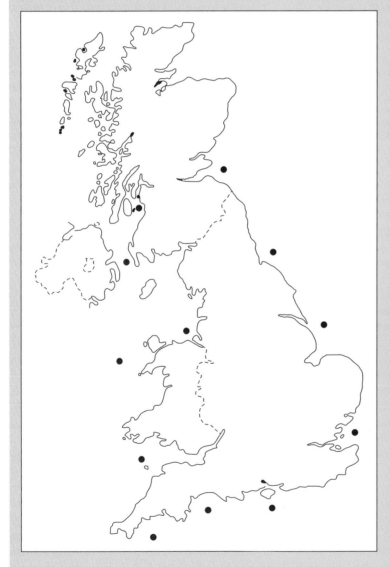

Box 3.1, Figure 1 Sites where, up until 1998, the United Kingdom disposed of sewage sludge on the seafloor. (Redrawn from various sources).

(continued)

Box 3.1 (Continued)

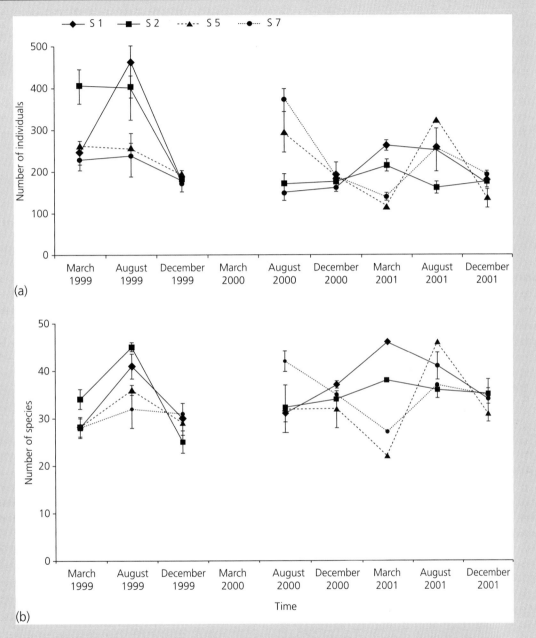

(a)

(b)

Box 3.1, Figure 2 Changes in the mean (± SE) number of individuals of macro infauna (> 0.5 mm) (a), and the number of species recorded (b) from 5 replicate 0.1 m² samples taken from inside (S1 and S2) and outside (S5 and S7) the Tyne sewage disposal site in the March, August and December 1999–2001 following the cessation of sludg e disposal in December 1998. Source: Birchenough, S.N.R. and Frid, C.L.J. (2009). Macrobenthic succession following the cessation of sewage sludge disposal. *Journal of Sea Research*, 62: 258–67.

(continued)

Box 3.1 *(Continued)*

However, at-sea disposal had environmental impacts. These ranged from smothering the seafloor and development of local anoxia to elevated levels of disease and metals in fish. The extent of these effects varied between sites: within the Thames estuary, conditions were so dynamic that no impacts were detected; at the Glasgow site, in deep waters off of Garroch Head, the waste accumulated had a profound but localised impact on seabed ecology. The Tyne site in Newcastle was intermediate in terms of the hydrography, and at that site there was a measureable impact on the environment but the effect on the biota was positive: the added organic matter acted as a food resource, stimulating productivity (Box 3.1, Figure 2).

Thus, while the public perception that putting sewage sludge in the sea was negative, in reality the measureable impacts were generally small, reversible (albeit over decadal scales at sites like Garroch Head) and reduced atmospheric pollution and the emission of greenhouse gases. On this occasion, societal pressure resulted in policy that was probably not the best practical environmental option (Section 7.4.5).

3.2.3 Other treatment regimes

While the classic four-stage sewerage treatment regime developed in Victorian Britain to deal initially with the problems of sewage in London was taken up around the world, its adoption is not universal. In many rural areas, the capital costs of such a plant were prohibitive and so small-scale solutions were used, in particular, septic tanks. A septic tank receives effluent and develops a microbial ecosystem that deals with the waste. Settled material including microbial biomass is periodically removed from the base of the tank while a liquid treated effluent continually flows out. Septic tanks are biological treatment ecosystems and toxic chemicals or overloading with BOD can cause their collapse.

For larger communities, waste is piped to large lagoons that effectively act like large, open septic tanks. Waste flows into these waste stabilisation lagoons, a variety of biological processes occur and a cleaner liquid flows out of the lagoon to the river or sea. Lagoons may be arranged as a series of ponds and reductions in BOD of up to 80% can be achieved. Typically these lagoons become centres for birds who feed there and scavengers such as rats. However, they are extremely cheap to construct and operate and so find favour in economically challenging situations such as in less economically developed countries.

3.2.4 The challenges of a mixed waste stream

The addition of water to human waste to propel it through a sewerage system also dilutes it. While this might dilute toxic components in the waste, it also increases the volume of material that treatment plants need to deal with and so raises their capital costs. More problematic is the addition of inorganic material into the waste stream; this includes industrial wastes that are often licenced to be discharged to the sewer, litter and material washed in from the roads/storm sewers, microplastics in the form of synthetic fibres (Section 4.2.2) and household chemicals and pharmaceuticals (Section 5.2). It is naïve to assume that because a product is used in the home it is somehow benign: many toilet-cleaning products contain metal salts and biologically active organics. Chlorine bleach, frequently used to clean sanitary fittings, can react with organic materials to form organochlorides (see Section 3.4).

When most sewerage systems were designed, no thought was given to the challenges of mixed waste streams as to a large extent the problem only arose subsequently. In many regions—most of Europe, the United States and Australia, for example—new sewerage systems separate human waste and domestic waste-water from industrial effluents and storm water. These are often supplemented by public education programmes to try and reduce the incidence of toxin-forming chemicals entering the domestic waste stream.

3.2.5 Other biological oxygen-demanding wastes

A variety of other wastes, particularly those from the food processing, paper manufacturing and the brewing sectors along with some organic chemical wastes including pharmaceuticals, also have a high oxygen demand. These wastes tend to be much easier to deal with than sewage as they contain the oxygen-demanding component but without the often highly variable mix of other waste compounds like metals, detergents and cleaning agents. Therefore, these wastes can usually be treated in facilities similar to a sewerage works and produce relatively clean end products; for example, methane to power the process, and contaminant- and pathogen-free sludge for use as a fertiliser.

Agricultural animal wastes tend to be similar to sewage in the challenges they pose for treatment with the organic, oxygen-demanding elements but they also contain veterinary drugs (Section 5.2.3) and other chemical contaminants. They are often stored on farms and leachate can enter local rivers and the groundwater.

One group of wastes frequently overlooked are those from aquaculture. Aquaculture facilities on land will usually be required to have effluent treatment facilities, as would any other industrial complex, but offshore aquaculture facilities do not. For example, a raft of salmon cages moored in a sea loch/fjord could easily produce as much organically rich waste as a small town. The fish themselves excrete and produce faeces but there is also an input from the food waste: high-protein pellets that are added to the cage as food but not all of which are consumed. These farms often release toxic waste and pharmaceutical wastes too and may present biosecurity risks (See Chapter 5).

3.3 Metals

Metals are a natural constituent of the environment and are transported through a range of physical, chemical and biologically mediated processes.

Seawater contains all the naturally occurring elements and therefore includes large amounts of metals. However, while some metals are essential for biological processes, many are toxic; in fact some of those that are essential for normal biological activity are toxic at high concentrations. Iron and copper, for example, are used in the respiratory pigments (haemoglobin and haemocyanin, respectively) of marine organisms, and a magnesium atom is at the heart of each chlorophyll molecule. A variety of metals are required in trace amounts by animals and plants but these are readily available in seawater. Pollution by metals is said to occur when human activity raises concentrations above the natural levels (i.e. contamination) and these impact on the biota.

Metals do not break down in the environment; once released they may be carried around the biosphere being diluted and dispersed by physical processes (atmospheric movements, rivers and sea currents and precipitation/chemical binding). Most anthropogenic inputs of metals occur in estuaries and nearshore waters and this is also where most of the impacts occur. Metals will often bind to clay mineral particles and so are deposited and become buried within the sediments where these particles accumulate. Concentrations of metals in sediments are often much higher than in the water column. This has major implications for pollution management, because even when the sources are eliminated and concentrations in the water decline, the sediments remain as a reservoir of contaminants that can become a secondary source of pollution if conditions change; for example, if the sediments are disturbed by dredging or burrowing organisms. For example, it is possible for discharges to a degraded estuary to be controlled so that the water quality improves. This kind of scheme is initially seen as a success as the chemical monitoring data show a decline in toxic metals and some biota begin to return. However, this recovery may then stall as the toxicity, of legacy contaminants, in the sediments limits the species of bottom-dwelling biota that may return, and the activity of burrowing organisms may release these accumulated contaminants from the sediment.

Living organisms deal with metals in a variety of ways: some can be regulated (excreted), but for many there is no excretory pathway and they accumulate in the tissues (Section 2.3). In order to reduce the toxicity effects caused by metals, some organisms have the capacity to form metallothioneins (Section 2.3.1), essentially making the metal biologically inactive. These proteins bind to the metal and so prevent it from interfering with normal metabolic processes. Once the capacity to produce metallothioneins is exceeded, further exposure to metals will cause a metabolic/toxicity effect.

In marine organisms, mercury is generally regarded as the most toxic metal, followed by cadmium, copper, zinc, nickel, lead, chrome, aluminium and cobalt. From a regulatory point of view, lead, mercury and cadmium are of most concern and are the subjects of strict national and international controls due to their high human toxicity.

3.3.1 Mercury

Mercury (Hg) is familiar to most people in its pure, inorganic form as a silvery liquid. Mercury enters the environment naturally from the weathering of mercury-bearing rocks, from the atmospheric emissions of volcanoes and hydrothermal vents in the deep ocean. Mercury is used in thermometers, light bulbs and batteries and a number of industrial processes. Mercury is used as an anti-fungicide (it is often present in foot-odour treatments, for example, and seed dressings). It is used as a preservative in some medicines and vaccines and has been widely used in dental amalgams (the material used in tooth fillings). Metallic (liquid) and inorganic mercury are not highly toxic but the organic form, methylmercury, is highly toxic and bioaccumulates (see Section 3.3.9). Methylmercury is formed by bacterial action on mercury in anaerobic conditions (i.e. marine muds) and as a detoxification mechanism in some higher organisms.

Mercury affects DNA and cellular processes involving it, brain and nerve cell function, and in mammals it affects reproductive processes and embryo development. Mercury is the only pollutant, other than pathogens, to have caused large-scale human mortality, the most dramatic of which was the Minimata incident (see Section 3.1).

Open ocean levels of mercury are typically 0.5–3 ng l^{-1}. In coastal waters concentrations may be 10 times higher whereas in industrialised areas concentrations can exceed 500 ng l^{-1}. Due to the potential of methylmercury to bioaccumulate, the levels of mercury in human food are closely monitored. Generally, fish for human consumption must contain less than 0.5 ppm (parts per million dry weight) mercury, although a higher level of 1.0 ppm is allowed in some species where the lower limit would exclude most of the catch from market. In such cases, some stocks of higher level trophic predators such as tuna or swordfish for example, consumers are advised to limit their weekly intake, and pregnant and nursing mothers are advised not to consume these fish.

3.3.2 Cadmium

Cadmium is highly toxic and while the levels required to cause death (i.e. in LD_{50} tests) are lower than for mercury, cadmium often causes sub-lethal effects at lower concentrations than mercury. Cadmium causes cancers, growth defects and reproductive anomalies in humans and vertebrates. Crustaceans seem to be particularly sensitive whereas molluscs are often able to accumulate cadmium with no obvious adverse effects (Section 2.3.1). In 1912, people in the Toyama Prefecture in Japan began to complain of spine and joint pain, a disease that became known as Itai-itai (meaning *it hurts, it hurts* in Japanese). The source was traced back to contamination of rivers by cadmium released by local mining operations. Although the contaminated rivers were used as a source of drinking and washing water, the main exposure route was via the irrigation of the rice crop with contaminated water, which led to metal-contaminated rice and human exposure via rice ingestion.

Cadmium is used in batteries and a range of microelectronic devices. It is also released from fuel combustion and the smelting of zinc and copper (as it is often a secondary component of ore-bearing rocks). Although some cadmium was discharged into rivers and the sea via wastewater outfalls, these have come under increasing control. The main input routes that remain are via the atmosphere and road run-off (car tyres, brake pads and paint can be sources). Therefore, cadmium inputs may be diffuse and not necessarily close to the sources.

3.3.3 Copper

Copper was the first metal to be mined by humans and was widely used in the ancient world, occasionally in ornaments and jewellery but most widely as bronze, its alloy with tin. The Bronze Age followed the Stone Age and represented the first widespread use of metals for tools, household wares (cooking pots, vessels and cutlery), jewellery and ornaments. Metalworking gave those with this technology considerable advantages over those tribes without, and the skills learnt in working bronze were subsequently used to work the more difficult, but harder and hence more useful iron. Copper is an excellent conductor of heat and electricity and its main modern applications are in electrical components and wires, in heat exchangers, pipework and plumbing.

In the past, copper sheets were used to cover the hulls of wooden sailing vessels, providing both physical protection from attack by wood borers such as the ship-worm (*Torredo*, a bivalve mollusc). It was also used to discourage growth of algae, barnacles and other fouling organisms on the hull, through its toxicity. Widespread use of copper sheeting by the British navy is thought to have contributed to the greater speed and manoeuvrability of British naval vessels compared with their French and Spanish counterparts during the Napoleonic wars. Similarly, the danger of a wooden ship suffering catastrophic hull failure due to ship-worms was reduced with the use of copper sheeting. So, merchants preferred their cargo to be carried in such vessels and led to the term *copper-bottomed* being used to refer to a sound investment.

With the advent of iron and then steel ships, physical protection was no longer needed but the growth of fouling organisms on ship hulls remained a problem because it slowed a vessel, requiring more fuel and increasing the time and costs of maintenance. Protecting the hulls with paints containing toxic compounds, known as anti-fouling paints, became common practice and many of these paints contained copper-based compounds. By their very nature these paints are formulated to release toxins into the water to produce a zone of toxic conditions close to the hull in order to kill or deter settling propagules and larvae. As such, anti-fouling paints are one source of copper into the marine environment.

Copper is an essential element for most living organisms including plants and animals. It is a key part of haemocyanin, the respiratory pigment used by a wide range of molluscs and arthropod species. At high levels copper is toxic, with many marine organisms showing toxic effects in the 1–10 µg dm^{-3} range. Surprisingly, this includes species (some bivalves and isopods) that utilise copper in haemocyanin and therefore have some mechanisms for regulating it (Section 2.3.1). Oysters do appear to be able to sequester large concentrations of copper into blood cells (leucocytes) without any adverse effects (see Box 2.1, Chapter 2).

In addition to leaching from anti-fouling paints, copper enters the marine environment from industrial effluents and in domestic waste-water. Copper dissolves from copper water pipes and is contained in many household chemicals and so forms part of the chemical cocktail that is urban waste-water (see Section 3.2).

3.3.4 Zinc

Over 11 × 10^6 t of zinc are produced worldwide each year, yet few things are made solely of zinc. Zinc is, however, a key component of many metal alloys and is widely used to galvanise iron and steel to prevent rusting; for example, shiny silver nails are galvanised with a coating of zinc. Zinc oxide is used in a wide range of products including plastics, inks and pharmaceuticals. Zinc is essential as a micronutrient for plants and animals, including humans. Excess zinc can cause cancers and neurological disorders in humans.

Zinc reaches the sea via urban wastewater discharges (that collect low levels in a variety of effluents) and industrial discharges, primarily from the metal manufacturing and processing sector. The levels of zinc in the open ocean are typically around 0.1 µg dm^{-3}, while in coastal waters such as the North Sea it can be 70 µg dm^{-3} suggesting considerable inputs from anthropogenic sources (although inputs form natural weathering would also raise concentrations near riverine inputs).

3.3.5 Nickel

Nickel, like zinc, is widely used in a variety of metal alloys to reduce corrosion. The US five-cent coin is known as a nickel as it was originally formed of 25% nickel and 75% copper. Nickel is also found in a number of modern batteries including those in hybrid cars (nickel-metal hydride batteries) and in rechargeable nickel-cadmium batteries. It is also a used as a catalyst, for example in the manufacture of hydrogenated vegetable oils.

Nickel is an essential nutrient for some plants, including some species of bean and baked beans (produced from haricot or navy beans) which contain high levels of nickel. Nickel is carcinogenic and can trigger allergic reactions in some people. It has been shown to cause toxicity to some algae (more than 600 µg dm^{-3}), crustaceans (around 150 µg dm^{-3}), and estuarine fish (more than 38 µg dm^{-3}). In the open ocean concentrations of nickel are of the order of 0.1 µg dm^{-3}, in the North Sea away from the coast around 0.3 µg dm^{-3}, but concentrations can exceed 1.0 µg dm^{-3} near the coast or in estuaries.

3.3.6 Lead

The use of lead has a long history. It was used by our ancestors in middle America and by the Greeks and Romans. These civilisations were able to extract and use lead, but the Romans also used lead in alloys, pewter for example, in drinking vessels and other tableware. The Romans produced 'white lead', a pigment and protective coating used in paints up until the mid-twentieth century when it was replaced by titanium dioxide (see Section 3.3.7). White lead and lead acetate-based cosmetics were used widely in western Europe between 1500 and the late 1700s as a skin whitener, and until the 1900s in Japan. Their use caused damage to the skin, hair loss, muscle paralysis and, in some cases, death.

Metallic lead used in pipes and traditional batteries is valuable and most is recycled. Lead that enters the environment does so in some industrial waste streams but it mostly enters via the atmosphere. Until the 1990s a lead-based additive was mixed into petrol/gasoline to improve its combustion and large volumes of the lead were therefore emitted with the exhaust gases. Atmospheric washout is the main route through which anthropogenic lead enters the marine environment. Since regulations were introduced banning lead as a fuel additive, marine inputs have declined dramatically.

In the marine environment, lead is not particularly toxic but it can produce neurological and behavioural effects and birth defects in mammals. The impacts of lead on the health and development of children has meant that strict controls now apply to the use of lead in pipes carrying drinking water and additions of lead to fuel.

One of the few demonstrated cases of a metal being biomagnified up the food chain (see Section 3.3.9) involved the death of around 2400 dunlins (a small wading bird) in the Mersey estuary in northwest England in 1979. Autopsies showed that these birds contained up to 10 ppm (wet weight) trialkyl-lead (an organic form, more toxic than inorganic or metallic lead) in their livers. It was later shown that the bivalve *Macoma balthica* on which these birds were primarily feeding had concentrated the lead, without any detrimental effects, that had been discharged from a local factory.

3.3.7 Iron

Iron has a low toxicity and is an essential element for plants and many vertebrates and invertebrates that use haemoglobin as an oxygen carrier. In the open ocean, far from terrestrial inputs, iron may in fact be a growth-limiting nutrient for phytoplankton (with the supply of iron in windborne dust transported off continents being a major ecological factor). It is therefore ironic that iron-rich wastewater discharges cause pollution in coastal waters.

Iron-rich effluents arise from the processing of titanium ores to make titanium dioxide (TiO_2). Titanium dioxide is the white pigment used in brilliant white paints and in plastics to colour them and reduce their photo-degradation (Section 4.2.2.2). Iron is a common element in the Earth's crust and it readily oxidises, and iron oxides are a common feature of mine water; for instance, the water pumped out of mines while they are operational, in leachate from mine spoil-tips (often with other metals) (Figure 3.4) and in the water that flows out of abandoned and flooded mines.

Figure 3.4 (see also Plate 3) The foreshore at Lynemouth in Northumberland in 1997 when the colliery was in full production and minestone was deposited on the foreshore. The large cobbles are stained orange and the pool is rust-coloured; both are the result of the iron-rich material in the minestone. (Photo: C. Frid).

The ecological impacts of iron are more physical than toxicological, with iron oxides precipitating and coating surfaces or adding to the cost of an organisms filtration/respiration. At sites of severe contamination, as evidenced by the degree of discolouration on rocks and shells, the zone of ecological impact is small and the effects quickly dissipate.

3.3.8 Tin

Metallic (inorganic/ionic) tin is not particularly toxic, however tin is extremely widely used and so enters the environment from a variety of sources. Tin is used in cans and containers for food and drink, in metal processing/plating, construction and within electrical applications. Organo-tin is also used in the glass industry, in perfumes and soaps, in the manufacture of plastics (as a stabiliser), in agrochemicals and in biocides.

Although tin is not particularly toxic it gained prominence as a marine pollutant following the use of the organo-tin compound, tributyltin (TBT) in marine anti-fouling paints (see Box 3.2). A worldwide ban on the use of TBT in anti-fouling paints was agreed by the International Maritime Organisation (IMO) in 1999 with a target date of 2003 for the last application and 2008 for removal from vessels. In many cases these target dates were missed but most maritime nations have now enacted legislation banning the application of TBT-containing paints, with a legally binding ban being enacted in 2008 by the International Convention on the Control of Harmful Anti-fouling Systems on Ships. The phase-out is occurring as vessels are repainted, but TBT will remain present in the water and surface sediments for several decades.

3.4 Persistent organic pollutants

Advances in chemical knowledge and technology coupled with the greater availability of organic- (oil) based material as a feed stock, led to the development, in the late nineteenth and early twentieth century, of a range of novel artificial compounds. Early examples included artificial dyes, the plastic Bakelite and the synthetic fibre rayon. Many of the early chemical formulations linked organic molecules to inorganic substances; for example, halogens. Many of these organic molecules share similar properties including persistence in the environment where they resist degradation. This has led to their characterisation as persistent organic pollutants (POPs).

In 1874, dichlorodiphenyltrichloroethane, (or 1,1,1-trichloro-2,2-bis(p-chlorophenyl) ethane), popularly known as DDT, was first synthesised. DDT is an example of an organic molecule containing a halogen atom, in this case chlorine, and is part of a family of compounds known as halogenated hydrocarbons. In 1939, the insecticidal properties of DDT were realised and it was developed commercially and widely used during the Second World War and, in the post-war years, to exterminate or deter lice, flea and mosquito populations (the carriers of typhus and malaria). It was also used to

---

Box 3.2 TBT, anti-fouling and transgender snails!

Ships' hulls quickly become colonised by marine fouling taxa such as microbial slimes, algae, barnacles, tube-worms, sea squirts, bivalve molluscs, hydroids and bryozoans and a range of other attached taxa. These fouling taxa may accelerate corrosion/degradation and even low levels of fouling can increase the drag on a vessel and so reduce speeds and increase fuel consumption. Heavy fouling can cause a 60% increase in drag, with a 10% loss of speed and a 40% increase in fuel consumption. This pushes up operating costs and adds to the pollution burden in the form of increased emissions of carbon dioxide and sulphur compounds.

Traditionally the hulls of wooden sailing ships were often covered in sheets of copper which slowly dissolved through contact with chemically active seawater (Section 3.3.3). The liberated copper ions adjacent to the hull prevented the settlement and growth of most fouling marine organisms. With the advent of iron and steel hulls the need for protection from boring organisms disappeared and the maritime industry looked for alternative solutions to the fouling issue. Antifouling paints were developed to combat the problem of marine fouling on vessels and man-made structures (piers, oil rigs, etc.). At their simplest, the paints contain toxins that kill or deter settling organisms. However, in order to be active the toxin needs to leach out of the paint and so over time the paint becomes less effective and needs to be replaced. Early paint formulations using copper as the biocide were typically active for two to three years. Repainting a large commercial vessel is expensive both in terms of the cost of the paint but also in docking fees and lost revenue. In the late 1960s, paint companies developed a new more effective active compound in the form of an organic tin compound, tributyltin (TBT).

TBT was more biologically active, as most organometal complexes are, and hence effective, but could also be linked to a new type of paint known as a self-polishing co-polymer (SPC) formulation. In a traditional paint, although only a thin layer, the active compound leached from the very surface layer only. This means that most of the, expensive, biocide remained in the paint and so did not contribute to the antifouling. In the SPC paints, the surface layer of the paint dissolves continually exposing a new 'polished' surface from which the toxin can leach. TBT SPC paints allowed vessels to dock less frequently, for example, every five years, representing a major saving for commercial vessel operators. These paints were also made available to the recreational boat market where again the reduced need to repaint the vessels ensured these paints rapidly came to dominate the market.

In the late 1970s and early 1980s oyster farms in Arcachon Bay (France) suffered a series of crop failures due to abnormal shell growth and low spat fall. Arcachon Bay contains, in addition to the oyster farms, large numbers of moorings for recreational craft. Analyses showed that the TBT leaching from these moored craft was building up in the waters of the bay and causing the mortality of oyster larvae and abnormal shell development through interference with oyster endocrine systems. At around this time work in the United Kingdom showed that populations of the dog whelk, a predatory gastropod mollusc common on rocky shores, were also declining with no settlement occurring in some areas for many years. Again, there was a spatial correlation with recreational craft use.

Experiments with the dog whelks showed that TBT was interfering with their hormonal systems and causing female dog whelks to develop male reproductive organs in addition to female ones. The condition was termed *imposex*: the male genitalia grew in front of the female genital opening preventing successful mating and so effectively making the individual infertile. These findings resulted in first national (in France and the United Kingdom), then European and subsequently international legislation banning TBT from use on vessels of less than 25 m in length and static structures. It was argued that for vessels more than 25 m in length, the economic costs of switching away from TBT would be high and lead to greater impacts from increased fuel consumption and because these vessels actually spent most of their time on the high seas or in transit, they did not contribute to the build-up of TBT in coastal waters where most of the impacts had been observed.

Following the partial ban on TBT, the observed impacts on dog whelks and oysters were reversed, although over decadal time frames. However, the impacts were also observed in offshore species of molluscs and pressure grew for an extension of the TBT ban to cover all uses. This ban was passed by the IMO on 1999 and legislated for internationally in 2008.

---

prevent insects spoiling food (grain and rice) when in storage. DDT is extremely toxic to insects but has a low toxicity to mammals, including humans, it is relatively cheap to make, easy to distribute and use and has a period of efficacy in the environment.

There is no doubt that the use of DDT saved thousands if not millions of lives directly (by preventing insect-borne disease) or indirectly by improving crop yields and the quantity and quality of stored food. However, the widespread and indiscriminate use quickly led to the development of resistance in insect populations. Its chemical stability meant it built up in the environment, accumulating in organisms and it became biomagnified. DDT was found to be highly toxic to fish if it entered watercourses and it had severe impacts on birds. It affected bird breeding through the production of thin-shelled eggs which broke in the nest (Section 2.2.4).

The increasing resistance of target insect populations to DDT and the concerns surrounding its wider impacts led to restrictions being placed on its use from 1972 in the United States and shortly thereafter in many other developed countries. However, the effectiveness and low cost of DDT mean it has an important public health role in many less economically developed countries.

Following the success of DDT, the chemical industry sought to develop other organohalides as pesticides both to be more effective/specific and to offer alternatives to DDT as it became less effective and more widely regulated. These compounds included lindane and the 'drins' (e.g. aldrin, dieldrin, endrin) and a range of other halogenated hydrocarbons were developed including the polychlorinated biphenyls that were used widely in electrical applications.

Lindane, or gamma-hexachlorocyclohexane, like DDT is an organohalide with insecticidal properties that has been used to control insects (e.g. lice/fleas and scabies). Like DDT it is long-lived in the environment and bioaccumulates, although it is quickly excreted once exposure ends. Lindane is a neurotoxin that is carcinogenic and may be an endocrine disruptor (Section 5.3). Lindane is banned in over 50 countries but pharmaceutical use continues in some developed countries and insecticidal uses continue in some less developed regions.

The 'drins' collectively describes a range of insecticides based around various chlorinated organic molecules. The most prevalent were aldrin, dieldrin and endrin. The 'drins' are all highly lipophilic but poorly soluble in water. They persist in the environment and bioaccumulate. The drins were phased out of use in the 1970 and 1980s except for very limited and specialized uses.

Polychlorinated biphenyls, or PCBs, are a range of compounds comprising multiple chlorine atoms attached to a double benzene ring. They are extremely stable and so found widespread use in high-temperature mechanical and electrical applications, for example, in the cores of transformers, in capacitors, as cutting fluids, as coolants and insulators and in flame retardants. However, they are biological active and although toxicity varies between the different PCB compounds, they are carcinogenic and cause endocrine effects. In 1968 in Japan, 400 000 birds died after eating food contaminated with PCBs. PCBs have been linked to reduced reproductive success in some populations of seals, although other populations with high PCB levels appear unaffected. PCBs are found in high concentrations in some seafood and so represent a human health threat (e.g. in salmon; see Section 2.3.3). Although the manufacture of PCBs ceased in the latter part of the twentieth century, large quantities are still present in operational engineering and electrical equipment. The end of life incineration of this equipment/material, if not carefully controlled, poses a continued risk of atmospheric releases and subsequent deposition into the marine environment.

The term *dioxins* is frequently used for the chemicals in the polychlorinated dibenzoparadioxins (PCDDs) and polychlorinated dibenzofurans (PCDFs) families of compounds. Strictly speaking, dioxin is 2,3,7,8-tetrachlorodibenzoparadioxin (TCDD). Dioxins are produced naturally (by forest fires and volcanoes), but are mainly produced as by-products of industrial processes (e.g. smelting, herb-/pesticide manufacture and the bleaching of paper) and the incineration of plastics and other organic materials (i.e. by waste incinerators). Waste industrial oils often contain high levels of dioxin contamination. Dioxins have a global distribution as a result of atmospheric transport but are not highly soluble in water, although they are found throughout the marine food chain due to their persistence and bioaccumulation in lipids. Dioxins are extremely toxic as they are carcinogenic and they bioaccumulate and biomagnify (they have an estimated half-life in the human body of 7–11 years).

Persistent Organic Pollutants (POPs) is the collective term used for a range of organic pollutants, including the halogenated hydrocarbons, that have a long environmental presence, a tendency to bioaccumulate and known impacts on human and ecosystem health. There are a very large number of compounds that can be classed as POPs. Many are synthesised while others arise as by-products of industrial processes or are produced during the disposal of other substances; that is, incineration. POPs are typically hydrophobic (water hating) and lipophilic (fat-loving). In the aquatic environment they therefore tend to associate with particulate material, especially organic particles or clay minerals, and once ingested tend to accumulate in lipid-rich tissues such as fat stores. In this way POPs can biomagnify up the food chain. Given their propensity to be stored in fat reserves they often show no impact until organisms start to use these reserves, for example in winter, during migration or the breeding season. The metabolism of the fat reserves liberates the accumulated POPs and there is a sudden increase in those circulating throughout the body and exerting metabolic effects.

The most widely reported impacts of POPs are on predatory birds and marine mammals. For example, there are accounts in the scientific literature of high body burdens of POPs (mainly PCBs) and increased incidence of reproductive abnormalities in the ringed and grey seal populations in the Baltic, beluga whales in the St Lawrence Seaway, harbour seals in the Wadden Sea and Californian sea lions. However, none of these studies was able to establish causation. For example, the grey seal population in the Baltic suffered increased reproductive disorders, and while tissue levels of PCBs were high they were lower than in the populations of grey seals on the Farne Islands where reproductive rates are so high that the population is subject to occasional culling.

In addition to a range of endocrine disrupting effects, such as the impairment of reproductive functions, POPs are also often characterised as carcinogenic. At least some have also been implicated in reducing the immunocompetence of top predators, leading to increased susceptibility to pathogens and altered patterns of behaviour. In 2001, the United Nations brought forward the Stockholm Convention on Persistent Organic Pollutants that set out to eliminate or severely restrict the use of most hazardous POPs. The convention covers 13 substances or groups of substances which have become known as the 'the dirty dozen' (Table 3.1).

Table 3.1 Substances originally regulated under the UN's Stockholm Convention on Persistent Organic Pollutants that entered into force in 2004. A further 11 substances/groups of compounds have been added subsequently. In all cases where elimination is prescribed, there are exceptions for certain, limited, uses.

Compound/group	Fate under the convention
Aldrin	Elimination
Chlordane	Elimination
Dieldrin	Elimination
Endrin	Elimination
Heptachlor	Elimination
Hexachlorobenzene	Elimination
Mirex	Elimination
Toxaphene	Elimination
Polychlorinated biphenyls (PCBs)	Elimination
DDT	Restriction
Section 1.01 Polychlorinated dibenzo-p-dioxins ('dioxins') and polychlorinated dibenzofurans	No Unintentional Production
Polychlorinated biphenyls (PCBs)	No Unintentional Production
Hexachlorobenzene	No Unintentional Production

Given the vast range of compounds that comprise POPs and the diverse range of behaviours of marine organisms, it is difficult to make simple conclusions about the impacts of POPs beyond identifying them as potentially highly biologically significant. A detailed study of bottlenose dolphins resident in Sarasota Bay in Florida showed that a wide variety of POPs were present in the group but the range and concentrations of the different groups of compounds varied markedly between individuals. The main drivers of these patterns were the age and sex of the individual.

Migratory humpback whales in the southern hemisphere feed, primarily on krill, in the Southern Ocean in the summer and migrate to tropical regions in the winter to calve and mate. During the migration the adults do not feed and examination of the tissue concentrations of a variety of POPs shows that blubber and other lipid- and protein-rich tissues are metabolised for energy during the migration. The concentration of metabolically active POPs increases dramatically during the fasting period but does not simply correlate with the amounts of blubber used, so the physiological basis for metabolic activity is complex.

3.5 Dredge spoil

Dredging, the removal of material from the seafloor, refers to two classes of activity: aggregate dredging and channel dredging. In aggregate dredging the purpose is to use the material removed from the seabed as aggregate; for example, in the construction industry. Whereas in channel dredging, the purpose is to alter the dimensions of a harbour, river or estuary to facilitate navigation by vessels. The material dredged often has to be disposed of in channel dredging. The term *dredging* is also used for certain types of fishing gear, particularly those used to harvest shallowly buried shellfish such as clams, but this application of the term is not considered further here.

3.5.1 Aggregate dredging

The construction industry uses vast quantities of sand, gravel, shingle, cobbles and crushed rock each year as building materials, as bulk materials in concrete or as the foundation materials for buildings, roads and railways, for example. It is estimated that the global demand for aggregates in 2015 was of the order of 48×10^9 t (valued at US$106.4 billion), but this figure fluctuates with the economic cycle. In 2010, a total of 37.4×10^9 t of construction aggregates were produced. Sand is the largest component of this, followed by crushed stone and then gravel, with 44.9% of the demand in the Asia–Pacific region. Recycled materials (i.e. from demolition, road planeing) meet only a small proportion of this demand due to limited availability and quality issues. The new construction materials have to be obtained from the environment either from quarries or watercourses.

In aquatic environments, natural physical processes erode rocks into smaller particles while moving these particles around. These processes produce sediments and tend to sort them into deposits of similar size. Water currents (river flows, wave action, tides) move particles and the stronger the flow the

larger particles that can be moved. This means that at any given location the sediment comprises particles that are too large (heavy) to be moved by the flow regime at that site (recognising that the critical flows may occur aperiodically and be associated with floods or storm waves). This process is known as sorting because it sorts sediment particles into deposits of particular and similar sizes. When landforms change these aquatic deposits can become naturally uplifted or buried, and these historically aquatic sediment beds can then be mined or quarried on land.

Quarries and mines are unpopular with local residents being sources of noise, dust and heavy traffic, and they tend to leave a legacy of a scarred and defaced landscape. The availability of suitable geological deposits in sites where quarrying and mining are acceptable is therefore limited. The alternative is to exploit the sediment beds currently in the aquatic environment by dredging the sediment. In small rivers this may simply involve a tracked front-end loader or excavator driving into the river or working from the bank and digging out the sediments. In deeper areas and in the marine environment, dredging occurs from vessels.

Dredge boats consist of large holds to contain the sediment and a means of lifting sediment into the holds (Figure 3.5). Older vessels or those operating in deeper waters use either a bucket dredge or a large grab. Bucket dredges comprise a large steel bucket that is pulled through the sediment deposit before being lifted back to the surface and tipped into the hold, while grabs usually comprise a pair of jaws that are lowered in the open position to the seafloor where they take a bite out of the seabed before being lifted back to the vessel and emptied into the hold. Both bucket and grab dredges are deployed from cranes on the dredge vessel. The lowering, raising and emptying process is slow and so restricts the rate of aggregate extraction, thus in shallower water they have been superseded by suction dredgers. Suction dredgers deploy a dredge head to the seabed on the end of a large diameter pipe. Powerful pumps on the vessel send a jet of water down a hose to the dredge heads to fluidise the seabed and then the water sediment slurry is pumped back up the large pipe and into the hold. The excess water is deposited into the hold along with the sediment and is allowed to overflow back

(a)

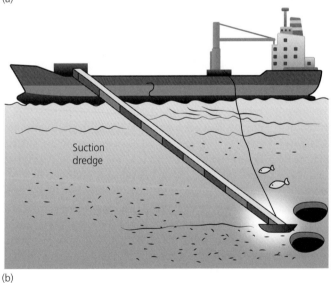

(b)

Figure 3.5 Sea floor dredging by (a) bucket dredger and (b) suction dredger. Both vessels will generate a plume of fine sediment from the overflow of water from the dredge holds. The bucket dredge takes relatively small bites out of the seafloor and produces a limited plume of sediment at the seafloor. The suction dredger uses a jet of high-pressure water to fluidise the seafloor before pumping the sediment up to the vessel. It therefore creates larger pits or trenches and considerably more sediment is disturbed to form a plume at the seafloor.

into the sea; in doing so it carries the finer particles along with it. The volumes of water used in suction dredging mean that this technique releases a large plume of fine material from the vessel.

The most dramatic impacts of dredging are the physical loss of habitat: what was once seafloor with its associated biota is now in the hold of the vessel and will in due course be brought ashore, washed free of salt and used in construction. Depending on how the vessel is operated, the dredge site may comprise a series of relatively small pits (mobile bucket/grab dredging), a pattern of linear dredge paths (mobile suction dredging) or large excavated pits (anchored dredging) (Figure 3.6). Monitoring by the UK government has shown that small isolated pits quickly fill in and disappear, particular in dynamic coastal seas, but large pits and dredge tow paths can take years or even many decades to recover physically and ecologically. Where dredging operations are regulated this has led to licencing conditions that limited the depth to

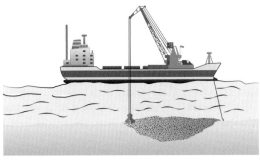

(a)

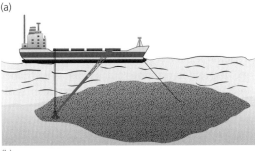

(b)

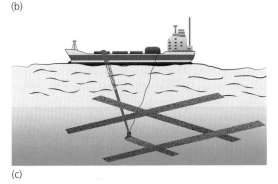

(c)

Figure 3.6 Dredging can be carried out either with the vessel anchored (a and b) producing a pit in the seafloor or with the vessel underway and towing the suction head (c), yielding a series of trenches or channels in the seafloor.

which aggregate beds can be dredged (to ensure the same physical habitat remains for colonisers) and restricted the total area that can be exploited to ensure that there are nearby sources from which the colonists can migrate.

The effects of the dredging extend beyond the footprint of the dredge pit; at the seabed, fine material is dispersed as plume while the overflow from the dredge vessel introduces fine particles into suspension at the surface. These fine particles have ecological impacts in the water column and where they are, eventually, deposited. In the water column, the fine material increases turbidity (reduces water clarity) shading phytoplankton, lowering productivity and altering species composition. The fine particles may clog the gills and feeding structures of filter-feeding zooplankton and fish. Larger organisms (fish, marine mammals, birds) may also show behavioural avoidance of turbid waters.

Fine material settles to the seafloor, smothering the benthos, clogging gills and filter-feeding apparatus and altering the characteristics of the seabed sediments. These attributes are a critical feature for burrowing benthos and fish spawning grounds. The effects may be widespread and may affect a number of sites, as settled particles may be remobilised during a storm and so be moved on to affect further sites.

3.5.2 Channel dredging

As the name implies, channel dredging is the dredging of shipping channels in the approaches to ports and harbours and is primarily concerned with facilitating navigation and not with the dredged material. Channel dredging uses essentially the same techniques as offshore aggregate dredging. When a harbour or port is constructed, the building works often alter the path of water flow so that even if initially a deep-water area, existing breakwaters and jetties create sheltered areas where sediments begin to accumulate. Together with the continual increase in commercial vessel size, this means that most harbours and ports undergo maintenance dredging (capital dredging is the creation of new shipping channels) to maintain sufficient depth and width of the shipping channels.

The dredge technology is essentially the same for channel dredging and aggregate dredging, although the nature of the sediment does introduce some major differences in operational procedures. First, the material in estuaries, ports and harbours is usually fine sand and muddy silts. This material has little to no value as construction material. So traditionally, the dredged material, known as spoil, was simply taken out to sea and dumped on the seabed. This meant that, in addition to the ecological impacts of extracting the seafloor and the plume of fines produced at the site, a further area was impacted by the direct deposition of large volumes of fine sediment. The spoil dump sites often showed reduced levels of biological diversity and productivity compared with the surrounding unimpacted sites because the seafloor was continually buried under large volumes of new waste. Over time, the fine material would be dispersed by water movement but as maintenance dredging is an ongoing process, the dumping and smothering are regularly repeated, disrupting any initial community recovery.

Port maintenance dredging can be traced back hundreds of years. In order to save time and money, initially the spoil would be dumped just outside the port. This led to the joke that in many cases, where tidal streams were strong, the spoil would be carried back into the port before the dredge boat got back, so being a dredger captain was a job for life. In more recent years, dredging operations required environmental licencing that specifies the disposal site, which is usually three to five nautical miles offshore.

Although the smothering effects of dredge spoil are ecologically significant at the locality, within coastal seas these effects are extremely minor and could be seen simply as moving some sediment around the system. However, the growth of human populations and industry in the vicinity of the ports has created the opportunity for a secondary and more serious pollution problem. Many waste products, including sewage (Section 3.2.1), metals (Section 3.3), persistent organics pollutants (Section 3.4) and litter (Section 4.2), end up in estuaries, ports and harbours. Some of these are washed out to sea but a proportion becomes incorporated into the sediments either by simple burial or by binding to the fine mineral particles (e.g. clay minerals will scavenge metals and POPs).

Dredging disturbs these legacy pollutants and potentially makes them biologically active again by mixing them into the water column where they are exposed to oxygen and the biota. Legacy contaminants have resulted in two further changes to operational procedures. First, to prevent wider exposure at the site dredging operations are now often conducted with no direct overflow of water and fines. This is difficult for suction dredgers, due to the volume of water pumped. Usually this involves the spoil being placed in a hold and the water overflowing into a further hold/tank where it is allowed to settle before the supernatant is pumped off. Second, physical barriers (silt curtains) or bubble curtains may be employed to retain any disturbed sediment at the dredge site (Figure 3.7a).

There have also been changes in the disposal regimes. Highly contaminated spoil can no longer be disposed of at sea and has to be brought ashore for disposal as contaminated waste, meaning incineration, placing in a containment landfill or chemical decontamination or glassification (i.e. being encased in an inert glass block). Less heavily contaminated materials can still be disposed of at sea but must be buried under uncontaminated material (so-called capping). Thus, the dredge spoil is removed from the contaminated estuary, carried out to sea, dumped on the seafloor, and then sediment from a clean site is dredged and disposed of on top of the contaminated spoil (Figure 3.7b). Over time, this should see the legacy of 200 years of industrial waste disposal in estuaries made safer as it is either buried in the estuary under modern, clean sediments, removed and buried at sea under a cap of clean sediment or removed to shore and cleaned up. The last two options are expensive and the costs are not being met by the original historic polluter, so while this may be a good environmental outcome it must be seen as failure of regulation.

In some environmentally sensitive areas the disposal of spoil at sea is seen as ecologically or socially undesirable. In Europe, continued efforts to reduce human impacts in the North Sea (Section 6.2) led the United Kingdom to declare that 'where possible', material from maintenance dredging would be used for 'beneficial' purposes. Since the 1990s the dredging industry has been working to get dredge material to be seen as a potentially valuable resource rather than a waste for disposal. However, in many industrialised regions the contamination of the dredged material with other wastes limits its possible applications. Uncontaminated material

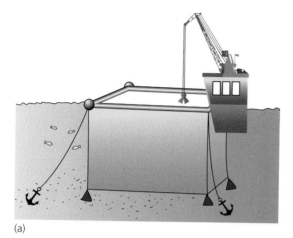

(a)

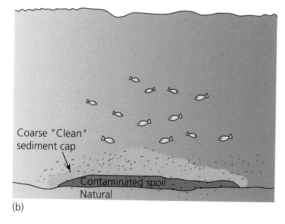

Coarse "Clean"
sediment cap

Contaminated spoil
Natural

(b)

Figure 3.7 (a) A dredge vessel operating inside a silt curtain to contain the plume of fine material. In low-energy environments, the curtain is often a physical barrier suspended from a float ring, an alternative arrange is a curtain of bubbles released from a ring of hoses on the seafloor. The latter design, a bubble curtain, also mitigates noise/vibration and so is also employed around pile-driving operations to limit noise impacts (see Chapter 4). (b) To prevent pollutants leaching from contaminated spoil—for example, material dredged from estuaries and ports that contains legacy contaminants—the deposited spoil can be covered (capped) with coarse, clean material. The use of coarse sediment limits the potential for erosion to remove the cap and hence expose the contaminated sediment.

Table 3.2 Some potential 'beneficial' uses of sediments derived from maintenance and capital dredging programmes. (Modified from Paipai 1994)

Beneficial use	Examples/notes
Sediment cell maintenance	Beach recharge—the adding of sediment to an eroding beach as part of shoreline management/protection
Construction and other engineered uses	Land reclamation for port/airport development, residential development, redevelopment of confined disposal facilities
Manufactured construction material	Dredged material mixed with cement, manufacturing of bricks
Replacement fill	Civil engineering applications such as filling voids, adding bulk material to earthworks/embankments, etc.
Shoreline stabilisation and erosion control	Various forms of shoreline protection including beach nourishment, coastal realignment, muddy shore profile engineering, offshore berms (e.g. feeder berms, hard and soft berms)
Amenity	Beach nourishment, derelict land restoration, recreation (e.g. hills for walks and picnic areas), landscaping (parks and other commercial and non-commercial landscaping applications)
Capping	Clean dredged material can be used to cap contaminated dredged material in offshore disposal sites
Habitat restoration	Enhancement and/or creation (e.g. saltmarshes, mudflats, wetlands, bird islands, gravel bars, oyster sandy/gravely beds)
Aquaculture	Construction and lining of ponds/bunds
Agriculture (e.g. cotton fields)	Soil 'improvement'
Horticulture (e.g. orchards, ornamental plant nurseries)	Soil 'improvement'
Forestry	Soil 'improvement'
Strip mine reclamation and solid waste management	Landscaping and soil 'improvement'

from maintenance or capital dredging can be used for beach recharge, land claim, landscaping, capping of contaminated disposal sites and a variety of other uses (Table 3.2).

The export of the mined minerals occurs from a number of ports along the Queensland coast inshore of the Great Barrier Reef (Section 6.7; Figure 3.8a) and these have been the subject of various capital and maintenance dredging projects. The spoil is generally disposed of at sites in nearshore waters. Increased riverine inputs of sediment from land clearing and mining combined with the input of dredge spoil have resulted in dramatic increases in the amount of fine sediments in the inshore system in recent decades and this is associated with increases in turbidity (Figure 3.8b).

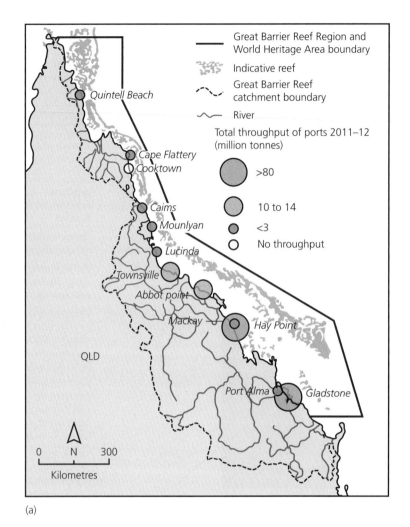

(a)

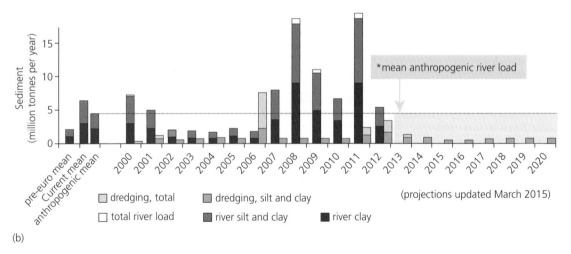

(b)

Figure 3.8 (a) The Great Barrier Reef (GBR) World Heritage Site and areas used by commercial shipping and the major ports. Most of the tonnage is coal exports. (b) Estimated quantity of sediment (total and fines: silt and clay) entering the GBR region from riverine sources (ten major rivers) and port dredging. The bars on the left indicate mean levels pre-European settlement (pre-euro), the current (1986–2009) mean (Current), and the mean non-dredging anthropogenic (anthropogenic) contribution. © 2014 Commonwealth of Australia. Both redrawn from the Great Barrier Reef Marine Park Authority (2014). Licensed for use under a Creative Commons By Attribution 3.0 licence: see https://creativecommons.org/licenses/by/3.0/au/legalcode.

Many species of coral are sensitive to the presence of these fine materials in the water and the turbidity/shading effects can limit photosynthesis by symbiotic zooxanthellae and clog the coral's feeding tentacles. The combined impacts of fishing, tourist visitors, nutrient and sediment loads and global warming (coral bleaching) are causing widespread degradation of the Great Barrier Reef, and in 2015 the UNESCO effectively placed the Australian government on two-years' probation under threat of declaring the World Heritage Site in danger if management measures were not strengthened to protect the region from damaging human activities. The initial response to this has been a ban on the disposal at sea of dredge from capital dredging (i.e. new ports or port expansion), but the disposal of spoil from ongoing maintenance dredging is set to continue.

3.6 Minestone and power station ash

The Earth's crust contains many useful substances and while some of them can be accessed from the surface, deposits of many extend underground. The process of extraction is generally known as mining or quarrying. In both cases, in order to access the geological deposits containing materials of interest other material has to be moved. The overlying material is either removed in a quarry or open-cast mine or the material is removed to provide a system of shafts and tunnels to reach the deeper deposits. In addition, the target mineral is rarely able to be extracted in isolation and requires some post-excavation sorting. Minestone or spoil are the generic terms for the solid waste generated from the overburden of rock removed; material removed during tunnelling and the waste rock sorted at the surface from the target minerals.

Quarries and open-cast mines often initially stockpile the spoil and then, as the active quarry face proceeds across the landscape, the spoil is used to backfill the exhausted area. This process minimises the overall impact on the landscape by partly refilling the void left by the excavated mineral. However, in mines accessed by tunnels, the spoil has to be stored or disposed as refilling an exhausted mine is uneconomic and potentially dangerous.

At many traditional mines the minestone/spoil was simply piled into mounds around the workings. While these are unsightly a common presumption was that this was after all 'just rock' and rock from that locality and therefore 'natural'. The high levels of metals in some rivers in Cornwall in the United Kingdom are due to leaching from spoil tips around tin and copper mines dating from Roman times 2000 years ago (Box 2.1). This shows that while spoil may be 'just rock', it can be a significant source of contaminants and that these impacts can persist for long periods of time. In 1966 the unnatural and unconsolidated nature of these mine spoil heaps was dramatically and tragically demonstrated when, following heavy rain, a spoil heap at the Merthyr Vale Colliery in South Wales collapsed and the ensuing mud/rockslide engulfed the village of Aberfan including the primary school. A total of 144 people lost their lives in the disaster including 116 children between the ages of 7 and 10 and 5 teachers. The subsequent enquiry resulted in a number of changes in legislation around the disposal of minestone and the maintenance of minestone tips. At coastal sites, depositing minestone into the sea was seen as a viable and cheap disposal option. Again, the simple justification was that it was after all 'just rock' and essentially natural.

Minestone disposal has two main impacts in the marine environment; the first is dependent on the geochemical nature of the material and the second is a physical consequence. Minestone from metalliferrous mines tends to be high in metal-containing minerals, and as is the case with the Roman mine tips, this can be leached from the minestone into the environment over time. Clearly, metal concentrations in the minestone are lower than in the main ores (otherwise the minestone would not be discarded), but this can still be a significant source of contamination or chemical change. Waters that have percolated through coalmine waste are often extremely acidic and rich in iron salts, making them potentially polluting in the receiving environment. Similar issues arise where abandoned underground mines flood over time, with the result that eventually the mine water overflows from the mine and into surface watercourses. When this happens it can cause unexpected and sudden influxes of metal-rich acid waters into surface streams and rivers with considerable effects. Because these mines have been closed and abandoned, often many years before, the application of the polluter-pays principle to remediation is difficult or impossible. Therefore, state agencies tend to be left to deal with the matter. Such effects are more severe in freshwaters because in the marine environment the buffering effect of the saline waters and the greater dilution and mixing capacity tend to minimise everything but very localised effects.

The most prevalent impact of minestone disposal in the marine environment is the physical impact because the minestone differs physically from the natural sediments in the disposal areas. During the operational phase of disposal, the continual addition of the waste acts to smother and bury the seafloor and associated biota. On the north-east coast of England prior to changes in legislation in 1992, minestone from local coalmines was disposed of at sea. Some colliers close to the coast simply pushed the stone off the cliff and on to the beach below, while others located up to 2 km inland had conveyor systems that moved the stone from mine to beach. Collieries located close to the estuaries of the rivers Tyne, Wear, Tees and Blyth (amongst others) conveyed the waste to a dock where it was loaded onto barges that were taken out to sea where the waste was dumped in designated offshore spoil disposal sites. At the peak of coal mining operations in the middle of the twentieth century, tens of square kilometres of the seafloor were subject to dumping; unfortunately there are no records of the amount dumped or the areas affected.

Once in the sea, the waste is subjected to a variety of processes. Waves and currents sweep small and light particles away, increasing turbidity in the process. The larger particles accumulate at the disposal sites altering the granulometry, and hence ecology, of the local seafloor sediments. The ecology of infaunal organisms is influenced by the nature of the sedimentary environment. The combination of the physical disturbance from the continual arrival of new material and the changes in granulometry (and associated physical parameters) have resulted in altered conditions for the benthos. However, as minestone is, after all, 'just stone' these effects resulted in altered communities but not catastrophic changes, and studies in the 1990s showed that following closure of the collieries the seabed recovered, albeit over several decades (see Box 3.3).

Coal is the most abundant fossil fuel on the planet (and so is relatively cheap); it is also easy to handle and store. The burning of coal is credited with driving the Industrial Revolution in Europe and North America, initially fuelling steam engines but subsequently, following the introduction of the steam turbine in 1906, used to generate electricity. Electricity-generating power stations provide large quantities of readily useable energy to consumers and industry. Coal is burnt to generate heat, which is used to boil water to produce the steam that turns the generators producing electricity. In order to make the combustion of the coal most efficient it is usually crushed to a dust and injected into the furnace in a stream of hot air. This ensures that the new fuel injection does not cool the furnace and that the energy is liberated almost instantly (the small particles have a high surface-area-to-volume ratio). As anyone who has burned coal in a domestic fireplace or stove will know, coal is not pure fuel; it contains mineral impurities (particularly silica and aluminium compounds). On combustion, some of these compounds are voided in the waste gases but some remain as ash in the grate. The same is true of coal combustion in power stations, and ash is a residue from energy production. As the coal was crushed prior to combustion, the ash particles are also very fine and operationally fall into two types: furnace bottom ash (fba) which remains in the furnace and fly ash. The latter is so fine it is initially entrained in the hot gases which are, in most power stations, later collected onto electrostatically charged precipitators at the base of the chimney to reduce the levels of particles emitted in the 'smoke'. Together the fly ash and fba are referred to as *pulverised fuel ash* or pfa. The pfa has been subject to high-temperature combustion and so is chemically fairly inert. The chemical composition of the pfa varies depending on the source and nature of the coal; coal from different deposits contains different mixes of other minerals. For example, pfa contains many of the

---

Box 3.3 Minestone and sea defences: Beneficial use of inert waste

At Lynemouth on the north-east coast of England, coal was mined from at least the nineteenth century, but increased technology allowed for deeper and more ambitious mining. This was partly to provide the coal to fuel the technological advances of the twentieth century. A new mine was established in 1909 that sent a shaft vertically down 150 m, and tunnels were extended out under the North Sea eventually reaching 16 km offshore. Not surprisingly, the mine had a problem with the ingress of water and at one time was pumping out three tonnes of water for every tonne of coal raised.

In the 1970s, an aluminium manufacturing company bought land adjacent to the mine and built a smelter and a dedicated coal-fired power station: the smelter used large amounts of electricity to refine the aluminium from the bauxite ore as this was more economical than buying electricity from a power company via the grid. The mine supplied coal to the power station via a conveyor from the pit head without the need for trains or lorries to move it over land. In the 1980s, following decreasing demand for coal in Europe, the easy availability of cheap coal from other parts of the world, and the disruption to mines by industrial action, most of the underground coal mines in the United Kingdom were not considered economic. The link with the aluminium smelter made the colliery at Lynemouth an exception.

However, in 1992 new pollution control regulations were enacted prohibiting the disposal of solid waste to sea. The colliery at Lynemouth bulldozed its minestone on to the foreshore and in doing so created an artificial raised beach that protected the mine and the power station from flooding by the sea during storms. Between 1993 and 1994 the mine had ceased operation during the privatisation of the UK mining industry, and during a storm in 1994, the sea broke through and flooded the power station to a depth of 30 cm.

Ecological surveys of the artificial beach showed that although it was impoverished compared to the unimpacted sites nearby, it still contained a functioning infaunal community. The sediments were coarser and more poorly sorted (more variable in size) than nearby beaches, and the ridge of minestone at the top of the shore was leaching iron-rich and acidic water onto the shore when it rained. Earlier ecological studies at other coastal mine disposal sites in north-east England showed similar changes, and following closure of the mines these sites recovered.

The marine waste-disposal legislation included an exemption that allowed waste disposal when there were beneficial effects (socio-economic or ecological). The mining company was able to argue that the minestone placement on the shore was beneficial in that it provided flood protection for the power station and that the ecological impacts of this were limited and ultimately reversible. If the mine had not used the spoil in this way then a concrete seawall (at a cost of many millions of pounds) would be required, and this would also alter the ecology of the shore and would do so permanently. The UK and European regulators accepted this was the best practical environmental option and the mine continued to operate.

Unfortunately, a major flood in the mine in 2004 resulted in its sudden and premature closure. This therefore ended the placing of minestone on the foreshore and the aluminium smelter and power station had to be protected by a new sea wall of concrete and large granite boulders imported from Norway (at a cost of £2.4 million).

---

metals and some of the radioactivity present in the original coal.

The pfa has a potential use as an inert filler in the construction industry and in the production of cinder blocks. In the United Kingdom and United States, typically around 40–50% of the ash is used in this way, but the demand for ash varies, being higher when the economy (and hence construction) is booming. The fraction of the ash not used has to be disposed. Placing an inert fine dust into landfill sites is inefficient and costly so most ash has been placed into heaps on site; these are initially unsightly but can be landscaped with a top covering of soil and planting. However, as many power stations are located on the coast, in order to use the sea as a source of cooling water (see Section 3.7), disposal of the ash to sea was investigated as a potential disposal option. Disposal of ash into the sea falls under the Dumping at Sea provisions of the London Dumping Convention (see Chapter 7) and as such it is regulated.

On the north-east coast of the United Kingdom a number of coalmines disposed of minestone at offshore disposal sites by dumping from barges. In the

1960s when Blyth power station was designed, it was decided to transfer ash to a purpose-built ship for disposal at sea. An area approximately 12 km offshore was designated as the receiving environment. The power station ash, unlike minestone, is a uniformly fine powder and, as it has been burnt, contains zero organic matter. When deposited in the sea it produces a plume of fine material that reduces light penetration, potentially clogs gills and feeding structures, and then settles to the seafloor as a blanket of sterile sediment free of organic matter. Monitoring of the disposal site showed that while there were no obvious toxic effects from the waste (the ash does contain metal residues from the coal), the smothering effects were severe. At the peak of the operation 42 km^2 were turned into an abiotic desert. No other power station in the United Kingdom therefore ever employed marine disposal of its ash.

In Europe, under the provisions of the 1992 North Sea Ministers' Agreement, solid waste dumping was banned and all power station ash is now utilised or disposed on land. Ecological studies at the former Blyth dump site show that as organic matter began to accumulate, organisms began to colonise the area, but even a decade after the cessation of dumping both species richness and the abundance of organisms remained lower in the former disposal site than at control sites.

Power station ash and minestone are solid wastes that are generally non-toxic. Around half the ash produced from power stations is used in the construction industry and in some cases minestone has also been used beneficially, for example as a construction aggregate (see also Box 3.3). These cases illustrate how one industry's waste can be another's raw materials. With solid wastes, such as ash and mine spoil, the costs and difficulty of moving material from the location where it was produced to the site of demand can be high and prohibitive; it is cheaper to get new stone of known quality and specification from a nearby quarry than to transport spoil of variable size and composition from a distant mine site. While ash and minestone from coalmines contains low levels of a variety of metals they are not generally regarded as highly toxic. Minestone from metalliferous mining operations can be a source of contaminants and, as demonstrated by the legacy from metal mining in Britain by the Romans, these can continue to present problems many years after the original mining operations ceased.

3.7 Heat

Traditional power-generating plants use an energy source—coal, oil, gas or a nuclear reactor—to heat water to steam which are used to drive turbines that generate electricity. In order to make this process efficient, the steam needs to be rapidly cooled after it has passed through the turbines so that it turns back into water to begin the process again. The cooling is achieved by passing the steam through a heat exchanger/cooler that transfers the heat (energy) to the coolant (which is usually water). A variety of other industrial processes also require a cooling step and these use essentially the same process. The water used as a coolant can be recycled, in which case a large reservoir must be maintained to allow it cool before reuse or it can be drawn from the environment. The advantage of drawing water from the environment is that being cool, it can hold/absorb a lot of heat and the water can be returned to the environment without the need for extensive tanks/lagoons on site.

For power stations and industrial activity located on rivers, a continually renewed supply of water is available, but the volume required may represent a large portion of the river's flow. This means that the water returned to the river needs to be cool enough that it does not affect the ecology. In practical terms this usually means that the returned water needs to be more than 2–3°C above the ambient temperature, and so cooling towers are used to cool the water prior to its return into the environment. Power plants on coasts and estuaries usually return the cooling water without any further cooling, at typically 9–10°C above ambient temperature. The cooling towers are the dominant visual features of inland power plants but are absent from coastal plants. The heat energy added to the environment by the warmed cooling water potentially constitutes thermal pollution.

Thermal pollution causes a number of impacts. Warm water carries less dissolved oxygen than cooler water and so increases the risks of hypoxia. Warm water also increases the metabolic rates of invertebrates and fish, increasing their food and oxygen demands (at a time when less oxygen is available). Many toxicants are more potent at higher temperatures and warm conditions may cause chemical reactions between contaminants to occur at higher rates than under ambient conditions. Warmer conditions may also interfere with biological rhythms such as reproductive cycles by masking or replacing natural cues.

The vast volumes and dynamic nature of coastal waters, for example, waves and tides mixing and dispersing waters, mean that the effects of thermal pollution in coastal seas have rarely been reported. The warm water tends to float on top of the cooler receiving waters so that it is in contact with air at the sea surface and quickly becomes reoxygenated as it cools and is mixed by waves and currents into the water column. The continuous point-source additions set up a mixing zone of warm, low-oxygen (stratified) water that gives way to normal conditions after mixing is complete. In semi-enclosed estuaries and bays where water exchange is restricted, ecologically significant warming can occur (Box 3.4). However, the most severe effects of cooling water discharges that have been reported are due to the hypoxic effects produced in rivers and stationary bodies of water such as docks.

In addition to thermal pollution, extracting water for use in cooling can cause additional impacts on the ecology of the source water for example by the entrainment (drawing in) of fish and other organisms. In order to prevent the growth of attached organisms within the cooling system, the water is often treated by adding chlorine that can react with organic matter and other compounds present in the water. This means that in addition to being heated, the effluent water may contain a cocktail of chlorine compounds and organic matter from the deceased biota (Section 5.7.1). The effluent pipes from the power station often prove to be popular feeding grounds for birds and fish attracted by the dead and moribund organisms. These organisms therefore risk exposure to any toxicants produced in the effluent.

Box 3.4 *Mercenaria* **fishery in Southampton Water: The dependence on power station emissions**

The story of the American hard-shelled clam *Mercenaria* in Southampton Water, a large estuary on the south coast of England, brings together the introduction of an invasive non-native (alien) species and a power station effluent. The city of Southampton sits at the head of the estuary (Box 3.4, Figure 1) and had a power station dating back to the nineteenth century that was closed down in the 1950s due to its negative impact on the city's air quality. It was replaced by a larger, cleaner power station on the opposite bank of the estuary at Marchwood. Both power stations used cooling water drawn from, and returned to, the estuary. Amongst the natural local fauna was the large bivalve mollusc, the soft-shelled clam *Mya arenaria*. Populations of *Mya* were dramatically reduced and even extirpated by the severe cold winters of 1947 and 1962–63.

In 1925, a barrel of American hard-shelled clams, *Mercenaria mercenaria*, were imported to Southampton to be tested as a possible new bait for catching eels. The clams were initially placed on the mudflats adjacent to the town power station. *Mercenaria mercenaria* naturally occur on the east coast of North America, from Nova Scotia in the north to as far south as the Yucatan Peninsula in Mexico. Within its natural range *Mercenaria* breeds when water temperatures exceed 22–23°C. Following its introduction in 1925, *Mercenaria* spread slowly around Southampton Water, but then in the 1950s it expanded more rapidly possibly occupying the niche vacated by the demise of the native *Mya* population.

Seawater temperatures in Southampton Water would rarely naturally exceed 18°C, but the effect of the warm water from the power stations, plus the additional cooling water discharges from the Fawley Oil Refinery and Power Plant in the lower estuary allowed *Mercenaria* to breed. Interestingly, it appears that the local population now breeds at temperatures 3–4°C lower than the native populations. Whether this is the result of an evolutionary shift or simply plasticity (the ability to express a variety of responses such as changing breeding temperature) is unclear.

In the 1960s, a fishery for *Mercenaria* developed which involved the dumping/discarding of undersize individuals and this deliberate spreading by fishers may have also helped to spread the clam around Southampton Water

(continued)

Box 3.4   (*Continued*)

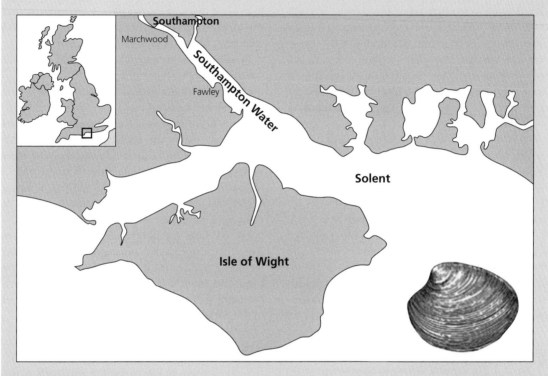

Box 3.4, Figure 1 Southampton Water on the south coast of England. The town power-generating stations were replaced by a larger-scale coal-fired plant at Marchwood and then an additional oil-fired plant at Fawley. Cooling water from these contributed to raising sea temperatures in Southampton Water allowing the non-native hard-shelled clam *Mercenaria* (inset) to breed.

and into nearby estuaries and harbours. By the 1980s the fishery was suffering from over-fishing and the closure of power station in 1983 meant that the remaining population failed to breed. *Mercenaria* is a long-lived species and a new, gas-fired, power station opened on the site in 2010, and while it discharges a much lower volume of water than its predecessor it will be interesting to see if any of the few, now, ageing clams begin to breed again.

3.8 Acidity

The combustion of fossil fuels, particularly coal and heavy fuel oil, in addition to producing large amounts of carbon dioxide also release a range of other pollutants into the air. These pollutants include sulphur compounds that react with water to form sulphuric acid. The sulphuric acid then returns to the Earth's surface in the form of acid rain. The phenomena was first described in the eighteenth century, following widespread industrialisation.

The quantity of acidic wastes deposited in northern Europe and North America had major impacts on freshwater ecosystems and forests. It produced acidic soils and degraded lake and stream ecosystems. Acid rain also fell over the seas and oceans. During the 1970s these changes became critical and the 1980s saw major efforts in North America and Europe to reduce atmospheric emissions.

While acid rain had extremely severe impacts on freshwater systems and upland soils, there was no discernable impact on marine waters in the regions affected. Freshwater contains low levels of mineral ions and so has a very limited ability to undergo chemical buffering. In contrast, seawater contains a wide range of ions that provide a high chemical buffering capacity and the pH of seawater generally remains close to the natural level of pH 8.2; that is, slightly basic. In fact the additional sulphur, as sulphite or sulphate ions, is utilised as a nutrient by marine plants as it is required for the synthesis of 3 of the 21 amino acids.

The sulphur dioxide and other acidic compounds emitted from industrial complexes are mixed and dispersed in the atmosphere so that the acid rain produced falls over a large area. Over time this produces drastic changes in the freshwater and terrestrial systems and the dispersed nature of the inputs did not exceed the seas buffering capacity. Of greater concern has been the emissions from vessels (Figure 3.9). The heavy fuel oil used to power large commercial vessels often contains high levels of sulphur compounds which when burnt are liberated as acidic gases. In fact, during refinery manufacturing, heavy marine fuels are often preferentially loaded with the sulphur fraction to prevent its addition to fuels used in trucks and automobiles, a case of deliberately sacrificing the marine environment

to protect urban environments. With the controls imposed on terrestrial sources of sulphur dioxide, attention has shifted to marine fuels, and controls were enacted by, amongst others, the United States, Canada and the European Union early in the twenty-first century to limit the sulphur content of marine fuel oils. These controls have now been incorporated into global standards.

The widespread acidification of soils and the dramatic impacts of acidification on lakes in North America and northern Europe led to a raft of controls on the emissions of acidic flue gases. These generally required the fitting of flue gas desulphurisation technology. The most widespread approach used limestone (calcium carbonate), the quarrying of which has its own set of environmental impacts, to absorb the sulphur dioxide producing calcium sulphate and water. The calcium sulphate, or gypsum, could then be used in a variety of ways including in the construction sector as a constituent of plaster. For power stations and industrial plants located at the coast, and particularly those using seawater for cooling, an alternative flue gas desulphurisation process was proposed. Instead of passing the gases over limestone or bubbling them through a limestone solution, the warm seawater from the cooling

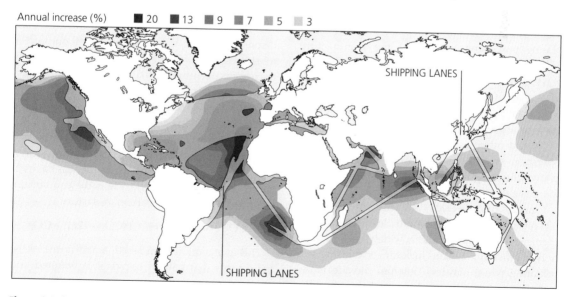

Figure 3.9 The annual increase in sulphur emissions along marine shipping routes associated with the growth in marine shipping. Source: http://www.solaripedia.com/13/119/1072/shipping_lanes_sulfur_emissions_map.html.

water could be sprayed into the flue gases. Reactive compounds such as sulphur dioxide would then be absorbed into the seawater and so removed from the gaseous emissions in a process known as *seawater scrubbing*. This, of course, results in the cooling water effluent being acidic. Laboratory trials showed that while the warm, acidic seawater had negative effects on a range of marine organisms, these effects only occurred at high levels of exposure and, once discharged, any effects would have been localised close to the discharge point. Seawater scrubbing has been employed at power plants in the United States and oil refineries in Norway. Thus, while this option is commercially viable, in the end, concerns about hot acidic seawater and the possibility of unexpected interactions with other substances (pollutants, organic material, etc.) present in the seawater have limited its widespread application.

Sulphuric acid-based inputs from combustion are not the only source of acidic residues in the marine environment. A number of industrial processes produce acidic wastes and the naturally basic and well-buffered nature of seawater make it an attractive disposal option because it will chemically neutralise the waste as well as diluting and dispersing it. As noted earlier, probably the largest and most difficult to control sources of acidic waste are from mine wastewaters and leachate from mine spoil. Other major sources of acidic wastes include metal plating/processing plants and the manufacture of titanium dioxide (TiO_2). Minor sources include waste liquor from food-pickling plants. Small-scale discharges of these wastes would usually be dealt with simply by dilution and possibly, particularly for food wastes, neutralisation by liming. The acidity would otherwise interfere with the biological operation of any sewage/wastewater treatment plant addressing the biological oxygen demand of the waste (Section 3.2.2).

Titanium dioxide (TiO_2) is a white pigment that has replaced, on safety grounds, lead-based white pigments in paints and and is widely used in a variety of other applications including stabilising and colouring white plastics. Titanium dioxide is extracted from its ore by two industrial processes: the chloride process and the acid-iron process. The former is the more widely used but requires a higher

purity of ore than the acid-iron process. One of the reasons the chloride process is more widely used is that although it uses large amounts of energy and some highly toxic chemicals, there are less onsite waste-disposal problems because some of the reagents are recycled. The iron-acid process is less costly and can use lower grade ore, so is more prevalent in more economically challenged locations. It produces a waste stream containing sulphuric acid and iron salts. In Europe, plants using the iron-acid process discharge iron-rich acidic waste to estuaries, via pipelines. Locally the precipitation of iron can cause impacts but the acid component appears to be quickly neutralised by the seawater.

In the 1970s and 1980s, a Belgium titanium dioxide factory dumped a liquid iron-acid waste from a specially designed vessel into the southern North Sea. The vessel would steam around inside a designated disposal area with the liquid waste discharged form a number of pipes located close to the propeller so the waste was actively mixed into the seawater immediately after disposal. Studies of the impacts of the operation could find no measureable increase in acidity or the iron content of the area even immediately after a disposal operation. In spite of this lack of a measureable effect, the practice was banned under European environmental legislation and has now ceased.

It is ironic that there is so little impact of the disposal of acidic wastes to sea when one of the most pressing environmental concerns in the early twenty-first century is that of ocean acidification (Section 8.3). The seas and oceans have absorbed over 50% of the carbon dioxide emitted by fossil fuel combustion. In doing so, the oceans have significantly slowed the rate of global warming and hence mitigated the impacts on terrestrial ecosystems. However, one of the reasons the seas have absorbed so much carbon dioxide is the equilibrium chemical balance of the carbonate system:

$$CO_2(g) + H_2O \leftrightarrow H^+ + HCO_3^- \leftrightarrow 2H^+ + CO_3^{2-}$$

This is an equilibrium reaction and is one of the main ways that natural seawater is buffered and any increase in acidity (i.e. hydrogen ions) pushes the reaction to the left. However, more carbon dioxide gas dissolving in the sea drives the reaction to the right, producing more carbonic acid (H_2CO_3), which in solution dissociates to hydrogen ions and a carbonate ion. At the time of writing the global ocean pH was about 8.2, a decrease of 0.1 pH unit from the long-term average. This is a seemingly small change (although the pH scale is a logarithmic one), but if atmospheric carbon dioxide levels continue to increase this change will accelerate, raising concerns about the ability of calcifying taxa to secrete calcium carbonate skeletons (see Section 8.3 for further discussion).

3.9 Synthesis

- The problems arising from sewage contamination are amongst the earliest accounts of pollution. Treatment technology is well developed to address the oxygen-demanding nature of the waste, the nutrient load that arises from the breakdown of the organic material, and the public health issues of pathogens and parasites. However, many coastal schemes utilise the marine environment to provide at least some of this treatment, thereby reducing capital and infrastructure costs.
- The addition of toxic chemicals and the presence of pharmaceuticals, along with potential to react with the organic wastes to produce novel compounds, means that sewage can contain many biologically active compounds. Not all of these compounds are removed by treatment and there is increasing concern about these biologically active materials (see Chapter 5).
- One of the seminal points in human history was the development of metalworking and the widespread industrialisation of the last 300 years has resulted in greatly increased releases of metal pollution from mining, manufacturing, and disposal operations. Seawater contains all naturally occurring metals and the impacts of metal pollution are usually restricted to near the sources. Some metals, however, are highly toxic and can be concentrated by the food chain, placing humans at risk. Examples include mercury build-up in tuna. The use of metals, and in particular organic forms in anti-fouling preparations, continues to pose a threat to marine systems.

- Like some metals, persistent organic pollutants (POPs) can bioaccumulate and are very persistent and will eventually gain a global distribution, DDT being a classic example. There is a wide range of manufactured chemicals now in circulation and the number of POPs is continually expanding. Many are carcinogenic and disrupt metabolic and reproductive processes. POPs have become a major focus of international pollution regulation.
- Extracting minerals from mines and obtaining sediments from the seafloor (whether for use as aggregate or to deepen shipping channels) result in sediments being redistributed. This causes physical impacts on the seafloor and turbidity in the water column. These effects can alter patterns of productivity and species distributions. Where mining generates metalliferous wastes or dredging occurs in regions with sediments contaminated by historic pollution, the movement of sediments can stimulate the release of these legacy contaminants and toxicity effects can occur.
- Thermal pollution and acid wastes only have local impacts, although the effects of thermal pollution can be locally significant when the distribution or biological functioning of species is altered due to the changed temperature regime. Looking ahead, global warming and the associated ocean acidification, linked with the atmospheric pollution from burning fossil fuels, mean that temperature and acidic effects will have global reach and the effects are not yet fully known.

Resources

For an interesting account of the evidence of health impacts of air pollution in the ancient world, see;
http://www.livescience.com/14420-ancient-egyptian-mummies-lung-disease-pollution.html.
For a comprehensive catalogue of the sources and effects of heavy metals in the marine environment, see;
http://www.marbef.org/wiki/heavy_metals.
http://www.who.int/mediacentre/factsheets/fs225/en/.
For information of the impacts and management of dredging in the Great Barrier Reef Marine Park, see;
http://www.gbrmpa.gov.au/media-room/latest-news/corporate/2015/independent-report-brings-together-knowledge-on-dredging-and-disposal.

Bibliography

Bengtson-Nash, S.M., Waugh, C.A. and Schlabach, M. (2013). Metabolic concentration of lipid soluble organochlorine burdens in the blubber of southern hemisphere humpback whales through migration and fasting. *Environmental Science & Technology*, 47, 9404–413.

Birchenough, S.N.R. and Frid, C.L.J. (2009). Macrobenthic succession following the cessation of sewage sludge disposal. *Journal of Sea Research*, doi:10.1016/j.seares.2009.06.004

Hong, S., Candelone, J.-P., Patterson, C.C., and Boutron, C.F.(1994). Greenland Ice Evidence of Hemispheric Lead Pollution Two Millennia Ago by Greek and Roman Civilizations. *Science*, **265**, 1841–843. DOI: 10.1126/science.265.5180.1841

Jones, K.C. and de Voogt, P. (1999). Persistent organic pollutants (POPs): state of the science. *Environmental Pollution*, **100**, 209–21.

Kennish M.J. (1998). *Pollution Impacts on Marine Biotic Environments*. London: CRC Pres.

McCook, L.J., Schaffelke, B., Apte, S.C., Brinkman, R., Brodie, J., Erftemeijer, P., Eyre, B., Hoogerwerf, F., Irvine, I., Jones, R., King, B., Marsh, H., Masini, R. Morton, R. Pitcher, R., Rasheed, M. Sheaves, M., Symonds, A., and Warne, M.S.J. (2015). *Synthesis of Current Knowledge of the Biophysical Impacts of Dredging and Disposal on the Great Barrier Reef: Report of an Independent Panel of Experts*. Townsville: Great Barrier Reef Marine Park Authority.

Nriagu, J.O. (1983). *Lead and Lead Poisoning in Antiquity*. London: Wiley, London.

Paipai, E. (1994). Environmental enhancement using dredged material. PIANC-Bulletin, **85**, 5–20. https://www.iadc-dredging.com/ul/cms/terraetaqua/document/1/3/0/130/130/1/terra-et-aqua-nr92–01.pdf.

Reijnders, P.J. (2003). Reproductive and developmental effects of environmental organochlorines on marine mammals. In: J.G. Vos, G. Bossart, M. Fournier, and T. O'Shea (eds). *Toxicology of Marine Mammals*. London: Taylor & Francis, 55–66.

Yordy, J.E., Wells, R.S., Balmer, B.C., Schwacke, L.H., Rowles, T.K. and Kucklick, J.R. (2010). Life history as a source of variation for persistent organic pollutant (POP) patterns in a community of common bottlenose dolphins (*Tursiops truncatus*) resident to Sarasota Bay, FL. *Science of the Total Environment*, **408**, 2161–172.

CHAPTER 4

Ongoing issues

4.1 Introduction

Although our understanding of the sources, fate and impacts of many contaminants are now relatively well known, legislated for and regulated by national and international bodies and conventions, a number remain problematic. Some of these contaminants are produced in such large quantities and/or are so persistent that they are now prevalent throughout the marine environment. All are known threats that have either been in large part ignored, took time to manifest, or have been challenging to manage. These constitute the ongoing issues in marine pollution management.

Our excessive production and disposal of vast quantities of plastic waste with exceptional longevity in to the environment is apparent all around us (Sections 4.2.1 and 4.2.2). Plastic can now be found from the abyssal plains to the intertidal and from pole to pole. Plastics are a persistent waste that have long been recognised as a problem (Section 4.2.3) however they have continued to enter the oceans in large quantities. The promises of the biodegradability of new materials are often deceptive and simply mean that they fragment quickly rather than truly biodegrading (Section 4.2.5.3). Far from sight and mind, the litter in the sea has received far less attention than litter on land, except for the occasional entanglements of charismatic marine megafauna. While terms such as the *Great Pacific Garbage Patch* are alarmist, they are ultimately based in fact and have helped to raise awareness of the problem. However, the challenges of marine plastic pollution may have just begun with the discovery that much of the plastic waste in the sea remains present as smaller and smaller particles that may be finding their way into our food chain (Section 4.2.5).

Although the impacts of adding extra nutrients to the marine environment are known, they are proving difficult to the manage and mitigate (Section 4.3). Nutrient additions are therefore an ongoing challenge that will continue to grow as human populations increase. Similarly, detergents will continue to be produced and added to the sea in larger quantities (Section 4.4).

The sources and impacts of oil pollution are known and regulated but some additions continue to occur (Section 4.5.1). The economic benefits of hydrocarbon exploitation are large and are presumed to outweigh the environmental costs and so production continues. More than a century of use has led to many of the toxic compounds contained within oil—for example, the toxic polyaromatic hydrocarbons—accumulating in coastal and estuarine sediments around the world. Technological advances such as the load-on-top method and double skins for tankers have reduced accidental and operational discharges from ships. However, accidents continue to occur and small amounts of oil are released through routine operations and the quantities are cumulatively significant (Sections 4.5.2 and 4.5.3), and so remains an ongoing challenge.

Some contaminants are of such high potential risk to human health that they cannot be ignored; for example, radioactive wastes which, despite being stored far from human habitation, are highly toxic and if disturbed could be released (Section 4.7). The generation of energy from nuclear sources is in some ways 'clean' in that it does not release atmospheric carbon dioxide and it is portrayed as a cost-effective means of generating energy. However, despite technological advances in nuclear energy, the industry continues to produce highly

Marine Pollution. Christopher L. J. Frid & Bryony A. Caswell.
© Christopher L. J. Frid & Bryony A. Caswell 2017. Published 2017 by Oxford University Press.
DOI 10.1093/oso/9780198726289.001.0001

radioactive waste with very long residence times and it must be treated and stored in such a way that does not damage the environment nor affect human populations. There is also renewed interest in further developing nuclear technologies to supply more of our energy needs, particularly in China.

The threats to biosecurity from non-native invasive species have been ongoing for centuries. Non-native invasive species introductions (NISs), some of which have been deliberate, have increased as global trade has grown and consequently coastal marine community composition has become somewhat homogenised (Section 4.8). An improved understanding of the biosecurity risks, the vectors and the consequences of NIS mean they are now recognised as a major threat and successful control measures have been enacted in some areas. In other areas they are a major problem and are having considerable economic impacts on marine resource exploitation and natural ecosystems. It is hoped that new developments in shipping operations will reduce NIS introductions, although these procedures could also increase the additions of other pollutants.

4.2 Plastics and litter

Discarded manmade materials composed of plastic, metal, cardboard or textile products find their way into the marine environment in large quantities (Section 4.2.1). The United Nations Environment Programme defines marine litter as 'any persistent, manufactured or processed solid material discarded, disposed of or abandoned in the marine and coastal environment'. Millions of tonnes of litter end up in the oceans every year that originate from both land-based and ocean-based sources (Figure 4.1). The land-based sources include landfill sites near to coasts and rivers and the disposal of litter by the public on beaches, piers or harbours. This litter may be transported to the sea by winds, rivers or in sewage and storm water run-off. Ocean-based sources include the accidental or deliberate disposal of waste, equipment and cargo overboard from commercial shipping (ferries, cruise liners, military and recreational vessels) and offshore platforms such as those used for oil and gas exploration (Section 4.5). Coastal and offshore aquaculture operations also introduce litter into the marine realm.

Fishing vessels regularly lose gear and other equipment and in some sea regions large quantities of lost gear are found (Section 4.2.3).

Litter inputs can be diffuse or point source and litter can be transported considerable distances before and, although less so, after sinking (Figure 4.1). Estimates of the proportions of plastic litter in the North Sea suggested that two-thirds of the plastic sinks, another 15% floats in the water column and the remainder is washed up on beaches. Litter degradation and overgrowth by marine organisms creates a loss of buoyancy causing it to sink. The break-up of litter into smaller and smaller pieces is leading to an increasing amount of sub-millimetric particles in the sea (Section 4.2.2). A considerable proportion of litter may become ingested or attached to marine animals (Section 4.2.3).

4.2.1 How much litter is present in the marine environment?

Ascertaining how much litter is actually present in the ocean is not a simple task because the global inputs to the sea are not well-documented and it is challenging to adequately sample litter from such a large area. In 1997, the US Academy of Science estimated that approximately 6.4×10^6 tonnes (t) of litter were entering the oceans each year. Some of the best data on litter inputs come from that collected by beach clean-up programmes (Section 4.2.1). In 2015, the total volume of litter collected from shorelines by the International Coastal Clean-up (ICC) programme was 18 060 t from 25 188 km of coastline (Figures 4.2a-c).

Most litter is composed of plastic, and in litter collected from the marine environment, plastic constitutes around 70% on shorelines (Figure 4.3), between 68% and 99% is present in the water column and 23% to 89% occurs on the seabed of the continental shelf (Table 4.3). Of the litter that has accumulated in seafloor canyons and seamounts, 41% is plastic and 34% is fishing gear. The remainder of the macrolitter is glass, wood, paper/cardboard and metal, although its composition varies greatly between the different regions.

The degradation rates for marine litter (Table 4.1) are thought to range from weeks to hundreds of years. Short-term and accelerated weathering

How plastic moves from the economy to the environment

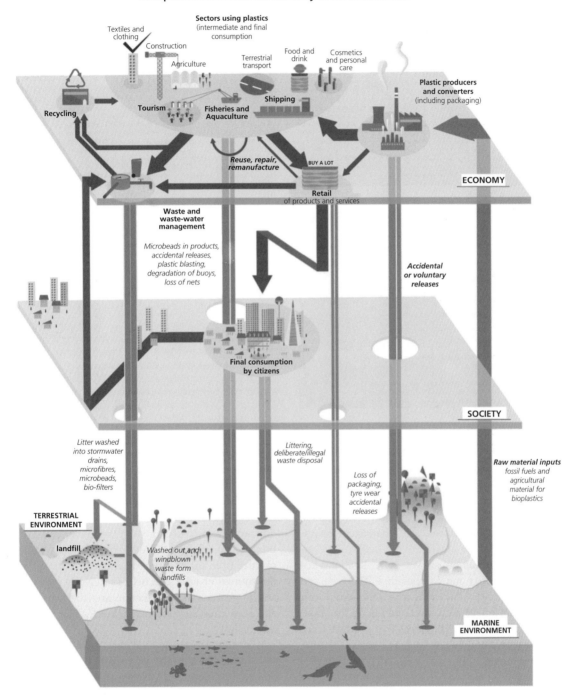

Figure 4.1 The main sources and pathways through which litter enters the marine environment beginning with production. Sinks include beaches, coastal and open ocean waters and the sediments below each. Transport may be wind-blown or transported through waterways, the sewerage system and/or stormwater drainage. Once within the water column, it may eventually sink and or accumulate on the seafloor, become buried in sediments or be ingested by marine organisms. Credit: Maphoto/Riccardo Pravettoni and GRID-Arendal (http://www.grida.no/graphicslib/detail/how-plastic-moves-from-the-economy-to-the-environment_e3d5#).

(a)

(b)

(c)

Figure 4.2 (see also Plates 4, 5 and 6) (a) Shoreline litter. © Greg Martin, Surfers against sewage 2016. (b) Common terns (*Sterna hirundo*) on the shoreline at Seaforth Nature Reserve, Liverpool. © Paul Scott 2016 (sgpscott@flickr/Facebook). (c) Mesoplastics and microplastics on the strandline of a beach in Cornwall, UK. © Tracey Williams (creator of the Lost at Sea Facebook group: https://www. facebook.com/groups/LostAtSeaGroup/).

experiments have shown that the least degradable materials are plastics. Ultimately, though it is not certain how long plastics take to degrade because this has not yet been observed in the environment. However, the very slow degradation rates of plastics means they are considered a persistent pollutant (Chapter 1).

Litter accumulation shows large spatial variability in quantity and composition. The type and quantity of macrolitter on beaches, the seafloor or floating in the watercolumn depends on the proximity to cities or industrial areas that produce large volumes of waste. The level of beach use, the nature, extent and proximity of offshore maritime activities such as fishing grounds and shipping lanes are also important. Local and regional hydrodynamics such as tides, storms and ocean current patterns also affect litter distribution.

The identity of litter items can be used to determine the use category and thus its source. Linking litter to the source is straightforward for some items like fishing gear, sewage debris or litter left by tourists, but others are indistinguishable. Ocean circulation models and lost cargo (e.g. the rubber bath toys; Box 4.1) can be used to determine the likely routes of litter from origin to destination. Understanding the sources of litter can help to develop measures and targets for litter reduction and the legislation and regulation surrounding activities that produce marine litter.

It is now largely agreed that most marine litter, around 80%, originates from land. Most beach litter is composed of plastic, glass or metal bottles, lids and other food packaging that originates from shore-based activities. This predominance of land-based litter has been shown for shorelines in

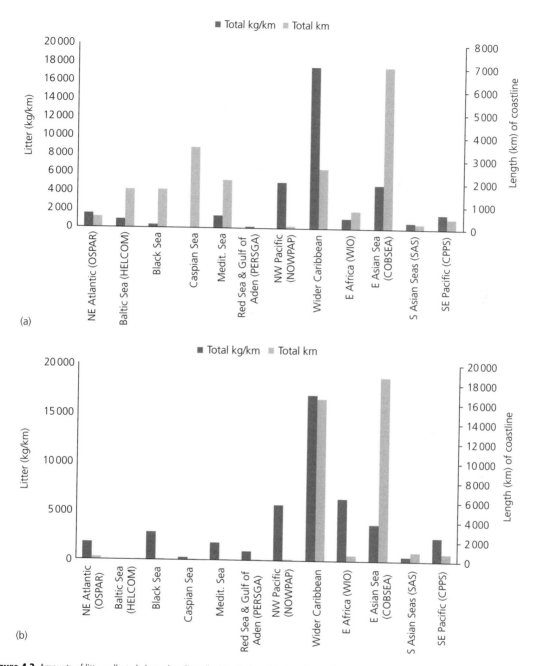

Figure 4.3 Amounts of litter collected along shorelines (kg litter/km) and the length searched (km) during UNEP International Coastal Clean up events between (a) 2005 and 2007, and (b) 2015. Differences in collection effort between regions may be responsible for some of the patterns observed. Data from: UNEP 2007 and Ocean Conservancy (2015, 2016). Note: data from the HELCOM region in 2015 do not include Poland, Denmark, Finland, Estonia or Latvia due to changes in ICC participation and so the length of shoreline searched was reduced.

Table 4.1 Degradation rates of marine litter. Credit: United States National Oceanic and Atmospheric Administration (NOAA 2016).

Material	Degradation time
Paper towel	2–4 weeks
Newspaper	6 weeks
Apple core	2 months
Cardboard	2 months
Waxed milk cartons	3 months
Wool gloves	1 year
Plywood	1–3 years
Cotton fabric	1–5 months
Cigarette butt	2–10 years
Painted wood	13 years
Plastic bags	10–20 years
Polystyrene cup	50 years
Tin can	50 years
Aluminium can	200 years
Plastic beverage holder	400 years
Plastic bottles	450 years
Disposable nappies	450 years
Monofilament fishing line	600 years

various parts of the Mediterranean (e.g. Greece, Malta, Gozo), the United Kingdom, South China, Australia and central and southern America. In some regions litter can mostly be attributed to ocean-based sources such as fishing, for example, in the North Pacific, the North Sea and the Baltic Sea (Section 4.2.3).

4.2.1.1 How much litter is present on our shorelines?

Data on shoreline litter are regularly collected by coordinated clean-up activities; for example, the International Coastal Clean-up programme coordinated by Ocean Conservancy, a non-governmental organisation (NGO) and other international shore-based programmes such as those coordinated by the World Wildlife Fund for Nature and Clean Up the World. Regional governments, national NGOs and charities also conduct litter surveys on beaches such as those of the UK's Marine Conservation Society (Figure 4.4), Legambiete (Italy) and the Mediterranean Information Office for Environment, Culture and Sustainable Development (MIO-ECSDE). Litter

is usually recorded as the number of items or weight of items within a specified length or area of shoreline. Some of the shoreline monitoring programmes have been running for more than a decade; for example, the International Coastal Clean up (more than 25 years), the United Kingdom Marine Conservation Society's Beach Watch programme (more than 18 years), the German North Sea beach litter monitoring programme (more than 20 years) and the OSPAR programme (over 10 years).

The weight of litter per kilometre of shoreline increased substantially between 2002 and 2015. The ICC programme, coordinated across 100 coastal countries, collected a total of 4.0×10^3 kg of marine litter from 21 000 km of shoreline in 2002, 1.02×10^6 kg from 20 183 km in 2005–07, and 18.06×10^6 kg from 25 189 km of coastline in 2015. The areas with the greatest amounts of beach litter in 2005–07 and 2015 (Figure 4.3a-b) were the wider Caribbean, the North-West Pacific and the South-East Asian Seas. In the other regions the amounts of litter found per km of coastline were similar to those previously collected, although in almost all cases more litter was collected in 2015 (but in many cases the length of shore sampled was less).

In 2015, 60% of items found on beaches were, in order of quantity, cigarette butts, plastic bottles, food wrappers, plastic bottle caps, straws/stirrers, plastic bin bags, glass bottles, plastic grocery bags, metal bottle caps and plastic lids. Most of the litter was made of plastic and the overall composition has changed little over the last decade. The exact composition of beach litter varies between regions depending on local sources. For example, beach litter in the north-east Atlantic and the Red Sea contained a greater proportion of fishing nets, line and rope; the Black Sea, Baltic Sea, south-east Pacific and coastlines of east Africa and south Asia contain more clothing, and the Caribbean and Mediterranean seas contain high proportions of cigarette butts.

The long-term data from the annual beach cleaning events in the United Kingdom by the Marine Conservation Society, like the ICC data, show a gradual increase in beach litter since 1994. Three times as much litter was collected from beaches in 2015 compared with 1994 (Figure 4.4). Macrolitter accumulation on the shores of remote high-latitude regions shows that the marine litter supply fluctuated throughout the 1980s and 1990s, and it has increased through time in most areas, for example, in the Gulf of Alaska, the Bering Sea, South Georgia and the South Shetland Isles (Table 4.2).

Table 4.2 Litter accumulation on the shores of isolated islands at high latitudes in the northern and southern hemispheres through time. Data from: Erikkson et al. 2013 and references therein.

Year	Location	Quantity (items km^{-2} d^{-1})
1985–86	Yakutat, Arctic	0.04
1986–87		0.05
1987–88		0.03
1985	Middleton Island, Gulf	1.59
1986	of Alaska	2.77
1987		2.71
1972	Amchitka Island, Bering Sea	0.53
1973		0.78
1974		1.36
1982		1.61
1987		0.85
1991	Bird Island, South Georgia	0.30
1992		0.30
1993		0.93
1994		0.70
1995		4.53
1987–88	Livingstone Island, South	0.07
1990–91	Shetland Isles	0.15
1993–94		0.34
1994–95		0.55
1995–96		0.84
1996–97		0.32

4.2.1.2 How much litter is present in the sea?

The amount of macrolitter present in the sea may be determined by visual seafloor surveys conducted by divers, snorkellers or using manta-tows in waters less than 10 m deep. These types of surveys are coordinated by programmes such as PADI project AWARE. Visual surveys are conducted using strip transects, line transects or plot surveys of the number of items per area of sea or seafloor surface. The recommended sampling methods for visual surveys vary with litter density, visibility and seabed complexity (e.g. seagrass bed or reef is more complex than a flat, soft sediment).

Seabed surveys can also be conducted visually in shallow waters, but in deeper water the seabed is sampled using trawls and similar methods as those used for scientific sampling and demersal fishing. The swept area of the trawl is calculated from its location data and the litter is quantified within this area. Surveys of seafloor litter are made during occasional scientific cruises and established annual or subannual bottom-trawl survey programmes; for example, the International Bottom Trawl Survey, the Baltic International Trawl Survey, the Beam Trawl Surveys, the Mediterranean Trawl Survey (MEDITS, run by the European Commission) or the Black Sea programme. Many of these bottom-trawl surveys are national programmes that were later combined into larger datasets and were originally established for fisheries monitoring

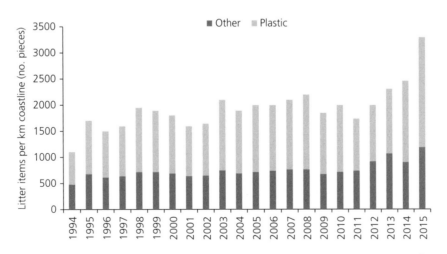

Figure 4.4 Beach litter collected (number of items per km coastline) from around 300 UK beaches (approximately 100 km of coastline) during the Marine Conservation Society's annual beach clean events between 1994 and 2015. Data from: MCS UK Beach Watch.

purposes. Some of the data from the bottom trawl survey programmes exceed decades; for example, the International Bottom Trawl Survey (more than 26 years), the Baltic International Trawl Surveys (more than 16 years) and the Mediterranean Trawl Survey (more than 22 years). The deepest regions of the oceans such as the continental slope, canyons and seamounts are sampled less regularly for litter because of the high costs of sampling. There is also generally less routine ecological monitoring in the deep sea and so the litter datasets are smaller/fewer for the deep sea. Litter may be sampled from the deep seafloor using trawls in soft seafloor areas and remotely operated vehicles for visual transects in rocky areas or on slopes.

Data on litter within the water column have not been routinely collected. There are no data exceeding five years' duration, unlike those from shoreline or seabed surveys, and there are large areas for which there are no data. Information on the amounts of macrolitter floating on the surface of the ocean is now assessed visually along line transects or within plots from ships, platforms or aircraft. The quality of data from visual surface surveys does depend on weather conditions and the skill of the observers. Identification of the type and therefore origin of the litter is obviously more difficult and for aerial surveys it is only reliable for objects exceeding 30–40 cm in size. Surface surveys may be conducted as part of scientific cruises, on ships of opportunity, from oil or gas platforms or during aerial megafauna surveys. Assessments of floating microlitter require different sampling methodologies (Section 4.2.2.3).

Litter can be found from the Arctic Ocean to the Southern Ocean and from the continental shelf down to the abyssal plains, in deep ocean basins and canyons (Table 4.3). Trawl net surveys show that considerable amounts of litter have accumulated on the continental shelves and most of this litter is made from plastic (Table 4.3). Within the northeastern Atlantic several zones have been identified where litter is accumulating, between hundreds and thousands of pieces per kilometre, on the seafloor; for example, in the Fram Strait, the Norwegian Margin, the southern Celtic Sea, the south-east coast of France, the north-east coast of Spain, in the Thyrenian Sea, off north-west Italy, and the Azores.

Table 4.3 Macrolitter in the sea, quantities and, where available, the proportion of plastic present. Litter from the canyons and seamounts, banks and deep-sea basins was 41% plastic. Data from: Morris 1980, Galgani et al. 2000, Kuriyama et al. 2003, Barnes et al. 2009, Pham et al. 2014 and Galgani et al. 2015.

Region	Type	Plastic	Year
Offshore: water column			
British Columbia	1.48 km^{-2}	92%	2011
North Sea	25–38 km^{-2}	70%	2011
Belgian coast	0.7 km^{-2}	95%	2013
South-west Malta	2000 km^{-2}		1980
Mediterranean Sea	10.9–52 km^{-2}	96%	2014
Ligurian Sea	1.5–25 km^{-2}		2003
Kerch Strait/Black Sea	66 km^{-2}		2007
North Pacific	459 km^{-2}	95%	2011
Chile	10–50 km^{-2}	80%	2009
Sorth China Sea	4.9 km^{-2}	68%	2011
Bay of Bengal	8.8 km^{-2}	96%	2013
Strait of Malaca	579 km^{-2}	99%	2013
Southern Ocean	0.032–6 km^{-2}	96%	2014
Offshore seabed: shelf			
Fram Strait, Arctic	3635–7710 km^{-2}	59%	2002–11
Kodiak Island, Alaska	11–147 km^{-2}	47–59%	1999
Norweigan margin	970 ± 380 km^{-2}		2007
West coast USA	30 km^{-2}	23%	2007–08
Oregon coast	149 km^{-2}	26%	1990
Celtic Sea	528 ± 247 km^{-2}	30%	1998
East Channel, UK	17.6 ± 6.7 km^{-2}	85%	1998
North Sea	156 ± 37 km^{-2}	48%	1994–98
Baltic Sea	126 ± 82 km^{-2}	36%	1996
NW Mediterranean	1935 ± 633 km^{-2}	77%	1998
Bay of Biscaye	142 ± 25 km^{-2}	74%	1994–97
France (Medit.)	76–146 km^{-2}	29.5–74%	1994–2009
Gulf of Lion	143 ± 19 km^{-2}	64%	1994
	40 ± 10 km^{-2}		2009
Bay of Seine	72 ± 5.8 km^{-2}	89%	1993
South-west Malta	97 ± 78 km^{-2}	47%	1994
Spain (Medit.)	4424 ± 3743 km^{-2}	37%	2009
Azores, Portugal	1439 km^{-2}	89%	2010–11
Gulf of Patras, Greece	89 ± 240 km^{-2}	79–83%	1999
East Gulf of Patras	188 km^{-2}	66%	2008
West Gulf of Patras	437 km^{-2}	28%	2008
Southern and western Greece	72 ± 437 km^{-2}	56%	1999

(Continued)

Region	Type	Plastic	Year
Adriatic Sea	378 ± 251 km^{-2}	70%	2005
Thyrenian Sea	5960 ± 3023 km^{-2}	76%	2009
Eastern Corsica	230 ± 72 km^{-2}	46%	1998
Sicily/Tunisia Channel	401 km^{-2}	75%	1994–98
ABC Islands, Dutch Caribbean	2700 km^{-2}	29%	2000
South China	693 km^{-2}	47%	2009–10
Iwate, Japan	1590 km^{-2}	43%	2011
Tokyo Bay	270–550 km^{-2} 185–338 km^{-2}	40–42% 48–59%	1995 2003
Monterey Canyon, California	632 km^{-2}	33%	1998–2011
Rio de la Plata	1509 km^{-2}	37%	1995
Offshore seabed: slope			
Hausgarten observatory, Arctic	3635 km^{-2} 7710 km^{-2}	57% 36%	2002 2011
North Faroe-Shetland Channel	0.3 ± 0.2 km^{-2}		2006
North-east Faroe–Shetland Channel	1.9 ± 1.0 km^{-2}		2006
Offshore seabed: canyons			
Dangeard and explorer canyon	720 ± 270 km^{-2}		2007
Nazaré canyon	420 ± 160 km^{-2}		2007
Lisbon canyon	6620 km^{-2}		2007
Setúbal canyon	2460 km^{-2}		2007
Cascais canyon	1060 km^{-2}		2007
Guilvinec canyon	3190 ± 2810 km^{-2}		2008–10
Whittard canyon	140 ± 40 km^{-2}		2010
Blanes canyon	3210 ± 1190 km^{-2}		2009–11
Offshore seabed: banks, seamounts, deep sea basins			
Anton Dohrn seamount	190 ± 100 km^{-2}		2005–09
Condor seamount	1460 ± 300 km^{-2}		2010–11
Josephine seamount	570 ± 330 km^{-2}		2012
Hatton Bank	190 ± 80 km^{-2}		2005–11
Rockall Bank	70 ± 50 km^{-2}		2005–11
Rosemary Bank	330± 230 km^{-2}		2006
Pen Duick alpha/beta mound	250 ± 170 km^{-2}		2009
Darwin mounds	970 ± 290 km^{-2}		2011
Algero–Balearic Basin	180 ± 150		2009
Crete–Rhodes ridge	120 ± 30		2009

Litter is also accumulating at high densities in part of the Caribbean, off the coast of South China and north-east Japan. Hotspots of litter accumulation in the deep sea include the Lisbon, Setúbal, Cascais, Guilvinec and Blanes canyons, the Condor seamount and the Darwin mounds. These areas have comparable amounts of litter accumulation to that on the shelves and beaches (Figures 4.3 and 4.4, Table 4.3). Litter becomes trapped in areas with low circulation and high sedimentation, and local seabed topography dictates where it accumulates (e.g. in canyons or other depressions). In the Gulf of Lion in France and Tokyo Bay, litter accumulation on the seafloor has decreased but it has increased in other areas such as the Gulf of Patras in Greece and at the Hausgarten station in the Arctic (Table 4.3).

Although less data are available on the quantities of floating litter, the amounts seem to be lower than for the seabed. This is presumably because ultimately macrolitter sinks and accumulates on the seafloor. Floating litter densities are high in the North Pacific, the Mediterranean and the Straits of Malacca. Floating macrolitter accumulates at ocean fronts and gyres (Section 4.2.1). However, much of the plastic now present in the oceans is so small that it has not been quantified by beach, benthic trawl or floating debris surveys (Section 4.2.1).

4.2.2 How much plastic is in the ocean?

Marine litter is dominated by plastic items (Section 4.2.1) that have very long residence times in the ocean and so essentially become permanent additions to marine environments. In order to tackle marine plastic pollution, we first need to know just how much plastic is present in the ocean. Based on the amounts of plastic production, waste generation and population densities, it was estimated that in 2010 between 4.8 and 12.7 × 10^6 tonnes of plastic were released into the oceans. These estimates are comparable with those for litter from the US Academy of Sciences in 1997 (Section 4.2.1). Data on floating plastics together with ocean circulation models suggest that 35 000 to 268 940 t of plastic may be floating in the oceans, and that these plastics are accumulating within the five subtropical ocean gyres (Figure 4.5). Size analyses of these floating plastics have shown that there are fewer plastic

Box 4.1 Lost at sea?

Shipping containers account for 90% of the non-bulk cargo transported around the world and approximately 120 million containers are transported by sea every year. It is estimated that there are 5–6 million shipping containers floating in the sea with on average 1679 lost at sea each year. These container losses occur during storms (42%), groundings and collisions (11%) or from the improper loading and securing of containers (Box 4.1, Figures 1a–d and Plates 7–10). The contents of these lost shipping containers are diverse and once liberated have led to numerous incidents around the world in which seemingly bizarre items have washed ashore; for example, the 5 million Lego® bricks that are still washing ashore in Cornwall despite having been spilt 20 years ago, the 60 000-odd pairs of trainers that have washed ashore along the western coast of North America or the 28 000 rubber bath animals that have circled half the globe since being lost overboard in the Pacific in 1992 (Box 4.1, Figures 1a–d and Plates 7–10). These bath toys are now being used by scientist's to track ocean currents and to understand the nature of the ocean gyres.

Containers are thought to regularly fall off of container ships along their regular transport routes, but there is no requirement within the shipping industry to report these losses. However, the containers represent a hazard to shipping and

(a)

(b)

(c)

Box 4.1, Figure 1 Shipping containers lost at sea (see also Table 4.4). (a) The *Svenborg Maersk* losing containers in extreme weather in northern France in 2014. © Paul Townsend 2014 https://www.flickr.com/photos/brizzlebornandbred/12852909293. Available under Creative Commons Attribution-No Derivs 2.0 generic licence (CC BY-ND 2.0, https://creativecommons.org/licenses/by-nd/2.0/legalcode). (b) The *MV Rena* grounded on a reef near Tauranga, New Zealand. © New Zealand Defence Force 2011 https://www.flickr.com/photos/nzdefenceforce/6386334175/. Available under Creative Commons Attribution 2 Generic licence (https://creativecommons.org/licenses/by/2.0/legalcode). (c) Sunken container in the Monterey Bay National Marine sanctuary. Credit: © 2011 NOAA / MBARI. Description: After seven years on the deep-seafloor, this sunken shipping container had been colonised by a variety of deep-sea animals, including crabs and deep-sea snails. (d) Lego® octopus washed up on a shoreline in Cornwall, UK, that may be one of the 5 million pieces of Lego® lost from the shipping containers transported by the *Tokio Express* in 1997. © Tracey Williams (creator of the Lego Lost at Sea Facebook page: https://www.facebook.com/LegoLostAtSea/).

Box 4.1 (*Continued*)

(d)

Box 4.1, Figure 1 (*Continued*)

the containers themselves and their contents introduce litter and potentially toxic chemicals into the marine environment (Table 4.4). The World Shipping Council and International Maritime Organisation (IMO) have developed new regulations, that came into force in 2016, to reduce losses by verifying container weights prior to loading. Additionally, the International Standards Organisation are producing new standards for securing containers on ships and the IMO is releasing new guidelines for securing cargo.

In addition to the ecological impacts of the container contents (Section 4.2.3), the container itself may release contaminants from its surfaces. Once the container sinks it may affect the benthic communities in a zone surrounding the container. The impacts of a sunken container lost, from the Med Taipei in 2004, in the Monterey Bay National Marine sanctuary are being studied by the Monterey Bay Aquarium Research Institute (MBARI). The study found that local sedimentation was to be affected, and in turn the communities inhabiting the seafloor were less diverse around the container. However, the container was also behaving like an artificial reef (Section 4.6), with sessile invertebrates attaching themselves to the container's surface (Box 4.1, Figure 1b). In deep water, shipping containers could persist for hundreds of years.

Table 4.4 A selection of shipping containers lost at sea, estimates of the number of containers lost and the nature of the litter.

Date, ship and location	Number	Contents
1997 *Tokio Express*, near Cornwall, UK. Extreme weather /sea conditions	62	5 million pieces of Lego® washed ashore in Devon, Cornwall, and Ireland. Plus unknown container contents
2016 *Blue Ocean*, Cornwall, UK. Extreme weather / sea conditions	1	Plastic bottles (containing H_2O_2), Lizard Peninsula, Cornwall, UK
2014 Unknown, Atlantic. Lost cargo	Unknown	Around 370 000 printer cartridges found in UK, Ireland, Hebrides, France, Portugal, and the Azores
2014 *Svenborg Maersk*, northern France Atlantic Extreme weather /sea conditions	520	Empty or dry food goods
1992 *Evergreen Ever Loyal*, Alaska, central Pacific. Faulty welding and improperly secured cargo	12	1 container of 28 000 rubber ducks reported from north-western US, Queen Charlotte Islands, north-eastern Atlantic, north-western Atlantic, south-eastern Pacific, north-eastern Australia, Indonesia, Hawaii
1990 *Hansa Carrier*, North Pacific. Extreme weather / sea conditions	21	61 280 pairs of trainers unknown debris from the others. Alaska, Oregon, Washington, Vancouver Island
2008 *MV Riverdance* grounded in Blackpool, UK. Extreme weather / sea conditions	90	Thousands of packets of biscuits, timber
1994 *Hyundai Seattle*, near the Aleutian Islands. Extreme weather and engine room problems	25–30	34 000 ice-hockey gloves, plus chest protectors and shin guards
2004 *Med Taipei*, Oakland, California. Faulty welding and improperly secured cargo	15	1159 steel-belted car tyres, furniture, clothing, plastic items, wheelchairs

(*continued*)

Box 4.1 (*Continued*)

Table 4.4 (*Continued*)

Date, ship and location	Number	Contents
2006 *Courtney L*, Virginia, US. Extreme weather /sea conditions	4	1 container had about 8000 bags of crisps washed up near North Carolina
1998 *APL M/V China*, Alaska, north Pacific Extreme weather /sea conditions	406	Clothes, sports gear, and electronics
2013 *MOL Comfort*, container ship sank near Yemen, Indian Ocean	4,500	Mostly electronics and clothes
2011 *MV Rena* grounded on a reef near Tauranga, New Zealand	900	Plastic beads, plastics for recycling, timber, meat; 21 containers of cryolite (toxic if ingested/inhaled)
2007 *MSC Napoli* beached in Branscombe, UK. Extreme weather /sea conditions	114	Motorbikes, car parts, hair products, and disposable nappies
2015 *Manoa*, San Francisco, US. Extreme weather / sea conditions	12	Packaging material

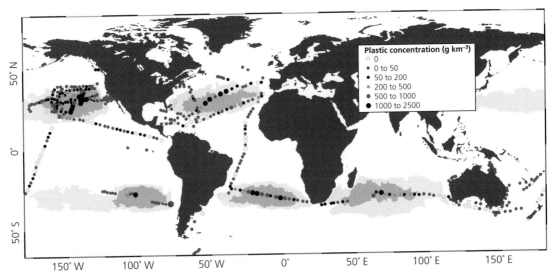

Figure 4.5 Distribution of floating plastic in the oceans shows concentration in the five main ocean gyre systems: the North Pacific subtropical gyre, the South Pacific subtropical gyre, the North Atlantic subtropical gyre, the South Atlantic subtropical gyre and the Indian Ocean gyre. Data are the mean plastic concentrations (g km⁻²) from 442 sites. The different shaded points indicate the different concentrations, and open circles = no plastics found. Grey-shaded regions indicate predicted accumulation zones based on ocean circulation models: dark grey = inner accumulation zones, light grey = outer accumulation zones and white areas indicate that no accumulation was predicted. © Andrés Cózar et al. 2014. Reprinted with permission from National Academy of Sciences and Andrés Cózar (Andrés Cózar et al. 2014. Plastic debris in the open ocean. *Proceedings of the National Academy of Sciences*, 111, 10239–244).

particles than expected, from fragmentation, in the less-than 5 mm size class. This discrepancy could mean that the plastics have fragmented into sub-micron-sized pieces, they have been transferred into food webs or they are sinking and accumulating on the seafloor.

4.2.2.1 The history and use of plastics

Plastics are synthetic or semi-synthetic organic compounds produced via the polymerisation of petrochemicals. Their polymeric structure confers both strength and flexibility, and these unique attributes have led to their widespread use throughout our daily lives. Natural 'plastic-like' materials exist, such as shellac, a resin secreted by female Lac bugs, natural gums derived from plants, trees and bitumen. These natural materials can be chemically modified or 'plasticised'. For example, the first plastics were produced from natural rubber by the Mesoamericans in the 1600s. The production of modern man-made plastics began in 1846 when a French chemist produced a type of plastic called galalith from milk proteins. In 1862, Parkesine was made from cellulose treated with nitric acid to produce nitrocellulose. Polystyrene, one of the most widely used plastics today, was discovered in 1839 and was made from the resin of a Turkish gum tree, and its widespread production began in the late 1920s. Cheap production costs meant that plastics became widely used. Their malleability and waterproofing properties gave them multiple possible applications, and in time, plastics replaced wood, glass, clay, stone, metal and animal products. One reason for the initial popularity of plastic products, aside from their versatility, was the metal shortages during the Second World War due to extensive weapons manufacturing.

In 2013, 299×10^6 t of plastic were produced throughout the world and their production has increased near exponentially since the 1950s. Plastics are now one of the most widely used materials in the world and production is predicted to rise to 540×10^6 t with a value of US654×10^9 by 2020. The largest producers of plastics today are China followed by Europe, producing 24% and 20% of global plastics, respectively. European plastic production is dominated by polyethylene, polypropylene, polystyrene and polyethylene terephthalate (Table 4.5).

Table 4.5 The proportions of the different plastics produced in Europe and their main applications. Europe is the second largest producer of plastic in the world. Data from: Plastics Europe 2015.

Plastic	Applications	Proportion
Polyethylene (PE); low and high density	Packaging, carrier bags, toys and milk bottles	29.6%
Polypropylene (PP)	Packaging, textiles, stationery, automotive construction	18.9%
Polyvinyl Chloride (PVC)	Clothing, building construction	10.4%
Polystyrene	Packaging, clothing	7.1%
Polyethylene terephthalate (PET)	Most plastic bottles	6.9%
Polyurethane (PUR)	Sponges and insulation panels	7.4%

In combination, these five plastics make up three-quarters of all plastic produced in Europe and the largest users are the packaging and construction industries.

Polyester, a type of polyethylene, and nylon, a polyamide plastic, have been produced in large quantities since 1935 and there is increasing evidence to indicate that synthetic fibres are also now prolific in the natural environment (Section 4.2.2). Synthetic fabric production is very high: in 2014; approximately 45 million tonnes of polyester were produced, and was almost twice the volume of cotton produced that year.

4.2.2.2 The nature of plastic pollution

Modern plastics are mostly synthesised from oil, gas or coal. They can also be produced by converting natural materials such as rubber, cellulose or collagen (the latter two being the dominant structural compounds in plants and animals). The production of plastics from fossil fuels uses 4% of the oil produced globally as feedstock and a further 4% is required to power the process. Plastics are shipped around the world as pellets (around 5 mm diameter) that are referred to as *nurdles*. Plastics contain a variety of other compounds for example for fire retardation, as fillers to bulk them out (usually chalk, wood, flour cellulose or starch), as plasticisers (oily compounds), colourants and stabilisers.

Plastic debris in the environment is categorised based on its size (and nature, if obvious) and is grouped into four main size classes: megadebris (more than 100 mm diameter), macrodebris (more than 20 mm diameter), mesodebris (2–20 mm) and microdebris (less than 2 mm). Alternative categorisations use 5–20 mm for mesodebris and less than 5 mm diameter for microdebris.

Macro- and megadebris include vehicle parts, containers from ships (Box 4.1), furniture and household appliances: almost all manufactured goods have been found in the ocean. Equipment used in the fishing industry such as crab/lobster pots, buoys, trawls, fishing line and large volumes of netting are a major problem in some areas (Section 4.2.3). The mesoscale debris often includes packaging, toys, cigarette butts, plastic bottle tops and similar objects that enter the ocean from land (Section 4.2.1), and the nurdles are used as plastic feedstock. These nurdles are regularly spilt during transport and handling on land, and may be washed or blown into watercourses or the sea. Large quantities of nurdles are produced; for example, in 1993, 27×10^6 t of nurdles were produced in the United States. They have been found in many oceans such as the Sargasso Sea (3500 nurdles km^{-2}), remote areas of the south-west Pacific and on New Zealand beaches in high concentrations (100 000 nurdles km^{-2}).

Meso- and microplastics can also be produced by the breakdown of larger objects. In the sea, most plastics are broken down by exposure to ultraviolet (UV) rays. This photodegradation is slow and in the sea it is slower than on land because the temperature changes are less extreme and seawater provides protection from UV rays. Different types of plastics degrade at different rates depending on their UV absorbance and the strength of the chemical bonds. Plastics containing chromophores (aromatic polymers: polyarylate and polyethylene terephthalate (PET)) have the greatest absorbance, and as the chromophores break down the plastic becomes bleached. However, impurities within plastics (Section 4.2.2) can also absorb UV and so they facilitate degradation. For plastics that accumulate on shorelines or in shallow waters, physical abrasion from waves and shifting sediments make them more brittle and prone to fragment. Plastics that are not exposed to sunlight—that is at depth in the water column or on the seafloor—degrade by thermal oxidation (changes in temperature) which is much slower. Although most plastics fragment into smaller and less visible pieces, this does not mean they have been eliminated from the environment (Section 4.2.2).

4.2.2.3 Microplastics

Much of the plastic present in today's aquatic ecosystems occurs as microplastics in rivers, floating in the ocean, in seafloor sediments, in shoreline sediments and within marine organisms (Section 4.2.1). A number of different size classifications are used to describe microplastics including less than 0.2 mm, less than 1 mm, less than 2 mm and less than 5 mm. Microplastics produced by the breakdown of larger objects are referred to as *secondary microplastics*. Alternatively, they may comprise *primary microplastics*; for example, the small particles that are produced as feedstock for the plastics industry (e.g. nurdles) or those being incorporated into many hygiene products as exfoliants (e.g. shower gel, toothpaste, face and body scrubs; Table 4.6), and within cosmetics. They are also used as scrubbers for paint stripping from vehicles or other surfaces. Microplastics are also used in medical and veterinary applications to deliver drugs to specific sites within the body. Other important microplastics now being found throughout the world's oceans are the polyester or nylon fibres from rope and clothing.

Microplastics can therefore enter the oceans through the breakdown of litter, industrial discharges (e.g. from plastic production, textile production, or use of scrubbers for cleaning) and through sewers and stormwater discharges (from washing clothes, hygiene products, cosmetics and perhaps

Table 4.6 Type and amount of polyethylene microplastics in samples of commercially available hygiene products. Data from: Gregory 1986, and Browne 2015.

Product	Microplastics by mass	Size of microplastic particles (diameter)
Hand cleanser	0.2–4.0%	63–500 µm
Facial cleanser	2.0–3.0%	63–500 µm
Cosmetics	0.5–5.0%	100–400 µm
Skin cleansers	47.0%	100–200 µm

medicines). Based on the sales figures for personal hygiene products, and assuming 10% of each product is microplastic, it has been estimated that 263 t of μm-sized polyethylene particles are released into the environment each year from the United States alone. In coastal waters adjacent to a polyethylene microplastic production plant in Sweden, concentrations of 0.120 particles ml^{-1} have been found.

Microplastics in the environment are measured using standard sampling methods for sediments (grabs and cores and sieving through a fine mesh) and the water column (plankton, manta and bongo net trawls with appropriate mesh sizes or from the continuous plankton recorder samples) followed by density separation to extract the plastics. Microplastic composition and type may be determined using fourier-transform infrared spectroscopy.

Small plastic fragments were first found in the environment in 1970 when they were picked up in plankton samples from the North Sea, and a few years later they were found throughout the northwest Atlantic. The amount of microplastics present in the water column and marine sediments ranges from hundreds to thousands of particles per square kilometre throughout the global oceans. The highest concentrations of floating microplastics occur in the subtropics and are associated with the large-scale convergence of currents, and are accumulating in large ocean gyre systems (Table 4.7; Figure 4.5). Ocean currents, wind-driven mixing, the formation of biofilms on microplastic surfaces, and the

Table 4.7 Microplastic concentrations in the five ocean gyre systems (with year measured). Microplastics being particles less than 5 mm diameter. Data from: Lusher 2015, Eriksen et al. 2014 and references therein.

Current system	Microplastic concentrations	
	Particles (km^{-2})	Particles (m^{-3})
North Atlantic Subtropical Gyre (2010)	20 328	0.004
North Pacific Central Gyre (2013)	85 184	0.017
South Pacific Subtropical Gyre (2013)	26 898	0.005
South Atlantic Subtropical Gyre (2014)	~100 000	0.020
Indian Ocean Gyre (2014)	~15 000	0.003

microplastics properties (e.g. density, shape and composition that all influence particle buoyancy) all affect their distribution within the environment.

Time series data from plankton trawls show that in the North Pacific central gyre (Figure 4.5), the concentrations of microplastics have increased by two orders of magnitude over the last 40 years. Data from the continuous plankton recorder have shown they are also accumulating in the Atlantic and increased four-fold between 1960 and 2000. In the North Atlantic gyre, concentrations are high but are considerably lower than in the North Pacific central gyre (Table 4.7), and have changed little over the last 30 years. Plastics removed from the stomachs of sea birds, fulmars and shearwaters in the northeast Atlantic show that between 1970 and 2012 microplastic composition shifted from predominantly primary to secondary microplastics.

Microplastics seem to also be accumulating in coastal waters. They are found in the highest concentrations close to shore; for example, north-east Pacific offshore waters contain 4–27 times less microplastic than the coastal north-east Pacific. Similarly, the North Atlantic, microplastics are accumulating off the Portuguese coast. Microplastic concentrations are high within enclosed water bodies such as the north-west Mediterranean (0.27 m^{-3}). Limited data are available from polar regions, but microplastic concentrations in the Bering Sea are between 2.0×10^7 m^{-3} and 0.004 m^{-3}. However, concentrations in the Arctic Sea ice are 100 fold higher (from 38 to 234 particles m^{-3}) than in the North Pacific central gyre (Table 4.7). No microplastics have yet been found in Antarctica.

Within marine sediments, between 1 and 40 microplastic particles (less than 1 mm diameter) occur per millilitre of sediment globally. Although not all regions have been sampled, high concentrations have been found in sediments near Portugal, the United Kingdom, Western Australia, South Carolina, South Africa, western and southern Chile, Japan, Mozambique and the Azores (Figure 4.6a). Sediment microplastic concentrations are highest near urban areas (Figure 4.6b), as also shown for the floating microplastics. In sediments, microplastics are dominated by polyester, polyvinyl chloride (PVC) and polyamide particles. In some instances, higher concentrations of microplastics occur subtidally, and hotspots occur at decommissioned sewage sludge disposal grounds (Figure 4.6c). This could be because of the high quantities of microplastics with in hygiene and cosmetic products (Table 4.6) and/or the artificial fibres released from laundered clothing (Figures 4.6c–d, Figure 4.7). During one wash, 1900 polyester fibres can be released from a single garment. These fibres are present in sewage effluent at 1 fibre l^{-1} and are concentrated in the sediments from sewage sludge disposal sites that typically contain more than twice the plastic fibres found at reference sites (Figure 4.6c). The microplastic fibres in sewage are dominated by

polyester (70%), acrylic (20%) and polyamide fibres (i.e. nylon) (16%).

4.2.3 The 'toxicity' of plastics

4.2.3.1 The physical effects

Plastics are not inherently toxic. They may leach chemicals that have chronic toxicity, although this is not yet well-established, and their impacts on marine life are largely physical. Marine animals become entangled in plastic debris and drown, suffocate or become strangled. Entanglement tends to occur with derelict fishing gears and similar large complex items. Some animals may become

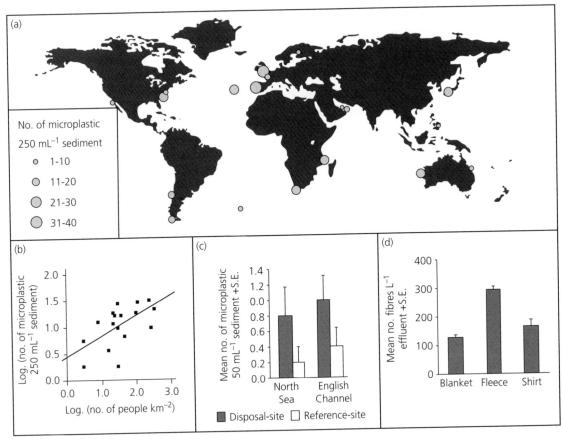

Figure 4.6 Sedimentary microplastics that mostly comprised artificial fibres. (a) Extent of microplastics in the sediments of sandy shores ($n = 18$). Circle size corresponds to the amount of plastic particles (less than 1 mm diameter) in 250 ml of sediment. (b) The relationship between population density and sedimentary microplastics. (c) The concentration of microplastics in sediments from past UK sewage sludge disposal sites compared with reference sites. (d) Numbers of polyester fibres discharged in domestic washing machine waste-water for different polyester fabrics. Reprinted with permission from (Browne, M.A., et al. 2011. Accumulation of Microplastic on Shorelines Worldwide: Sources and Sinks. Environmental Science and Technology, 45, 9175–179). Copyright (2011) American Chemical Society.

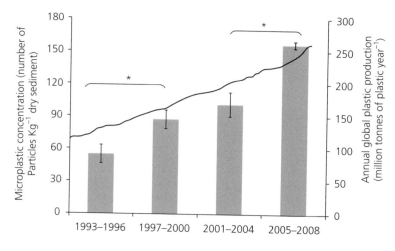

Figure 4.7 Microplastic concentrations in sediments collected from Groenendijk, Belgium through time. Including fibres, granules and film. Bars = number of fragments per kg dry weight sediment, line = global plastic production. *Indicates significant difference between groups at p < 0.05 when compared with Mann-Whitney U test. Reprinted from *Marine Pollution Bulletin*, 62, Claessens, M., De Meester, S., Van Landuyt, L., De Clerck, K. and Janssen, C.R., Occurrence and distribution of microplastics in marine sediments along the Belgian coast, 2199–204, Copyright (2011), with permission from Elsevier.

Figure 4.8 The Grassholm Island gannet breeding colony (40 000 birds) nests on a mass of nylon rope/netting. Inset box: an entangled juvenile. Reprinted from *Marine Pollution Bulletin*, 62, S.C. Votier, K. Archibald, G. Morgan and L. Morgan, The use of plastic debris as nesting material by a colonial sea bird and associated entanglement mortality, 168–72, Crown Copyright © 2010 Published by Elsevier Ltd., with permission from Elsevier.

permanently entangled (e.g. in six-pack can holders and similar smaller pieces of plastic) and continue to be mobile. As the entangled animals grow, appendages may become constricted or lacerated, and the attached plastic may impair feeding and swimming (Figures 4.8 and 4.9a–e) or cause mortality. At least 136 marine species have been reported entangled with plastic and this includes a wide range of sea birds (51 species), whales (11 species), seals (19 species), turtles (6 species), fish (36 species), crustaceans (8 species) and other mammals such as otters and dugongs. Data on the number of entangled animals are obviously better in areas near human habitation: those in remote areas of the open ocean may never be recorded.

Large species of sea bird—for example, albatrosses, gulls, penguins, petrels and gannets—seem to be particularly susceptible to entanglement. Being plunge divers, gannets become entangled while diving for food. At a gannet colony in Helgoland in Germany, 29% of birds were found to have died after becoming entangled in plastic litter, and in a Dutch gannet colony 6% mortality occurs due to entanglement (Figure 4.8). This has increased

Figure 4.9 (see also Plates 11, 12, 13, 14 and 15) Examples of marine megafauna entangled in macroplastic litter. (a) Caspian gull (*Larus cachinnans*) entangled in string at Ainsdale beach, Southport, United Kingdom © Paul Scott 2016 (sgpscott@flickr/Facebook). (b) Turtle entangled in fishing nets. Credit: U.S. National Oceanic and Atmospheric Administration. (c) Triggerfish entangled in fishing nets, at Acqua bella, Ortona, Abruzzo, Italy. © Claudio Stoppato 2013. (d) and (e) Steller sea lions with plastic packing bands entangling theirs necks causing lacerations in (d). © Alaska Dept. of Fish and Game, the activity was conducted pursuant to an NMFS Permit.

through time. On Grassholm Island in Wales, large quantities (18.6 t) of macroplastics have accumulated which the gannet breeding colony preferentially uses for nesting material (80% used synthetic fibre rope and 15% fishing nets; Figure 4.8). In the colony, 65% of the adult gannets become entangled, and 80% to 90% of the chicks become entangled, with a mortality rate of up to 75%. Fortunately, volunteer groups visit Grassholm each year to help release the birds and so the mortality is often lower than this.

Entanglement is often witnessed in seals who are naturally curious and often end up with plastic over their necks (Figure 4.9) or entangled in fishing nets where they drown. Estimates from Australia and New Zealand concluded that 1478 fur seals and sea lions die each year from entanglement in fishing gear. The high entanglement rates for seals may be due to overlap between their feeding grounds and productive commercial fishing grounds. On Farallon Island in California of the 914 sea lions, stellar sea lions and elephant seals examined 32% showed signs of entanglement and 68% had evidence of constriction from earlier entanglements.

Whales often become entangled with plastic caught in their mouths or tails and although some can free themselves, others cannot. A study in the North Atlantic found that 89% of entangled whales had become caught up in pot and gill nets that are set on the seafloor and connected to buoys at the surface. Turtles often become entangled and drown or have their locomotion or growth impaired (Figure 4.9) A study of stranded turtles on the shores of the Canary Islands found that 25% had become entangled in fishing nets.

The second main physical threat from plastics to marine life is ingestion. At least 36% of the 312 species of sea bird are known to have consumed plastic in some form. This often comprises pellets and bits of styrofoam, and in extreme cases cigarette lighters, toy soldiers, buttons, golf tees, pens and other plastic items. Marine animals consume plastic because they mistake it for food or perhaps because it is present in the guts of their prey (see Section 4.2.3). Visual feeders select for specific types, sizes and colours of plastic. Turtles consume plastic

bags, mistaking them for jellyfish. Planktivorous sea birds ingest more plastic than those that eat fish because while floating plastic can be confused with fish eggs or zooplankton, it is less likely to be mistaken for fast-swimming pelagic fish. Similarly, several pelagic fish species only consume white spheroids and others only consume transparent plastics, again perhaps mistaking them for fish eggs or other zooplankton.

The amount of plastics consumed by fulmars show regional variations in the plastic ingested. In the North Atlantic, 79% to 99% of fulmars contain plastic; in the North Pacific, 84% to 88%; in the Davis Strait separating Greenland from Canada, 36% of birds contained plastic. The composition of plastic in fulmars also varies between regions and is thought to reflect regional differences in waste composition (Section 4.2.2). In the North-East Atlantic the OSPAR commission advises that an acceptable ecological quality standard (EQS) is less than 0.1 g of plastic in fewer than 10% of sea birds. However, in the North Sea 58% of beached dead birds were found to contain more than 0.1 g plastic, and so in terms of plastic pollution parts of the north-eastern Atlantic are not of good ecological quality by this measure. There are fewer reports of seals containing plastics and where found, their consumption is thought to be indirect through ingesting contaminated fish.

Plastic ingestion by large mammals such as whales is not as well-documented because in many cases their bodies sink rather than wash ashore and so autopsies are rarely performed. However, the autopsies that have been performed on beached whales have shown that they do consume plastic. In 2008, two sperm whales beached in California were found to contain 70 kg of plastic, mostly fishing net, line and plastic bags (this represents 2% of the body mass of a 16 m whale). The whales died of gastric rupture followed by starvation. Beached pygmy sperm whales, beaked whales, killer whales, manatees, dolphins and harbour porpoises have also all been found containing plastic. This is a mixture of plastic packaging and fishing gear and it is not clear if ingestion is accidental or deliberate being mistaken for food.

There are many documented cases of turtles ingesting plastic; for example, a turtle found off Hawaii contained 100 000 pieces of plastic, including a toy truck. In Brazil, dissection of 38 stranded green turtles, *Chelonia mydas*, found 61% contained plastic, and it was the cause of mortality for 13%. Similarly, 56% of stranded turtles in Florida were found to contain plastic. Turtle feeding selectivity is reflected in their gut contents, consistently consuming white or colourless bags or plastic pieces. Common species of fish such as flounder and mullet, as well as sea snails, contain plastics, and in mesopelagic fish, plastic concentrations are 12 000–24 000 t y^{-1} in the North Pacific subtropical gyre. Most fish ingest only small pieces of plastic (see Section 4.2.3).

The ingestion of plastic may block internal passages such as the gut or gullet and thus inhibit feeding or breathing. By occupying space in the gut, plastics can reduce the feeding stimulus and create the false impression of being well fed. This also reduces the capacity of the gut and so only small meals may be consumed. Reduced feeding means reduced fitness: animals that consume less have lower potential to accumulate fat reserves. Consequently, there is less energy for flying, feeding, swimming and mating. This can affect behaviours such as migration and there is less energy available for the production of gametes (eggs and

sperm). Juvenile animals have smaller guts and so often die of malnutrition. For example, albatross chicks are fed regurgitated plastic pellets by the parents until their guts become full of plastic and they can no longer feed efficiently (Figure 4.10). Plastics may also block the secretion of gastric enzymes used to break down food, they can affect the production of steroid hormones, and interfere with ovulation. The effects of plastic ingestion vary between species and those that migrate are more susceptible to population-level effects than those that do not. For example, a study of Californian red phalaropes found that body-fat reserves were lower the more plastic birds consumed and this affected their long-distance migration to the breeding grounds.

4.2.3.2 Ghost fishing

Lost fishing gear is thought to represent approximately 10% of marine litter and its distribution varies according to where fishing activity is concentrated (Section 4.2.1). Data on discarded fishing gear are patchy, and the best are from Europe and parts of the United States (Table 4.8). It is estimated that approximately 1% of fishing nets are permanently lost at sea, although in some cases it may be as high as 30% (Table 4.8). In deep water, fisheries losses are higher. Most fishing gear is either dumped at sea during illegal and unregistered fishing activities (to

Figure 4.10 (see also Plate 16) Remains of albatross chick with gut full of plastic pieces presumably ingested while still alive. ©Sparkle Motion/Flickr 2014 https://www.flickr.com/photos/54125007@N08/14876384246/. Available under Creative Commons Attribution 2.0 Licence https://creativecommons.org/licenses/by/2.0/legalcode.

Table 4.8 Summary data of the amounts of fishing gear lost or abandoned at sea in different regions. Data from: Macfayden et al. 2009.

Region	Gear type	Gear quantity
Pearl and Hermes atolls, North Hawaii	Mixed	107 t (2002) 90 t (2003)
North Sea and north-eastern Atlantic	Bottom set gill nets	0.02–0.09% boat y^{-1}
English Channel and French North Sea	Gill nets	0.2% boat y^{-1} sole 2.11% boat y^{-1} seabass
Mediterranean	Gillnets	0.05% inshore hake 3.2% sea bream
Gulf of Aden	Traps	20% boat y^{-1}
ROPME Sea area	Traps	260 000 y^{-1}
Indian Ocean	Tuna longline	3% hooks/set
Queensland, Australia	Blue swimmer crab traps	35 boat y^{-1}
North-eastern Pacific, Bristol Bay	King crab traps	7000–31 000 y^{-1}
Newfoundland	Cod gillnet	5000 y^{-1}
Canadian Atlantic	Gillnet	80 000 y^{-1}
Gulf of St Lawrence	Snow crab traps	792 y^{-1}
New England	Lobster traps	20–30% boat y^{-1}
Chesapeake Bay	Traps	30% boat y^{-1}
Guadeloupe, Caribbean	Traps	20 000 y^{-1}

avoid being caught fishing illegally) or is lost when damaged during legitimate fishing due to entanglement or extreme weather. Fishing gear is expensive and legitimate fishermen will usually go to great efforts to retrieve lost gear.

Abandoned fishing devices are referred to as 'ghost nets' because they may continue to fish despite being lost or abandoned at sea. While some fishing gear cease to fish once lost, many continue to capture marine life. These may continue to be fully functional; crab/lobster pots, set nets such as drift or gill nets and long-lines are just as effective whether there is a line to a buoy at the surface or not. In addition, because they are never recovered and cleared, animals entangled in the gear begin to decompose and act as an attractant that lures other marine life that in turn becomes entangled. This cycle of capture and baiting continues until the net sinks, becomes 'balled up' by rolling over

the seabed or is physically degraded. Ghost nets can continue to fish for many years, and they are moved around the oceans by currents accumulating a range of animals. Ghost nets can reach a mass of 6 t. Some fishing gear can be very large, like drift nets that can hang in the water column unanchored for a period of time and could legally be up to 50 km in length prior to 1960. In 1992 the use of drift nets exceeding 2.5 km were banned by the United Nations (UN) for use in international waters.

In addition to ghost fishing and the general impacts of marine plastic pollution, lost or abandoned fishing gear can cause damage to the benthos when scraped across the seafloor by currents; for example, fishing gear that has a base anchored to the seafloor. The base itself can weigh up to 100 kg and so does considerable damage, similar to benthic trawling, when dragged by currents.

4.2.3.3 The impacts of microplastics

Microplastics span the same size range as sedimentary particles and planktonic organisms that are at the base of the food chain, and so they may be readily ingested by marine animals. In laboratory studies, microplastics are ingested by animals from all feeding guilds and nano-sized particles have even been absorbed into unicellular algal cells (Table 4.9). For several species these observations have also been verified *in situ* (Table 4.9). Observations from wild animals have also found large quantities of microplastics inside a range of vertebrates. Many species of surface-feeding sea birds—for example, procellarids (petrels, fulmars, prions, shearwaters), larids (gulls and terns) and suliforms (shags)—from across the world have been found with large quantities of microplastic pellets, beads, plastic fragments and polystyrene spheres in their guts. However, it is not clear whether sea birds are impacted by microplastics to the same degree as macroplastics (Section 4.2.3); that is, it is not clear whether mortality actually occurs.

Although, there is little data for mammals. Furseal scat has been found to contain microplastics and harbour seals have had microplastics (more than 100 μm) removed from their digestive tracts. Although effects have only been shown indirectly, based on contaminant loads in blubber, it seems

Table 4.9 A selection of taxa that are known to have ingested plastic in laboratory studies, the type of plastics ingested, their size and the organisms feeding mode. Data from: Wright et al. 2013 and Thompson et al. 2004. PS = polystyrene, PVC = polyvinyl chloride, PE = polyethylene, LDPE = low density polyethylene. *indicates species which have also been found to contain microplastics *in situ*.

Taxa	Plastic type	Feeding mode
Algae		
Scenedesmus sp.	Absorbs 20 nm positively charged nanoparticles	Autotroph
Chlorella sp.	Absorbs 20 nm positively charged nanoparticles	Autotroph
Benthic invertebrates		
Semibalanus balanoides barnacle	230 μm PVC	Suspension/filter feeder
Orchestia gammarellus amphipod	20–2000 μm	Detritivore
Nephrops norvegicus crustacean*	Nylon strand balls	Scavenger
Carcinus maenas shore crab	8–10 μm PS	Scavenger/predator
Crangon crangon brown shrimp*	300–1000 μm	Deposit feeder
Arenicola marina lugworm	20–2000 μm	Deposit feeder
Gaeleolaria caespitosa fan worm	3 and 10 μm PS	Suspension/filter feeder
Marenzelleria spp. mud worm	10 μm PS	Suspension/filter feeder
Holothuria floridana, H. (Halodeima) grisea, Thyonella gemata, Cucmaria frondosa holothurian	250–1500 μm spheres/ fibres, < 0.5 mm PVC fragments	Deposit feeder
Tripneustes gratilla sea urchin	32, 35 PE	Grazer
Dendraster excentricus sand dollar	10, 20 μm PS	Deposit feeder
Strongylocentrotus sp. sea urchin	10, 20 μm PS	Grazer
Mytilus edulis bivalve mollusc*	2–30 μm, 0–80 HDPE	Suspension/filter feeder
Mytilus trossulus bivalve mollusc	10 μm PS	Suspension/filter feeder
Placopecten magellanicus bivalve mollusc	20 μm PS	Suspension/filter feeder
Crassostrea gigas bivalve mollusc*	2, 6 μm PS	Suspension/filter feeder
Synchaeta spp. rotifer	10 μm PS	Deposit feeder
Zooplankton		
Lepas spp. gooseneck barnacle*	1410 μm	Suspension/filter feeder
Stombidium sulcatum ciliate	0.41–10 μm	Suspension/filter feeder
Tintinnopsis lobiancoi ciliate	10 μm PS	Suspension/filter feeder
Echinoderm larvae: sea urchins, sea stars, sand dollars, brittle stars, and sea cucumber	10–20 μm PS spheres	Suspension/filter feeder
Galeolaria caespitose polychaete larvae	3 and 10 μm spheres	Suspension/filter feeder
Acartia tonsa calanoid copepods	14–59 μm	Suspension/filter feeder
Parasagitta elegans arrow worms*	100–3000 μm PS	Predators
Vertebrates		
Pomatoschistus microps common goby	1–5 μm PE	Predator/scavenger
Gadus morhua Atlantic cod	2, 5 mm	Predator/scavenger
Oryzias latipes Japanese medaka	3 mm LDPE	Predator/scavenger
Dicentrarchus labrax seabass larvae	10–45 μm PE	Predator/scavenger

likely that filter-feeding whales could also ingest large quantities of microplastics. Microplastics have only once been reported from the digestive tracts of turtles. The risks of microplastic ingestion may be greater for juvenile and neonatal turtles that consume smaller food particles and have smaller gut volums.

A range of wild-caught fish species have been found that contain microplastics of 5 μm to 5 mm diameter which were mostly polystyrene. Fish found containing microplastics span ten different taxonomic orders and include whiting, cod, haddock, herring, anchovy, saithe, five-bearded rockling, Atlantic silverside, several species of lantern fish, perch, goby, bergall, horse mackerel, dragonet, solonette, searobin, thickback sole, John dory, hatchetback and gurnard. In the English Channel, 37% of fish, from ten species, have been found to contain on average two microplastic particles per fish. One-third of the mesopelagic fish sampled from the North Atlantic gyre contained plastic and the dominant species contained an average of six particles per fish. Many of the fish in the gyre are myctophids that feed in the surface water column at night and presumably the microplastics appear similar to their zooplankton prey. Fish larvae—for example, small sculpin *Myxocephalus aenaeus*—have also been found which contain plastics less than 0.5 mm. A squid stranding in British Columbia found 27% of the 30 Humboldt squid washed ashore contained plastic pellets of 3 mm to 5 mm in their stomachs.

Unicellular algae that have absorbed microplastics show inhibited photosynthesis and the release of damaging reactive oxygen species (Table 4.9). Zooplankton feeding has been found to decrease, and lugworms fed plastic have lost body weight due to reduced stomach volume after microplastic consumption. Microplastics have even been shown to be taken up from the gut into tissues and the animal's circulatory systems. In mussels microplastics have been found in the lymph system. Within animals, microplastics could cause blockages, abrasion to delicate tissues, and may leach toxicants into the body.

Some animals may be able to egest microplastics and so would not accumulate them. The extent to which this occurs is unknown, but plastics have been found in lugworm (*Arenicola marina*) and seal faeces, and both copepods and sea birds can regurgitate microplastics, although it's not clear if this occurs regularly in response to plastic consumption. Some suspension- or filter-feeding taxa may reject microplastic particles prior to ingestion through the production of psuedofaeces although this carries an increased energetic cost and so may negatively impact the animal.

Plastics contain chemical additives such as flame retardants, phthalate plasticisers, stabilisers and colourants which may cause toxicity when released during photodegradation. Experiments with plastic particles have found that both bisphenol-A (BPA) and phthalates readily leach from plastics: both are known endocrine disruptors and phthalate is a known carcinogen (Section 5.3). Toxicity tests with invertebrates and fish have shown that at high concentrations these two compounds affect development and reproduction in a range of invertebrates and fish. The brominated flame-retardants within plastics also have carcinogenic effects, disrupt endocrine systems and cause neurotoxicity.

In addition to leaching toxicants, microplastics may adsorb and accumulate other toxins that are present in the environment such as polychlorinated biphenyls (PCBs), hexachlorocyclohexane, dichlorodiphenyltrichloroethane (DDT) and its breakdown products, nonylphenol and polycyclic aromatic hydrocarbons (PAHs) such as phenanthrene. These compounds are highly toxic and include carcinogens, mutagens and endocrine disruptors. The small size of microplastics mean they have high surface-area-to-volume ratio and are highly adsorbent. As microplastics travel around the oceans they may concentrate toxicants, making them bioavailable once ingested. Plastics that have been used as scrubbers may have elevated concentrations of toxicants because of their use in cleaning applications. These microplastics are often reused which increases their toxicant load. Once contaminants become adsorbed to microplastic particles their degradation slows considerably and so association with microplastics may increase the environmental persistence of some contaminants. The adsorbency varies with the type of microplastic, for example, polyethylene is the most adsorbent, PVC and polypropylene are less adsorbent. This potential to accumulate contaminants is even being used as part of a sampling programme called the International Pellet Watch to monitor global patterns of marine pollution (see resources).

The contaminated plastics may cause toxicity not only to the organisms that consume them but also to taxa higher up in the food web. Toxicants bound to plastics have been found to be biologically available to those that consume them. For example, in shearwaters PCBs can be assimilated from plastics in their guts, and phenanthrene has been taken up from plastic in the gut of *Arenicola marina*. The shearwaters acquired contaminants directly from plastic consumption but also from contaminated prey. These contaminants may leach into digestive fluids before being transferred into other tissues (e.g. fat stores) where they may accumulate. Ultimately, therefore, these contaminated plastics also represent a considerable threat to predators that may accumulate a higher contaminant burden than their prey through biomagnification (Section 1.3). Plastic has been acquired through prey ingestion by shearwaters, Antarctic fur seals and sea lions, and a number of these pollutants are known to bioaccumulate (e.g. DDT and PCBs). Thus, they also represent a threat to human health, especially for those taxa consumed whole, like shrimp or bivalve molluscs such as mussels and clams. Microplastics have been found in the guts of both wild shrimp and farmed mussels purchased from retail outlets.

4.2.3.4 *Other impacts of plastics*

Other phenomena associated with marine plastic pollution include the high concentrations of plastic waste on the seafloor in many coastal regions (Table 4.3) which can have the effect of *smothering* the benthos or coral colonies, for example. The plastic creates a barrier which inhibits essential gas exchange and results in deoxygenation of sediments/corals in the immediate vicinity, and this physical smothering effect is not confined to plastic waste. Another impact of plastic pollution is the increased rafting or transport opportunities for sessile taxa. The attachment of organisms to very buoyant plastics means they can be transported long distances before the plastic becomes overloaded and sinks. A

variety of taxa have been found attached to these rafts: bacteria, algae, diatoms, barnacles such as the invasive species *Elminius modestus*, polychaetes, mussels, hydroids, bryozoans (60 different species) and tunicates. The presence of plastics has been estimated to have doubled the amount of floating debris in the ocean available for rafting, facilitating the spread of non-native invasive species (NIS; Section 4.8). The encrusting bryozoan, *Jelyella tuberculata*, is thought to have crossed from Australia to New Zealand on floating plastic pellets. This species has also been found attached to plastic washed ashore in Florida.

4.2.4 The regulatory framework

A lack of regional and coordinated global action on the issue of marine litter has meant that there have been large inputs into the marine realm, some of which are persistent. Recognition of the scale of the problem of in the early 1970s led to the advent of regulation (Table 4.10). In 1972, the London Convention prohibited the dumping of black-list substances at sea, and the later protocol (1996) somewhat confusingly took the opposite approach, banning all substances except those on a 'white-list'. This was followed by the International Maritime Organisation's introduction of the MARPOL (Marine Pollution) convention in 1973 in a step to reduce the international dumping of waste from ships at sea. Annex V of the convention regulates the release of garbage and prohibits plastic disposal at sea. However, the legislation surrounding dumping at sea was largely ignored and this led to the redrafting of MARPOL Annex V in 2013 to incorporate stronger obligations for governments and offshore operators. Under the revised Annex V the discharge of all wast including plastics is prohibited with a few operational exceptions and those that conflict with passenger health and safety. Additionally, responsibility was placed on governments to provide facilities for the disposal of waste in terminals and harbours, and all large ships (more than 100 t) and platforms were required to produce garbage-management plans and carry garbage-disposal/incineration logs.

In 1974, the United Nations began the Regional Seas Programme to focus on improving marine

habitat quality by creating sustainable management practices. Within each of the regions, national
legislation and regional programmes exist and the
Regional Seas Programme specifically aimed to coordinate efforts within the different regional zones.

This effort was followed by the Global Programme
of Action to Protect the Marine Environment from
Land-Based Activities in 1995 that sought to assist
states in taking action on marine pollution. Separate international legislation exists surrounding the

Table 4.10 Key conventions, protocols, programmes, directives and resolutions regarding marine litter pollution; listed in chronologic order.

Year	Conventions, protocols, programmes	Agency
1972	*London Convention on the Prevention of Marine Pollution by Dumping of Wastes and Other Matter* – 87 states are party – Prohibits dumping of black-list substances	International Maritime Organisation
1973/1978	MAPROL 73/8 *International Convention for the Prevention of Marine Pollution from Ships* Protocol added in 1978: *Prohibition of at-sea disposal of plastics and garbage from ships* – 139 states are party – Annex V regulates release of garbage and prohibits release of ship-generated plastic	International Maritime Organization
1974	*Regional Seas Programme* – 18 regions, 140 countries – Legal, administration, and financial framework for UN Agenda 21 – Facilitates coordination within and between sea regions on actions to protect the shared regions	United Nations Environment Programme
1992	*Basel Convention on the Trans-boundary Movements of Hazardous Wastes and Their Disposal.* – 183 states are party – Minimising hazardous waste production using life-cycle approach and minimise transfer of hazardous waste between countries especially those less developed – Includes hazardous marine litter from land-based sources	United Nations
1995	*Global Programme of Action for protection of the marine environment from land based activities* – 100 countries participate – Aims to prevent degradation of marine environment from land-based activities, to assist states in taking action to prevent, reduce or eliminate degradation and aid recovery – Partners from government, UN agencies, private sector, NGOs, and regional seas partners	United Nations Environment Programme
1995	*Code of conduct for Responsible fisheries* – Fishing should be conducted based on MARPOL Annex V to protect marine environment and prevent loss of fishing gears.	Food and Agriculture Organization
1996	*Protocol to the London Convention* – 47 states are party – Prohibits dumping of all except white-list substances. – Waste generated in normal operations at sea not considered waste, but prevents export of waste to sea for incineration/dumping in any area	International Maritime Organization
2004	A/RES/59/25 on sustainable fisheries: calls for regional and sub-regional administration to take action to address the issue of lost/abandoned fishing gear and other marine debris and collect data on amounts, ecological impacts, and economic costs	United Nations General Assembly Resolutions
2005	A/RES/60/30 calls for national, regional and global action to address marine litter problem. Asked IMO to review MARPOL Annex V	
2005	A/RES/60/31 on sustainable fisheries: stakeholders and governments should closely coordinate to address the issues of lost fishing gears	
2008	A/RES/63/111 identifies need to raise awareness and support improved waste-management practice	
2013	Review of MARPOL Annex V – In response to UNGA resolution A/RES/60/30 reviewed Annex V – Revised annex now prohibits discarding of all waste, including all plastics, into the sea with some exceptions (e.g. animal carcasses, cargo residues, cleaning agents), and under conditions where passenger safety is at risk or accidental losses occur. Obliges governments to provide adequate disposal facilities at ports/terminals.	International Maritime Organization

movements of waste, and targets specific wastes such as fishing gear that are a significant problem in some regions (Table 4.10).

In 2015, GESAMP, the joint Group of Experts on the Scientific Aspects of Marine Pollution, recommended urgent action to reduce the inputs of plastics and microplastics into the sea. Also, recognition of the scale of, and potential risks from, marine microplastics (Sections 4.2.2 and 4.2.3) has led to pressure from environmental groups to ban the use of microplastics in cosmetics and hygiene products. For example, the 'Beat the Microbead' campaign launched in 2012 and co-ordinated by the United Nations Environment Programme accredited Plastic Soup Foundation (see Resources). In 2015, the United States introduced the Microbead-free Waters Act, which is expected to apply to manufacturing from 2017 and all commerce from 2018. A number of other countries are beginning to follow the United States in limiting and perhaps banning the use of microplastics in these products; for example, Canada and The Netherlands. A number of cosmetic/personal hygiene product manufacturers and retailers are also making commitments to cease production or sales of these products.

4.2.5 The future of marine plastic pollution

Plastic pollution carries significant costs to the environment (Section 4.2.3), industry and human well-being and many of the costs have yet to be determined (Section 4.2.5). Recognition of the scale of the plastic pollution problem has stimulated action to reduce plastics in the environment. This can be achieved in a number of ways. By removing them from the environment (Section 4.2.5), by increasing plastic recycling and by developing schemes to discourage the use of plastics and promote reuse and recycling (Section 4.2.5). Alternatives to petroleum-based plastics are also being developed that can biodegrade in the environment.

4.2.5.1 The costs of plastic pollution and the economic incentives

There are economic as well as environmental costs associated with marine litter such as the impacts upon marine ecosystem services (Section 1.5)

including recreation, food supply and transport by sea. There are direct costs associated with, for example, clearing litter from popular beaches and harbours/marinas. Clean-up activities have been valued at €1.9 × 10^6 y^{-1} for shorelines and €2.4 × 10^6 y^{-1} for harbours/marinas in the United Kingdom (Table 4.11) for those conducted by the state. However, a considerable amount of shoreline litter collection is done by volunteers (Section 4.2.1) and the cost of their time has been estimated at €131 000 in the United Kingdom in 2009 for two clean-up programmes. High levels of litter can affect tourist

Table 4.11 Example economic costs of marine litter by sector. Data from: Mouat et al. 2010.

Sector and activity	Cost	Region
Tourism		
Beach cleaning	€1 900 000 y^{-1}	UK
Beach cleaning	€10 400 000 y^{-1}	Belgium and The Netherlands
Shipping		
Removal of litter from harbours and marinas	€2 400 000 y^{-1}	UK harbours
Rescue by emergency services due to entanglement	€830 000–2 189 000 y^{-1}	UK waters (2008)
Fisheries		
Fouling of vessels/gear	€11 700–13 000 y^{-1}	Scottish fishing industry
Repairing of gear	€245 700–273 000 y^{-1}	Scottish fishing industry
Dumped catches due to contamination (oil, fuel filters, detergents, paint)	€140 400–156 000 y^{-1}	Scottish fishing industry
Cleaning litter from gear	€7 720 000–13 000 000 y^{-1}	Scottish fishing industry
Aquaculture		
Removal of entangled litter from boat propellors, intake pipes, and farms	€156 000 y^{-1}	Scottish aquaculture
Farming		
Damage to livestock and property from marine litter blown onto land	€252 000 y^{-1}	Shetland Isles

visitor numbers and therefore revenue is lost due to the risk of injury from broken glass, metal or medical waste such as syringes, and it acts as an aesthetic deterrent. The amount of lost revenue may be very high; for example, in Goeje Island in South Korea, a litter pollution incident in 2011 caused a drop in visitors of 500 000 with lost revenue of €23–29 × 10^6.

There are costs associated with the damage done to shipping when plastics block motors or become entangled in propellers, as well as costs associated with rescuing passengers from entangled vessels. For the fishing industry costs accrue from damage to fishing gears, lost efficiency due to the fouling of trawls with litter (Section 4.2.3) and lost revenue due to contaminated catches and ghost fishing (Table 4.11). Some other impacts are more difficult to value. These include the effects on human health from injury during shipping entanglements, collisions or from the litter on beaches, for example (Table 4.11). Losses of cultural ecosystem services may also occur which might include the value of an area for recreation or the loss of inherent aesthetic value due to litter.

Who then pays the costs associated with plastic pollution? The international expectation is that the polluter pays the costs of pollution clean-up and damage (Sections 1.5 and 7.4.1). For marine litter, currently most of the costs are borne by stakeholders (e.g. the fisheries, aquaculture, tourism and shipping sectors). Although these stakeholders are responsible for a portion of the litter in the sea, the largest sources are from land. Land-based litter is produced through the feedstock manufacturers, the product manufacturers and the retailers of items containing plastic and finally the public who litter or flush non-biodegradable waste down the toilet (Section 4.2.1). Thus, the polluter may be the manufacturer, the public, or other industry sectors. The polluter-pays approach has not traditionally been employed to diffuse sources of pollution due to the problem of attribution (it is usually difficult to identify the polluter), but litter can be both a point source and diffuse input. While it might be possible to identify the producers of fresh litter, it is more difficult for weathered litter. Another challenge is that

the offshore sources of litter are more difficult to monitor or regulate.

The redistribution of the costs of litter pollution onto the polluters would provide incentives to use less plastic or switch to non-polluting or biodegradable materials. It can thus provide an important mechanism for reducing marine litter. One approach to managing the costs is the concept of *extended producer responsibility* (EPR) whereby the product remains the producer's responsibility, both financially and logistically, once it becomes waste. Such schemes have been introduced in Europe which involve the development of *producer responsibility organisations* (PRO), bodies established by polluters to meet their recovery and recycling obligations. In Europe, EPR is part of the Packaging and Packaging Waste Directive (94/62/EC) and often takes the form of paying a fee (tax) to the PRO based on the amount of waste produced. These fees are used to support recycling schemes and the contribution to recycling varies between countries. For example, only 10% of the costs are recovered in the United Kingdom while in other countries 100% of the costs of collection and recycling are recovered. These EPR schemes do not presently include the costs of removing litter from the environment.

The revised MARPOL Annex V requires that wastes generated offshore are disposed of at port disposal facilities and the relevant state covers the costs of disposal. In some countries these costs are passed onto the polluter by reclaiming costs through port fees. It is not yet clear how effective these procedures are and they vary between countries. In the Baltic states, for example, all ships are charged a standard fee for waste disposal, which is included in its port fees and this seems to be having some success.

In some countries, landfill taxes or levies are used to incentivise recycling, although this can result in increased illegal fly-tipping and therefore needs to be used in conjunction with greater penalties against this common practice. Twenty countries in Europe introduced landfill taxes after the introduction of the EU Landfill Directive (1999/31/EC) and increasing fees for landfill may help to solve the greater challenge of sustainability. Some schemes are investing these taxes in improving waste-management infrastructure and schemes. Some, but not all, landfill levies have seen accompanying decreases in waste sent to landfill and increased recycling and composting. A number of schemes are used which focus on consumers such as greater provision of litterbins and waste collection and fines for littering and fly-tipping. Deposit-refund schemes for packaging waste are often very effective; for example, in South Australia three-fold reductions in littering have occurred for some products like glass drinking bottles. Similarly, schemes to reduce the use of plastic bags (Section 4.2.5) have been very successful.

The solution to marine litter therefore may be a multi-targeted approach that disincentivises the use of plastic by producers and consumers (e.g. plastic bag schemes that target consumers and the EPR schemes that target producers), charges for collecting and disposing of waste in landfill and the development of schemes to reduce offshore wastes through, for example, port waste management and improved operational practices to minimise losses at sea (like containers and fishing gear). However, there also needs to be consideration of the potential toxicity of plastics in the marine realm. Ordinarily, regulations surrounding the licensing and discharge of pollutants would address toxicity (Section 7.4.2), but such regulation is not currently applied to plastic products because their effect is generally not one of toxicity.

4.2.5.2 Clean-up, reduction and recycling

The clean-up of plastic wastes from the environment is undertaken by local authorities, NGOs and volunteer groups. For intertidal litter such clean-up activities are numerous and often collect very large amounts of litter. Offshore clean-up is, of course, more challenging and costly to achieve although some collaborations with fisheries and diver projects (e.g. Section 4.2.1) remove litter from the oceans. For example, 'Fishing for litter' (coordinated by *Kommunenes Internasjionale Miljøorganisation*) involves fishermen bringing ashore the litter collected in their trawls during fishing. Programmes such as Ghost Nets Australia specifically target lost fishing gear. These clean-up activities, even those programmes that utilise volunteers, are costly (Section 4.2.5), inefficient and they do not clean-up

microplastics (although by removing macroplastics they could lower the amounts of secondary microplastics). A more effective means of reducing plastic pollution would be to tackle it at its source.

One option for reducing the amounts of marine and terrestrial litter is to use less by moving human consumption away from so-called disposable materials towards those made from biodegradable materials or those that can be more effectively recycled (e.g. metals, glass, natural fibres or alternative plastics (Section 4.2.5)). In the last decade there has been global action to reduce the number of lightweight high-density polyethylene (HDPE) plastic bags used. These bags comprise a considerable proportion (e.g. Section 4.2.1) of marine litter. In Bangladesh in 2002, the first complete ban on lightweight plastic carrier bags was introduced and similar approaches have now been adopted in China, Taiwan, parts of India, Morocco, Mexico, Tanzania, Papua New Guinea and China. Some countries have struggled to enforce these bans, but others have had great success. For example, in China, plastic bag use by supermarkets dropped by 40 billion bags (80% of the total) in one year. In Myanmar, a different approach was taken and production stopped and in Kenya, imports and manufacturing ceased. A levy or tax on the use of lightweight plastic bags by consumers has been introduced in some parts of Europe, the United States, Australia, Hong Kong, Malaysia, Indonesia, Israel and South Africa.

In 2014, the European Union introduced a directive to discourage the use of lightweight plastic bags, but member states were to derive their own means of doing so. These measures ranged from levies in Demark, the United Kingdom and Ireland, to outright bans in Italy and The Netherlands and in Germany the stores providing the bags are taxed a recycling fee. In the United Kingdom in 2015, the introduction of a tax (5 pence per bag) led to an 85% reduction (6 billion bags) in plastic bag use in its first year. Denmark has also shown great reductions with each person now using on average only 4 bags y^{-1} compared with 466 bag y^{-1} in other parts of Europe. However, there are many European states that still have not taken steps towards reducing their use of plastic bags. Similarly, in the United States and Australia, there is a lack of concerted national action across states: some have had success whereas others have not taken any action. Some countries with strong environmental profiles have not yet taken any action, for example, Canada and New Zealand.

Although generally the plastics recycling industry is growing, by approximately 7% y^{-1} over the last decade in western Europe, there are large regional variations and of the total produced, only a small proportion of plastics are actually recycled. For example, in the United States in 2013, 9% of all plastics in municipal waste were recycled, and of these only the HDPE and PET were recycled in any substantial quantity (around 30% of that disposed of). This pattern is true of most countries where HDPE plastic bottles form the bulk of the plastics recycled. Although most householders widely support recycling schemes, the diversity of plastic formulations and the lack of standardisation in production and labelling present a barrier to recycling. Plastic is ground up, melted down and recombined into new products that carry the recycled plastics label. Estimates of the proportions of 'mismanaged plastic waste' (e.g. that is littered, fly-tipped or disposed of in open landfill) with the potential to end up in the ocean show that very high quantities may originate from parts of China and Oceania. Between 2.5 and 10 × 10^5 t of plastic waste is estimated to originate from North America, Brazil, Alaska, India, Pakistan, South Africa and parts of North Africa. The smallest amounts are discharged from Canada, Greenland, Iceland, Sweden, Norway and New Zealand, which is presumably partly attributable to their lower population densities. Transient increases in local populations from tourism may explain the high amounts in some areas (e.g. Turkey, Egypt, Morocco, South Africa and Thailand).

4.2.5.3 Alternative plastic products

There is increasing interest in sustainable plastic products that can be broken down naturally and so do not persist in the environment. Materials called *bioplastics* that are partly derived from biological materials—for example, wheat, maize, tapioca or sugar cane—show great promise. The potential for many of these products has existed since the inception of industrial-scale plastic production (Section 4.2.2), but the costs were previously too high to make commercial production viable. However, being of biological origin does not necessarily mean bioplastics can be broken down by bacteria and fungi— that they are truly *biodegradable* (Table 4.12). In this section we only consider those bioplastics that are biodegradable.

Table 4.12 Commercial 'bioplastic' compounds. Data from: UNEP 2015, Hammer et al. 2012, and Australian Academy of Science 2016.

Bioplastic	Constituents and use	Biodegradability
Plastarch (PSM)	**Produced from:** maize starch **Uses:** Similar properties to polyethylene and polypropylene, thermoplastic (to 125 °C)	**Biodegradation:** in compost, soil, water. Incineration produces non-toxic smoke and residues can be used to fertilise crops
Polyhydroxyalkanoate (PHA)	**Produced by:** bacteria as food storage molecules. Microbes are fed with natural carbon; for example, waste effluent, plant oils and carbohydrates **Uses:** food packaging and medical implants; thermoplastic (to 180 °C)	**Biodegradation:** in the marine environment, non-toxic and could process other wastes; for example, sewage
Poly-3-hydroxybutyrate (PHBs)	**Produced from:** biomass (as for PHA) **Uses:** medical implants	**Biodegradation:** in terrestrial environments
Polylactic acid (PLA)	**Produced from** plant starches for example, maize, wheat or sugar **Uses:** films, fibres, cups, bottles.	**Biodegradation:** industrial processing required taking about 60 days
Polybutyrate (PBAT)	**Produced from:** petrochemicals and starch **Uses:** packaging, bags and wraps.	**Biodegradation:** by composting.
Polycaprolactone (PCL)	**Produced from:** petrochemicals and starch **Uses:** biomedical and modelling	**Biodegradation:** in terrestrial environment and some marine after 12 months

The greater production efficiency, requiring lower temperatures and so less energy, and the increased costs of crude oil have made bioplastics economically viable. Several companies are now producing these materials (Table 4.12).

Bioplastics currently comprise a small fraction, around 1%, of the total plastics market (estimated at US$5.8 billion by 2021), but have experienced considerable growth since 2000. The benefits of using bioplastics are that they reduce the use of non-renewable hydrocarbon raw materials, some are biodegradable without intervention, and the manufacturing processes can use up to 65% less energy than traditional methods.

Plastarch and polyhydroalkanoate (PHA) are bioplastics made from biological materials that are biodegradable in the environment and so represent the ideal solution: making plastic use sustainable and non-polluting. Plastarch production uses maize starch and PHA is produced from bacteria that are fed organic matter. The bacterial production of plastics (PHA and polyhydroxybutyrate (PHB)) if efficient and economic could have the added benefits of performing waste treatment (Table 4.12). PHB is similar to PHA yet has only been found to be biodegradable in terrestrial environments.

Polylactic acid (PLA) is a bioplastics made from the natural starches that are extracted and fermented to make lactic acid, and this is polymerised to form PLA. This bioplastic has similar properties to PET plastics and although it is fully biodegradable it requires very specific conditions and so in order to be recycled it needs to receive specialist industrial processing. Similarly, the bioplastics that contain petrochemicals in addition to starch products (e.g. polybutyrate (PBAT) and polycaprolactone (PCL); Table 4.12) only degrade in terrestrial environments and/or need composting conditions.

4.2.5.4 The challenges of microplastics

Perhaps the most challenging future aspect of plastic pollution is the introduction of billions of very small, almost invisible plastics into the environment and, potentially, our food supply (Section 4.2.3). To tackle this challenge, more information is needed on microplastic pollution. There is insufficient data on the amount of microplastics and their nature that could be used to identify their sources so that their

inputs can be managed. With regard to the risks to ecosystems and human health, the GESAMP assessment (Bowmer and Kershaw 2010) of microplastic pollution identified the following main areas of data deficiency: the effects of microplastic ingestion on individuals, populations and food webs; the effects of the toxicants and additives they carry, or those adsorbed from the environment; and the risks of entry into human diets. Secondary microplastic pollution may be minimised by introducing policies to reduce plastic use and manage disposal practices (Section 4.2.5). However, the management of primary microplastic disposal will require different mechanisms and legislation due to the differences in size and source. For example, primary micro plastics might be reduced by fitting filters to washing machines to block the release of artificial fibres into waste waters. These advances require the support of new legislation and regulations surrounding not only waste disposal but also to compel the microplastic production industry to release information on properties, composition and use (data which are not currently made available by the industry).

4.3 Nutrients

The release of plant nutrients into aquatic environments has been occurring for centuries. Plant nutrients include nitrogen, phosphorous and silica and the two of primary concern as marine pollutants are phosphorous and nitrogen. Discharges of nitrogenous wastes into the aquatic environment consist of nitrate (NO_3^-), nitrite (NO_2^-), ammonia (NH_3) and ammonium (NH_4^+) which exist in equilibrium in aqueous solution and/or are readily transformed by chemical and bacterially mediated nitrification and denitrification. Phosphorous occurs as phosphate (PO_4^{3-}). In freshwater ecosystems, phosphorous is generally the limiting plant nutrient whereas in the marine environment it is usually nitrogen.

The sources of anthropogenic nutrients include animal waste products (Section 3.2) that enter the sea either via run-off from farmland or aquaculture sites or from the discharge of treated or untreated sewage effluent directly into the sea or through stormwater overflow (which shares pipework with sewage effluent; Section 3.2). There has been sharp increases in nutrient releases over the last 100 or so years due to applications of fertilisers to arable crops which began in the 1800s. These applications usually far exceed what is required and the excess nitrogen and phosphorous accumulate in soils or are washed into local watercourses.

The increased intensity of livestock farming has contributed to the high levels of nutrients discharged via animal faeces that are washed off land by precipitation and enter the sea through rivers, streams and groundwater (Figure 4.11). The leaching of animal wastes from the food-processing industry (e.g. slaughterhouses and, canning plants) also introduces nutrients to waterways. Similarly,

excess food and waste from industry (Section 3.2.5) adds nutrients directly to the sea or, if operations are land-based, indirectly through rivers and streams.

Nitrogenous gases are released into the atmosphere during the combustion of fossil fuels or biomass and other industrial activities. These nitrogenous compounds become deposited into the sea through wet or dry deposition (e.g. through winds and dust or precipitation). This inorganic atmospheric nitrogen pollution takes the form of nitrous oxide (N_2O), nitrogen dioxide (NO_2) or ammonia (NH_3). The main sources of nitrogen dioxide (NO_2) are the combustion of fossil fuels and

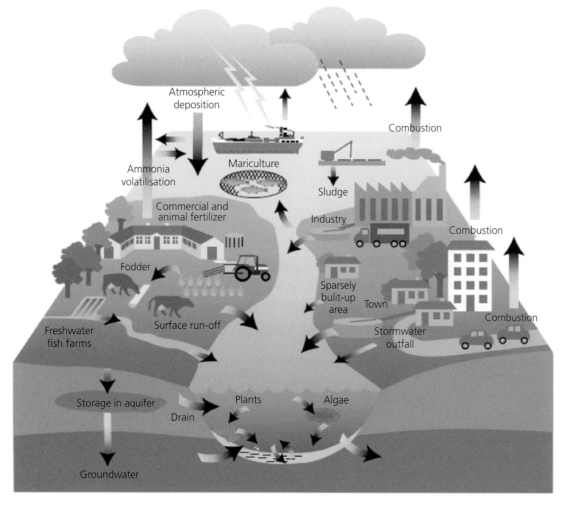

Figure 4.11 The aquatic nitrogen cycle with sources of nitrogen pollution. Source: 2003 AErtebjerg, G., Andersen, J.H. and Hansen, O.S. (eds). *Nutrients and Eutrophication in Danish Marine Waters. A Challenge for Science and Management.* National Environmental Research Institute. 126 pp. © National Environmental Research Institute.

industrial discharges. Nitrous oxide (N_2O) is also produced from fossil fuel combustion, the industrial production of nylon and gaseous emissions from agriculture. These agricultural sources include the bacterial breakdown of fertilisers applied to crops and animal wastes that release N_2O, whereas atmospheric NH_3 mostly originates from the industrial processes within which it is an end-product and/or by-product. Ammonia is a highly alkaline corrosive solvent that is very reactive and so has wide-scale industrial applications in, for example, the production of coke from coal, in metallurgy, the textile industry, within fuel cells and in the preparation of industrial chemicals (such as fertilisers, refrigerants, rubbers and plastics).

The main routes by which nitrogen and phosphorous enter the marine environment are therefore diffuse (Chapter 2) via the atmosphere or as the run-off of detergents, animal wastes, sewage and fertilisers applied to farmland. The two largest contributors of nitrogenous inputs into the oceans are agriculture and industrial emissions, and for phosphate they are agriculture and waste-water (containing sewage and detergents). In 2009, the European Union consumed 1.0×10^7 t nitrogen fertiliser and 4×10^6 t phosphate (P_2O_5) (European Fertiliser Manufacturers Association 2009). Global atmospheric emissions of nitrates from anthropogenic activities were approximately 10.6×10^6 t N_2O, 122×10^6 NO_2 and 49.3×10^6 t NH_3 in 2008 (Table 4.13). Approximately $1.0–1.7 \times 10^6$ t of sodium tripolyphosphate, the commonest phosphate additive, are used each year in detergents (Section 4.4).

Nitrate and nitrite cause excessive aquatic plant growth (*eutrophication*) that can limit light penetration through the water column and introduce large quantities of carbon into the system (from excessive algal growth) which accumulates on the seafloor causing enrichment and deoxygenation (Section 3.2). Decreases in oxygen (hypoxia or anoxia) have

Table 4.13 Annual atmospheric nitrogen emissions from anthropogenic activities from 1860–1993 and projected for 2050. Data from: Galloway et al. 2004 and EDGAR emissions database v. 4.2

	1860			1993			2050		
Gas	N_2O	NH_3	NO_2	N_2O	NH_3	NO_2	N_2O	NH_3	NO_2
Emissions ($\times 10^6$ t y^{-1})	1.4	7.4	2.6	9.18	47.2	36.2	10.6	49.3	122

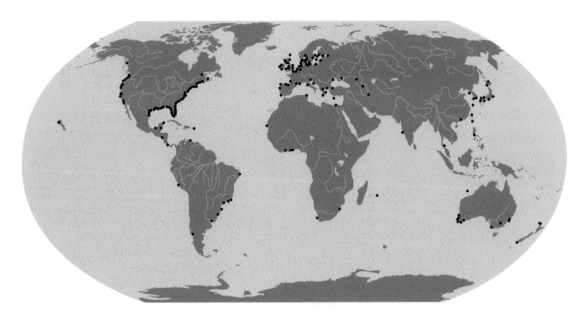

Figure 4.12 Global coastal hypoxic areas caused by anthropogenic factors as of 2009. Each point represents one hypoxic system. Credit: Rabalais et al. (2010). This work is distributed under the Creative Commons Attribution 3.0 License (http://www.biogeosciences.net/about/licence_and_copyright.html). See also interactive hypoxia map in resources.

severe effects on most marine organisms causing mass mortalities, reductions in community diversity, body size and fitness (Sections 3.2, 6.3 and 6.4). The number of deoxygenated zones in the ocean caused by anthropogenic activities now exceeds 500, covering an area of more than 240 000 km^2 (Figure 4.12).

Some of the algal growth comprises blooms of species characterised as harmful algal blooms (HABs) due to the production of toxins or physical characteristics (mucus, discoloration) that affect the ecology or use of the water body. These HAB species include diatoms, dinoflagellates and cyanobacteria. They produce substances that can be toxic to the fish and invertebrates that consume them, and therefore can also find their way into the human food chain through contaminated seafood. Mass mortalities of marine fish, cetaceans, turtles and sea birds have occurred due to consumption of HAB species or animals that have fed on them. In humans, these poisonings are referred to as paralytic shellfish poisoning (caused by saxitoxin from several species of dinoflagellate and cyanobacteria), neurotoxic shellfish poisoning (due to the brevetoxin from the dinoflagellate *Karenia* spp. and the alga *Chattonella*), amnesic shellfish poisoning (from domoic acid produced by diatom species) and diarrheal shellfish poisoning (from okadaic acid and dinophysistoxins produced by dinoflagellate *Dinophysis* spp.). Paralytic shellfish poisoning causes nausea, loss of vision, paralysis and potentially death, and amnesic poisoning causes memory loss and, potentially, brain damage and death. Diarrheal shellfish poisoning as the name suggests causes diarrhoea, vomiting and cramps but it is not usually fatal, and similarly, neurotoxic shellfish poisoning causes vomiting and neurological manifestations such as slurred speech but is also not generally fatal. Harmful algal blooms are monitored in countries where blooms are known to be common, for example, in the United States by the National Ocean and Atmospheric Administration (NOAA) and by the Centre for Environment, Fisheries and Aquaculture (CEFAS) in the United Kingdom.

Nutrients, particularly nitrogenous wastes, are an ongoing challenge because as human populations continue to expand, more crops are needed requiring more fertilisers and more agricultural

and human waste is produced. Changes in land use have facilitated the delivery of nutrients into the marine realm by increasing the amounts and rates of nutrient run-off from land into watercourses and eventually the sea. Naturally some of these nutrients will be absorbed into soils, leach into groundwater and support plant growth on land.

Nutrients can be removed from water by aquatic macrophytes; for example, seagrass or macroalgae or terrestrial wetland and saltmarsh plants. However, in many cases, the eutrophication kills off many of the plant populations that might help to buffer against eutrophication. Examples of this can be found at Chesapeake Bay, with the loss of extensive saltmarsh beds (Section 6.5), and in the Black Sea, with the loss of extensive red and brown macroalgal fields (Section 6.4). Artificial methods for removing excess nutrients include the use of *bioextraction* that involves cultivating plants and animals to remove nutrients. Animals such as the filter-feeding bivalve molluscs can remove substantial amounts of organic matter, and indirectly nutrients, from a water body and can then be harvested commercially as food. Oysters and mussels have been shown to be particularly efficient at removing nutrients and are the basis of remediation programmes in Chesapeake Bay where attempts are being made to restore the extensive historic oyster reefs (Section 6.5).

The EU Groundwater Directive 2006 states that nitrates should not exceed 50 mg l^{-1}. The 2013 European Commission report on nitrates in Europe found that for most member states from 2008–11 more than 10% of groundwater sampling stations exceeded 50 mg l^{-1}, with 10–20% between 25 mg l^{-1} and 50 mg l^{-1}. For marine waters, 1.4% of stations exceeded 25 mg l^{-1} and 75% were less than 2 mg l^{-1} (but data were lacking for marine waters). Water-quality guidelines for drinking water in Canada, Europe and Australia also use the 50 mg l^{-1} value. At high concentrations, nitrates can be toxic to aquatic animals at 20 mg NO_3–N l^{-1} for adults, and juveniles may be sensitive to 2 mg NO_3–N l^{-1} (Camargo et al. 2005), and they can be toxic to humans. Ammonia is highly toxic to aquatic life and is considered in the next section (Section 4.3.1).

4.3.1 Ammonia

Ammonia (NH_3) occurs widely in nature as a product of excretion, the breakdown of organic matter and the fixation of nitrogen by bacteria (Figure 4.11). In aqueous environments, NH_3 exists in equilibrium with its ionised form ammonium (NH_4^+):

$$H_3O^+ + NH_3 \leftrightarrow H_2O + NH_4^+$$

The amount of NH_3 or NH_4^+ available in an aqueous solution changes with pH and temperature. As they increase, the aqueous equilibrium shifts towards the left and more NH_3 becomes available; as salinity increases, NH_3 availability decreases. Ammonia is highly toxic to aquatic organisms whereas the ionised form (NH_4^+) is not. The toxicity of NH_3 to aquatic organisms therefore depends on the ambient environmental conditions, with toxicity approximately doubling for every 10 °C-temperature increase and increasing ten-fold with each pH unit. Under anaerobic conditions (low dissolved oxygen), NO_3^- is reduced to toxic NH_3 again, increasing the potential exposure of aquatic organisms to it.

Ammonia enters the environment through the discharge of animal wastes, the combustion of fuel and atmospheric industrial discharges (Section 4.3). It is also used in the food industry as an antifungal agent and in many household and industrial cleaning products such as bleaches and disinfectants. So, it is widely used by the health service. The leaching of wastes from landfill (containing 400 mg NH_3 l^{-1}) represents another source to aquatic environments. Overall, 90% of environmental NH_3 originates from the production of fertilisers and livestock waste. It is lost from the environment by bacterial nitrification, the rate of which is dependent on the temperature, pH and the availability of oxygen. Ammonia may be lost to the atmosphere through volatilisation or because it is taken up by plants. It does not accumulate in sediments or organisms.

4.3.1.1 Ammonia toxicity

Although NH_3 is toxic to humans at high concentrations if ingested or if it comes into contact with the skin, it is usually not toxic at the concentrations found in the environment. However, it is highly toxic to aquatic organisms and its toxicity varies with environmental conditions. So while moderate toxicity occurs at ambient water temperature and pH, as both increase it becomes more toxic. Changes in salinity and dissolved oxygen content may also affect NH_3 toxicity.

Ammonia is very toxic to most species of fish and high concentrations cause damage to fish gill epithelia that can lead to suffocation and death. It stimulates glycolysis and inhibits the Krebs cycle, resulting in the build-up of toxic metabolites and interferes with energy production, as adenosine triphosphate (ATP). By impairing NH_3 excretion or enhancing its uptake, osmoregulation is disrupted and the resulting electrolytic imbalances impact fish liver and kidney functioning. It diffuses across cell membranes, becomes ionised and interferes with glutamate receptors, resulting in increased transfer of Ca^{2+} and Na^{2+} into neurones which interact with intracellular enzymes leading to cell death. During exercise and stress fish produce more NH_3 and so become more sensitive to NH_3 in their environment. The detoxification mechanisms of NH_3 in fish are related to feeding and so underfed fish are more susceptible to its toxicity.

Juvenile fish are more sensitive to NH_3. For example, one-day-old flounder larvae have a median lethal concentration (LC50) of 0.27 mg NH_3–N l^{-1}, for 23-day-old larvae the LC50 is 1.02 mg NH_3–N l^{-1} and for adult fish it is 50–66 mg NH_3–N l^{-1}. For Dover sole exposed for 42 days at pH 6.9–7.9, the no observed effect concentration (NOEC) is 0.06 mg NH_3–N l^{-1}. Juvenile turbot exposed for 11 days had an NOEC of 0.11 mg NH_3–N l^{-1}. Some species of fish are tolerant to NH_3 because they either maintain NH_3 excretion or convert it to less toxic substances (e.g. glutamine or urea). For example, the marble goby, *Oxyleleotris marmoratus*, mud skippers, *Boleophthalmus boddaerti* and *Periophthalmodon schlosseri*, toad fish, *Opsanus beta* and elasmobranchs detoxify ammonia to urea. Rainbow trout, on the other hand, can actively excrete ammonia and the weather loach, *Misgurnus anguillicaudatus*, accumulates and volatilises NH_3 through its skin and mouth when exposed to air.

The toxicity of ammonia to invertebrates is better known for freshwater invertebrates and the effects include mortality, reduced fertility and other reproductive problems. The toxicity of ammonia to invertebrates varies from 0.057 to 19 mg NH_3–N l^{-1} and marine molluscs seem generally to be the least sensitive whereas echinoderms are the most sensitive (Table 4.13). In molluscs, NH_3 causes reduced respiration and feeding, the impairment of byssal secretions and mortality. As for fish, ammonia seems to be more toxic to juvenile invertebrates (Table 4.14). Ammonia inhibits the growth and photosynthesis of marine algae which are less sensitive than marine invertebrates and fish, perhaps due to their use of ammonia as a nitrogen source. However, some algae are sensitive such as the green alga *Nephroselmis pyriformis*. Long-term toxicity tests with benthic estuarine diatoms found that the lowest observed effect concentration (LOEC) for growth after 10 days was 0.2 mg NH_3–N l^{-1}.

Freshwater taxa are thought to be more sensitive to NH_3 than marine taxa as reflected by the different regulatory limits, but interspecific toxicity is highly variable. More freshwater toxicity data are available and so provide a more complete picture compared with the marine taxa. The acute toxicity of NH_3 to fish has been shown to vary up to a salinity of 10, after which NH_3 toxicity begins to increase with salinity up to 35. The toxicity to marine fish and crustacea is within the range for freshwater taxa. Freshwater molluscs are very sensitive to NH_3 compared with some marine mollusc species that seem more tolerant (Table 4.14).

The Environmental Quality Standard for ammonia in Europe is 21 µg NH_3–N l^{-1}, and although this is above the toxicity for most taxa (Table 4.14) it may not be adequate for juveniles or in situations where water quality is poor (e. g. in low oxygen conditions) and NH_3 toxicity is higher. Thus, Predicted No Effect Concentrations (PNECs) have been proposed as follows: 5.7 µg NH_3–N l^{-1} for short-term exposures and 0.66 µg NH_3–N l^{-1} for long-term exposure of marine biota to NH_3 (*cf.* for freshwater taxa 6.8 and 1.1 µg NH_3–N l^{-1}, respectively). Australian and New Zealand environmental guidelines predict that 1.7 mg NH_3–N l^{-1} will protect 80% of marine species (and less than 2.3 mg NH_3–N l^{-1} will protect 80% of freshwater species). In the United States, the marine guidelines for

ammonia specify that the 1-hour mean concentration should not exceed 0.233 mg NH_3–N l^{-1} and the 4-day mean concentration should not exceed 0.035 mg NH_3–N l^{-1}.

4.4 Detergents

Detergents have many uses in daily life in homes and businesses, in healthcare, in industry and as fuel additives. Sales of domestic cleaning products in the United States were US$52 × 10^9 y^{-1} and in Europe US$31.7 × 10^9 y^{-1} between 2010 and 2013 (for Germany, France, Italy, the United Kingdom, Spain and The Netherlands) (Statista 2016). Of these sales, 30% were disposable paper products, 21% were laundry detergents, 15% household cleaners and 10% dishwasher detergents.

Most modern detergents are composed of a suite of synthetic compounds with higher solubility in hard water compared with natural soaps. The major components of these synthetic detergents include *alkaline builders*, anionic, non-ionic or cationic surfactants and bleaches that break down organic matter. The alkaline builders are phosphate compounds that maintain the desired pH and remove calcium, softening the water and improving the effectiveness of the surfactants and bleaches. Alkaline builders are added to laundry detergents, dishwasher detergents at both domestic and industrial scales and for other industrial applications. The commonest alkaline builder is sodium tripolyphosphate. Surfactants, or surface active agents, lower the surface tension of a liquid. These compounds are usually *amphiphillic*, meaning they have both water repelling and attracting chemical groups. Surfactants perform many functions as emulsifiers, detergents, dispersants or foaming agents and so are present within most cleaning products. Detergents may also contain small amounts of other compounds such as the solvent mono ethanol amine, bleach activators like trichloroethylenediamine, and enzymes for breaking down proteins and fats. Some toxic surfactants such as sodium dodecyl sulphate (SDS) are also used in personal care products such as toothpaste, shampoo, bath foam and household cleaning products.

Table 4.14 Acute (24–48 h) and chronic (96 h) toxicity of ammonia to a range of marine taxa (LC50 for animals and EC50 for algae) at specified pH and water temperature. Note: if pH and temperature are not provided the data were not available; if tests differed in duration the number of hours is provided in parentheses. Data from: US Environmental Protection Agency 1989, Allan et al. 1990, Miller et al. 1990, Boardman et al. 2004, Wang 2015.

Taxa	LC50 or EC50 mg NH$_3$-N l^{-1}	
	Acute	**Chronic**
Fish		
Monocanthus hispidus	**pH 8.07, 23.4 °C** = 0.83	–
Sparus aurata	**pH 8.10, 27 °C** = 1.93	**pH 8.1, 17.5 °C** = 2.55
Sciaenops ocellatus	**pH 8.1, 25.5 °C** = 0.55	–
Pseudopleuronectes americanus	**pH 8, 7.5 °C** = 0.49	–
Dicentratus labrax	–	**pH 8.15, 17.5 °C** = 1.70
Menidia beryllina	–	**pH 8, 18 °C** = 0.98
Scopthalmus maximus	–	**pH 8.15, 17.5 °C** = 2.55
Cyprinodon variegatus	–	**pH 7.25, 25 °C** = 2.80 **pH 8.05,13 °C** = 2.10
Molluscs		
Potamopyrgus antipodarum	**pH 8.3, 20.4 °C** = 2.72	**pH 8.3, 20.4 °C** = 2.00
Crassostrea virginica	**pH 7.7–8.0, 20 °C** = 19.10	–
Mercenaria mercenaria	**pH7.7–8.2, 20 °C** = 5.36	
Anadara granosa	–	**pH 7.9** = 1.8 (168 h)
Argopecten irradians	–	**pH 8.1** = 5.3
Crustaceans		
Eucalanus elongates	**pH 8, 20.3 °C** = 0.87	–
E. pileatus	**pH 8.2, 20.5 °C** = 0.79	–
Latreutes fucorum	**pH 8.07, 23.4 °C** = 0.77	–
Mysidopsis bahia	**pH 6.8–9.2, 23.2 °C** = 1.02	**pH 8.0, 25 °C** = 1.70
Metapenaeus ensis	**pH 7.7, 25 °C** = 0.87	**pH 7.7, 25 °C** = 1.76
Penaeus chinensis	**pH 8.12, 25 °C** = 3.88	**pH 7.94, 26 °C** = 2.10
Echinoid larvae		
Strongylocentrotus purpuratus	–	0.057 (72 h)
Dendaster excentricus	–	0.03 (72 h)
Rotifers		
Brachionus plicatilis	–	**pH 7.5–7.9** = 8.6
Algae (EC50)		
Nephroselmis pyriformis	**pH 7.5, 20 °C** = 0.013 **pH 8.7, 20 °C** = 0.031	–
Phaeodactylum	**pH8–9.5** = 17 (1.5h)	–
Dunaliella tertolecta	**pH8–9.5** = 17 (1.5h)	–

Table 4.15 Ecotoxicity of a range of surfactants, commonly used in detergents, to aquatic taxa. Sodium dodecyl sulphate is synonymous with sodium lauryl sulphate. Data from: Swedmark et al. 1971, Davis and Gloyna 1967, Mariani et al. 2006, Persson et al. 1990.

Taxa	Detergent	Concentration	Endpoint
Algae and bacteria			
Vibrio fischeri	Sodium dodecyl sulfate	2.6 mg l^{-1}	EC50 (96 h) bioluminescence
Gloeocapsa sp.	Sodium dodecyl sulfate	50 mg l^{-1}	*N*-fixation
Dunaliella tertiolecta	Sodium dodecyl sulfate	4.8 mg l^{-1}	IC50 (96 h)
Laminaria saccharina	Mixture of dodecyl benzene-sulfonate, sodium lauryl ethersulfate and lauric diethanolamide	50 mg l^{-1}	Zoospore motility
Molluscs			
Crassostrea virginica		25 µg l^{-1}	Hatching
C. virginica larvae	Alkyl sulfonate	100–500 µg l^{-1}	LC50 (48 h)
Cardium edule	Octylphenol 11-ethylene oxide	10–30 mg l^{-1}	LC50 (48 h)
	Nonyl phenol 12-ethylene oxide	33–100 mg l^{-1}	
	Lauryl ether sulfonate-3 ethylene oxide	10–33 mg l^{-1}	
	Alkylbenzene sulfonate	20 mg l^{-1}	
	Dodecyl benzene sulfonate	15 mg l^{-1}	
Cardium edule	Lauryl ether sulfonate-3 ethoxylate	50 mg l^{-1}	LC50 (96 h)
	Nonylphenoxy 10-acetic acid	5 mg l^{-1}	
	Tallow alcohol ethoxylate	< 5 mg l^{-1}	
Mya arenaria	Dodecyl benzene sulfonate	70 mg l^{-1}	LC50 (96 h)
	Lauryl ether sulfonate-3 ethoxylate	50 mg l^{-1}	
	Nonylphenoxy 10-acetic acid	18 mg l^{-1}	
	Tallow alcohol ethoxylate	100 mg l^{-1}	
Mytilus edulis	Alkylbenzene sulfonate	> 100 mg l^{-1}	LC50 (96 h)
	Dodecyl benzene sulfonate		
	Lauryl ether sulfonate-3 ethoxylate		
	Nonylphenoxy 10-acetic acid		
	Tallow alcohol ethoxylate		
Mytilus edulis	Dodecyl benzene sulfonate	200 µg l^{-1}	Fertilisation < 50%
larvae	Dodecyl benzene sulfonate	50 µg l^{-1}	Growth and settlement
Pecten maximus	Dodecyl benzene sulfonate	< 5 mg l^{-1}	LC50 (96 h)
Crustacea			
Balanus balanoides	Dodecyl benzene sulfonate Nonylphenoxy 10-acetic acid	50 mg l^{-1}	LC50 (96 h)
Balanus balanoides nauplii	Dodecyl benzene sulfonate Nonylphenoxy 10-acetic acid	< 25 mg l^{-1}	
Carcinus maenas	Alkylbenzene sulfonate	> 100 mg l^{-1}	LC50 (96 h)
	Dodecyl benzene sulfonateLauryl ether sulfonate-3 ethoxylate		
	Nonylphenoxy 10-acetic acid		
	Tallow alcohol ethoxylate		
Crangon crangon	Octylphenol 11-ethylene oxide	33–100 mg l^{-1}	LC50 (48 h)
	Nonyl phenol 12-ethylene oxide	33–100 mg l^{-1}	
	Lauryl ether sulfonate-3 ethylene oxide	> 100 mg l^{-1}	
Gammarus tigrinis	Tetrapropylene alkylbenzene sulfonate	25 mg l^{-1}	LC100
Tigriopus fulvus nauplii	Sodium dodecyl sulfate	7.4 mg l^{-1}	LC50 (96 h)

(*continued*)

Table 4.15 (Continued)

Taxa	Detergent	Concentration	Endpoint
Paracentrotus lividus	Sodium dodecyl sulfate	3.2 mg l^{-1}	EC50 (96 h) egg fertilisation
Fish			
Pleuronectes flesus	Octylphenol 11-ethylene oxide	33–100 mg l^{-1}	LC50 (48 h)
	Alkylbenzene sulfonate	6.5 mg l^{-1}	LC50 (96 h)
	Dodecyl benzene sulfonate	1.5 mg l^{-1}	LC50 (96 h)
	Lauryl ether sulfonate-3 ethoxylate	< 5 mg l^{-1}	LC50 (96 h)
	Tallow alcohol ethoxylate	> 0.5 to < 1.0 mg l^{-1}	LC50 (96 h)
Gadus morhua	Alkylbenzene sulfonate	3.5 mg l^{-1}	LC50 (96 h)
	Dodecyl benzene sulfonate	1.0 mg l^{-1}	
	Lauryl ether sulfonate-3 ethoxylate	< 5.0 mg l^{-1}	
	Nonylphenoxy 10-acetic acid	6.0 mg l^{-1}	
	Tallow alcohol ethoxylate	> 0.5 to < 1.0 mg l^{-1}	
Dicentrarchus labrax	Sodium dodecyl sulfate	7.3 mg l^{-1}	LC50 (96 h)
Lepomis machrochirus	Alkylbenzene sulfonate	5–10 mg l^{-1}	Impaired egg and sperm development

Wide use of detergents has resulted in large quantities entering the oceans via rivers and estuaries from industrial discharges, as diffuse pollutants from land and in sewage outflows from domestic use. The phosphate alkaline builders, although not highly toxic, introduce large quantities of phosphorous into the marine environment and contribute to the problems of eutrophication (Section 4.3). Many of the surfactants that are included in detergents are, however, highly toxic (Table 4.15). Due to their amphiphillic nature, surfactants become concentrated near the air–sea interface and can bind to sediments. Surfactants are also able to interact with other hydrophobic pollutants that are relatively insoluble such as PCBs and PAHs (Section 3.8) and can mobilise them. Therefore, in addition to their innate toxicity, surfactants are able to release legacy contaminants such as PCBs from sediments.

The toxicity of surfactants range from μg l^{-1} to mg l^{-1} and they inhibit growth, interfere with reproductive and developmental processes and cause mortality of many taxa exposed for 48 to 96 hours. They inhibit algal growth (Table 4.15) and the motility of algal reproductive stages. The detergents used to clean-up oil spills are very toxic to algae (Section 4.4). They can impact the fertilisation, development and hatching of the larvae of bivalve molluscs and crustacea at μg l^{-1} (Table 4.15). Fish experience gill damage from detergents that can impair respiration, and mortality can result from reductions in surface tension caused by surfactants. The effects on aquatic taxa depend upon their ionisation state: many studies have shown that anionic detergents tend to be more toxic than non-ionic or cationic detergents, but the toxicity varies with taxa and detergent. Anionic and non-ionic surfactants can interfere with biogeochemical cycling by interfering with the nitrogen fixation of cyanobacteria such as *Gloeocapsa* spp. (Table 4.15).

A particularly toxic group of non-ionic surfactants used in detergents during the 1940s to 2000s that have now been phased out of use in most developed countries are the alkyl ethoxylates. However, because alkyl ethoxylates are bioaccumulating and were used in very large quantities globally for more than 50 years, they will continue to exist in the environment and organisms as a legacy contaminant and so represent an ongoing challenge. One widely used alkyl ethoxylate is nonylphenol ethoxylate that was widely used in oils as a lubricant, in laundry and dishwasher detergents. Nonylphenol ethoxylate is highly toxic to aquatic life (ranging from 17 μg l^{-1} median effect concentration (EC50) to 5.4 mg l^{-1} median lethal concentration (LC50)).

It bioaccumulates and functions as an endocrine disruptor (Section 5.3) and so is a threat to human health. Human exposure to nonylphenols occurs via contaminated drinking water and food where it occurs in concentrations of 0.1–19.4 µg l^{-1}. The daily intake has been estimated at around 7.5 µg d^{-1} for human adults. Nonylphenol ethoxylates were used in very large quantities domestically and the main sources into the marine environment were through waste-water discharge. Nonylphenol production by Europe, Japan, China and the United States in 2000–2002 exceeded approximately 260 000 tonnes y^{-1}. They are now listed as hazardous substances by the EC Water Framework Directive and the US EPA water guidelines recommend less than 1.7 µg l^{-1} in seawater.

Nonylphenol ethoxylates were replaced by alcohol ethoxylates that degrade more quickly in the environment. Alcohol ethoxylates have not yet exhibited any endocrine disrupting or carcinogenic effects. The no-observed adverse effect level (NOAEL) is 50 mg kg^{-1} which is well below the predicted local environmental concentration of 1.0 µg l^{-1} and the 9.8 µg l^{-1} found in sewage discharges. Detergents are regulated in Europe under REACH (Registration, Evaluation, Authorisation and Restriction of Chemicals) regulations on the basis of their degradability (EC 648/2004). Evidence for biodegradability must be presented which shows that for all surfactants used in detergents, more than 60% mineralisation should occur within 28 days. Regulations also consider the introduction of phosphates into the environment and their role in eutrophication (Section 3.2).

4.5 Oil and gas

Oil spills are probably the aspect of marine pollution that is uppermost in the minds of the general public. However, classic oil spills only account for around 12% of the oil entering the marine environment each year (Figure 4.13). To understand oil pollution we first need to understand oil. Oil is the partly decomposed and fossilised remains of marine plankton. These plankton lived in past oceans and after their death sank to the seafloor where conditions (low oxygen) were such that they only partly decomposed before being buried. These buried remains became further altered by heat and compression during the millions of years since their burial. Oil therefore only occurs in certain types of rock and in a limited number of localities. The oil also varies chemically between deposits due to both the nature of the marine plants from which it is composed, the conditions on the seafloor at the time of burial and the geological conditions it has been exposed to since. Oil from two oil wells only a short distance apart can therefore possess very different chemical and physical characteristics.

The material found in geological reservoirs is in the form of *crude oil*, and is a complex mixture of hydrocarbons of different chemical length, different extent and types of chemical bonds, with different co-occurring compounds like metals. Crude oil when extracted varies between a light-coloured fluid that pours readily at ambient temperatures (i.e. contains a predominance of low molecular-weight hydrocarbons) and material that is thick and tar-like that can only be poured (or moved in pipelines) if heated.

Crude oil is transported by pipelines and ships to refineries, where the different components of the crude mix are separated into useful products; for example, petroleum, lubricating oils, heavy fuel oil, wax, tar, organic raw materials (e.g. plastic feedstock (Section 4.2.2), organic chemicals) and so on. These products are then transported to end-users by pipeline, sea or by road or rail tankers.

4.5.1 The biological effects of oil

Crude oil is toxic if ingested, causing damage to the lipophilic components of membranes in particular. The toxicity varies, depending on its composition and source. Crude oil can also cause mortality through physical impacts such as coating the gills or other respiratory structures, interfering with the waterproofing and thermal protection of birds and mammals. The physical and chemical properties of the refined hydrocarbons vary but they are often more aggressive than crude oil. For example, the lighter petroleum and diesel fuel fractions are more toxic than crude oil and will also be more effective at stripping natural oils from feathers and fur, causing thermal and water-proofing problems for animals. However, these light fractions are also more volatile and will

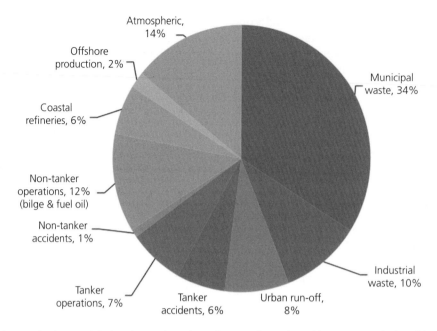

Figure 4.13 Sources of anthropogenic hydrocarbons to the marine environment. Non-tanker accidents represent oil released from the fuel tanks of vessels that are not transporting oil as cargo. Non-tanker operations are those that apply to all shipping. Data from: GESAMP (2007).

therefore only persist in the marine environment for a limited time, thus also limiting contact with the biota.

4.5.2 Sources of oil in the marine environment

The UN Group of Experts on the Scientific Aspects of Marine Pollution (GESAMP) estimated that, in 1993, approximately 2.3×10^6 t of oil entered the marine environment. A US Academy of Science report in 2002 estimated the figure to be about 1.3×10^6 t, but ranged from 0.47×10^6 t to 8.4×10^6 t. The biggest difference between the two estimates was the amount of oil released by natural seepage from geological deposits. Estimating the amount of oil leaking from, or being eroded from, oil-bearing rocks on land and under the sea is difficult and may have previously been substantially underestimated. New data on the number of seeps and their seepage rates, determined using synthetic aperture radar to quantify slicks from satellites (Figure 4.14), suggest that around 6.0×10^5 t enter from natural seeps every year, representing between 30% and 46% of the global estimates of the total oil inputs to the environment. However, there still remains some

uncertainty surrounding these new estimates due to lack of data on seepage rates.

GESAMP further categorised the sources of the oil (Figure 4.13), and it is interesting to note that oil-tanker accidents, the numbers of which vary greatly from year to year, account for around one-quarter of the total oil industry contribution (6% compared to 21% for the total oil industry inputs; Figure 4.13) and is smaller than the amount entering the seas from atmospheric fallout (14%). While these sources contribute similar total amounts, the nature and impacts are quite different. For example, atmospheric fallout occurs over very large areas and involves mainly material emitted from the incomplete combustion of hydrocarbons when used as fuel on land (i.e. in vehicles or by industry) and at sea. The effects of this fallout are likely to be minimal.

There are 600 locations where natural seepage occurs and while the volumes may be large, these are continuous discharges and so the local environment will respond, as it would to any point source discharge. There will be a strong effect zone, then a zone of more limited change, before there is a return to more 'normal' conditions. If the seep remains constant in its nature then the affected zone will also remain stable in extent and response. An equilibrium has been established and because these are natural seeps the changes are also natural and do not constitute an 'impact' or pollution. By contrast, an oil spill from a ship or well accident will release large quantities of oil in locations that have had no prior exposure to oil. The effects of these large spills are therefore often catastrophic (see Section 4.5.3).

Historically, tankers were also responsible for operational discharges of oil associated with the dual use of the tanks for oil as cargo and water as ballast on the return journey. When a tank is emptied some oil remains clinging to the sides of the tank, this 'clingage' can comprise 1500 to 2000 m^3 of oil on an average VLCC (Very Large Crude Carrier—a supertanker). Tankers simply loaded ballast water into these tanks and on approach to the port of laden this ballast water was pumped out, along with some of the oil that now contaminates

it. This was an operational discharge. In the late 1960s and early 1970s, partly driven by environmental pressure but also by the rapid increase in oil price in the early 1970s, oil companies revised their procedures. First, the water was directed to a slop tank and the natural separation of oil (floating on top) and water allowed the relatively clean water to be pumped out from the bottom of the tank. The new cargo was then loaded into the tank. This method of separation is referred to as the 'load-on-top' method. With pressure to recover more of the valuable clingage, the tanks were designed so that they could be washed with high-pressure water to remove this old cargo and prevent the contamination of the new load. The oily washing waters were then pumped into the slop tank for separation and again, only a limited amount of oily water was discharged. A further development was the use of crude oil washing: whereby high-pressure crude oil is sprayed around the tank which mobilises the clingage that can then be discharged. While this

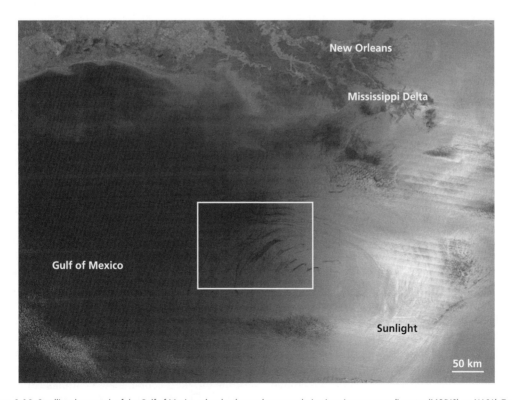

Figure 4.14 Satellite photograph of the Gulf of Mexico taken by the moderate resolution imaging spectroradiometer (MODIS) on NASA's Terra satellite on 13 May 2006. Inset box highlights oil slicks forming from natural seeps in the central Gulf of Mexico. Credit: National Aeronautic and Space Administration.

delays the discharge at port, the offloading of the extra oil recovered, when oil prices are high, offsets this cost.

In spite of the technical changes in oil-tanker operations, unfortunately in some circumstances oily water continues to be discharged at sea as an operational discharge. However, the load-on-top and crude oil washing procedures have greatly reduced the volume of oil released.

Following the grounding of the *Exxon Valdez* in 1989 (Section 4.5.4), the United States enacted new legislation that mandated changes in tanker design. These included the use of a double skin for the hull—so the oil tank would not be breached in the event of a minor collision that pierced the outer hull skin—and also the use of separate tanks for ballast water, keeping the cargo and ballast separate. While these measures reduce oil spills following minor collisions, a double skin would not have mitigated the *Exxon Valdez* oil spill where the ship ground hard on rocks in Prince William Sound, Alaska.

Both oil tankers and other ships release oil as part of their routine operations (Figure 4.13), and this includes oil contaminated water from bilge spaces and engine rooms, oil in cooling waters and oil spilt during fuelling operations, for example, from split pipes or the incorrect use of valves. Accidents involving commercial vessels can also release oil slicks. For example, in 1986 the bulk carrier *Kowloon Bridge* suffered steering failure and ran on to rocks south-west of Ireland. Unfortunately, the ship had taken on fuel only a couple of days before. As the vessel broke up 1200 t of heavy fuel oil was released causing widespread damage to wildlife and local fisheries.

While 13% of anthropogenic oil enters the marine environment from transport operations associated with moving crude oil from source to user, many of the sources of crude oil are located under the seafloor. Offshore marine oil exploration and production generates pollution accounting for a further 2% of the anthropogenic inputs of oil to the sea (Figure 4.13).

Exploration to discover new geological reservoirs containing oil and gas involves seismic surveys, which produce noise pollution (Section 5.7). In general, the impacts of such surveys are limited as they are one off events and are generally timed to avoid seasonal aggregations of marine mammals, birds or fish.

Once a possible reservoir has been identified, the next step is to drill a test well. Usually a vessel is either anchored over the location of interest or a rig is floated into place and jacked up on legs that extend down to the seabed. Once in place, a drill is deployed. The mechanics of the drilling involves a cutting head of hardened metal and industrial diamonds, and the continual lining of the bore hole with a liner, usually made of steel. The whole assembly is pressurised as the gas/oil in the reservoir is under pressure and if an unpressurised drill punctured the reservoir, the oil/gas would flow back out and up the borehole causing what is known as a *blow-out*. The drill needs to be cooled and lubricated because it becomes extremely hot during drilling operations. The drill is hollow and a fluid, known colloquially as 'mud', is pumped down the inside of the drill to emerge from perforations in the drill. This fluid pressurises the well and so acts to prevent a blow-out, and the mud cools and lubricates the drill head before flowing back up the shaft between the drill and the liner. In doing so it carries the rock cuttings back up to the surface.

On the surface, the cuttings are initially held in mesh trays to allow the drill mud to drip off and be recycled, and then the cuttings are typically dumped from the vessel/rig onto the seafloor adjacent to the drill site. The cuttings cause a direct physical (smothering) impact on the seafloor and because they are contaminated with a residual load of mud, this causes a toxic effect.

Drilling mud is far from the simple substance that the name implies. Rather, the drill mud is specifically formulated for the type of rock being cut and so the composition varies as the drill passes through the different strata. It contains substances that cool and lubricate, and these include detergents, organic solvents and metal salts, all of which have a degree of toxicity to marine organisms. For many years, these active agents were carried in a matrix of highly toxic diesel oil. The resultant oil-based muds were extremely toxic and the effects of the oil/mud contaminated cuttings could often be traced for up to 10 km down current of a drill site. These impacts were therefore occurring at every test well, whether or not it developed into a production well, and also at production sites when additional wells were drilled into the reservoir. Since the start of the twenty-first century, there has been considerable pressure on the oil industry to use water-based muds, which are far less toxic but still contain many toxic agents, and are more costly than oil-based solutions. Water-based muds are also unsuitable for some types of operations/rock types.

Around half of all the oil that enters the marine environment does so from land-based sources (Figure 4.13). These include industrial discharges including coastal oil refineries, chemical plants and a wide range of other manufacturing processes that use oils in some form. These discharges are generally point source and so should be regulated and controlled under national licencing regimes. A more significant problem is the quantities of oil entering as diffuse inputs via urban run-off (Figure 4.13). These inputs may enter the oceans via rivers or streams, surface water drains or urban wastewater treatment works. The discharges include uncombusted fuel oil deposited on to hard standings, spilt oil, oils from domestic sources (i.e. following a home lubricant oil change in an automobile) and illegal dumping of oil waste by unscrupulous waste contractors. The diversity and dispersed nature of these latter inputs make them difficult to control and educating users is an important element of reducing these sources.

4.5.3 The fate of oil in the marine environment

As noted earlier, oil is the partially decomposed remains of marine organisms and as such is essentially biodegradable. However, oil is considerably less labile than the newly produced organic matter, and both the hydrocarbons in the oil and some of the substances incorporated into it during geological transformation are toxic to many organisms. Thus, the hydrocarbons in the oil can be biologically degraded or transformed by microorganisms including 79 species of bacteria, 9 species of cyanobacteria, 103 species of fungi and 14 species of algae. However, degradation is dependent on the availability of oxygen and nutrients (primarily nitrogen and phosphorous). The process is slow and depends upon the fraction of oil, the species of microbe and the water temperature as well as the availability of limiting resources. The oil is converted to carbon dioxide and water and small quantities of a range of salts. Anaerobic degradation can also occur but there are less species of microbes capable of doing so and the process is far slower than in aerobic environments.

In practice, oil that is incorporated into marine sediments is not broken down significantly due to a lack of oxygen. Such oil can persist in anoxic environments for decades. The bacterial breakdown of slicks is limited to activity at the interface where the oil meets oxygen and nutrient containing seawater. Refined oil products and heavy crude oils are often too toxic or too refractile to be subject to bacterial action.

Oil residues in liquid effluents from urban waste waters or industrial discharges will often be in solution/suspension and will be dispersed and mixed with a large volume of water. Under these conditions, biological degradation will occur, albeit often slowly. In areas with regular inputs of oil, including natural seeps, oil-degrading bacteria will be present in the local microbial assemblage and able to act on any oil that becomes available. Oil-degrading microbes are a relatively rare component of the microbial assemblage in the wider ocean.

4.5.4 Oil spills and oil spill clean-up

Offshore oil exploration accidents (e.g. sub-sea wellhead equipment failures, blow-outs) and shipping accidents (tankers and non-tankers) tend to release large quantities of oil over a relatively short period of time. The oil, as it floats on water, tends to rise to the surface and, as it is hydrophobic, combines to form a slick. Oil slicks can be formed of crude oil or refined oils such as fuel oil or tar.

While the common perception of an oil spill is of a wrecked oil tanker, the largest spills, in terms of volume of oil released, come from incidents involving multiple oil wells damaged in conflicts (three of the top five incidents), two from blow-outs and one from a sub-sea well accident (Table 4.16). The collision of two laden tankers, the *Atlantic Empress* and the *Aegean Captain*, resulted in the seventh largest oil spill to date. The *Exxon Valdez* spill in Prince William

Table 4.16 The largest ten (at the time of writing, 2016) oil spills. Note: that estimates of the oil released during the Kuwait and the Gulf War are unreliable but are the best available estimates.

Spill / Tanker	Location	Date	Tonnes (x 10³) of crude oil
Kuwaiti oil fires	Kuwait	16 January 1991–96 November 1991	136 000
Kuwaiti oil lakes	Kuwait	January 1991–November 1991	3409–6818
Lakeview Gusher	United States, California	14 March 1910–September 1911	1,200
Gulf War oil spill	Kuwait, Iraq and the Persian Gulf	19 January 1991–28 January 1991	818–1091
Deepwater Horizon	United States, Gulf of Mexico	20 April 2010–15 July 2010	560–585
Ixtoc I	Mexico, Gulf of Mexico	3 June 1979–23 March 1980	454–480
Atlantic Empress / Aegean Captain	Trinidad and Tobago	19 July 1979	287
Fergana Valley	Uzbekistan	2 March 1992	285
ABT Summer	Angola, 700 nautical miles (1300 km) offshore	28 May 1991	260
Nowruz Field Platform	Iran, Persian Gulf	4 February 1983	260
Castillo de Bellver	South Africa, Saldanha Bay	6 August 1983	252
Amoco Cadiz	France, Brittany	16 March 1978	223

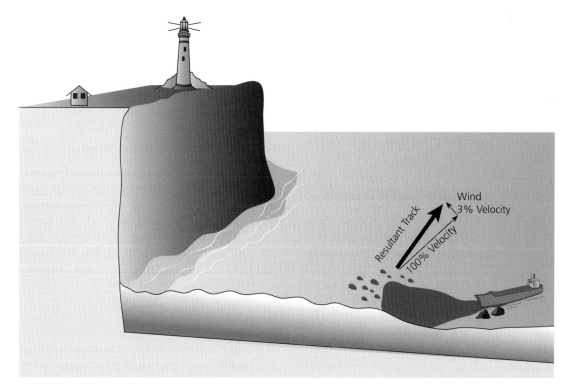

Figure 4.15 Oil slicks move in a predictable way. On stationary water (i.e. a lake) they are blown downwind at about 3% of the wind velocity. On moving water, with no wind, they move down current at the velocity of the current. On the open sea with winds and current the oil-spill track can be predicted by resolving the vectors of the current and wind effects.

Sound in Alaska in 1989 currently ranks as only the 28th largest spill in history, although it commanded global media attention at the time.

Over time, an oil slick moves under the influence of the wind and currents and its nature changes. As a rule of thumb, a slick moves on a track that is the result of two forces. It tries to move in the direction of the water flow (at the water speed) but it is also pushed downwind at a rate equivalent to 3% of the wind speed. The actual slick therefore moves under the influence of both water and wind velocities

(Figure 4.15). This predictability of the track of the slick allows its movement to be anticipated in advance, based on tidal predictions and the weather forecast and allows emergency measures and equipment to be mobilised to both the slick's predicted path and destination.

As the slick sits on the surface of the water it moves but also expands over a wider area as the oil spreads out into a thin layer. The volatile components evaporate from the surface while the soluble components leach out from below (Figure 4.16). In

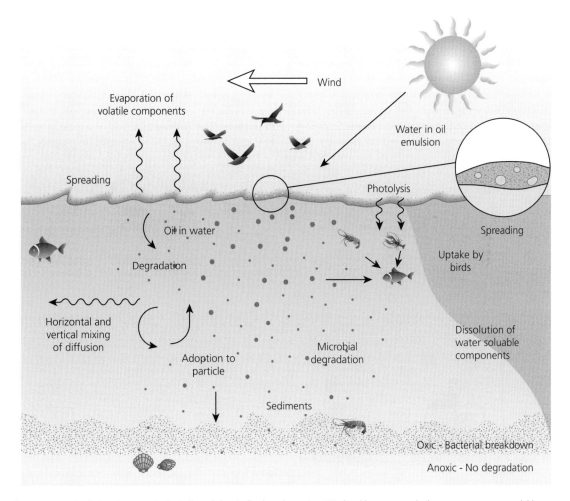

Figure 4.16 Oil split into the ocean is physically and chemically altered over time. Wind and heat cause volatiles to evaporate, more soluble fractions are leached into the water, physical mixing liberates droplets of oil to be dispersed in the water column while adding water to the slick, producing an emulsion. Light can stimulate the formation of cross-bonding which, with the loss of volatiles, makes the oil increasingly viscous and tar-like. As the oil becomes denser it sinks and on the seabed it becomes incorporated into sediments and sediment can become incorporated creating tar balls. Bacterial breakdown of oil requires nutrients and oxygen and so is restricted to the water column and surface sediments.

addition, the friction of movement at the oil–water interface and, in particular, wave action, may mix water into the oil, forming an emulsion (known colloquially as 'chocolate mousse'), and droplets of oil become dispersed throughout the water (Figure 4.16). The action of sunlight on the oil causes some photodegradation but can also cause more resistant polymers to develop and, combined with the loss of the volatiles, results in the slick taking on a more refractive tarry nature. Eventually the slick becomes heavier than the water and sinks, the oil mixing with seabed sediments to form 'tar balls', a mix of oil and sand, that are moved as bed load (Figure 4.16) or they become buried in sediments where further degradation is unlikely.

When an accidental oil spill occurs a response is usually initiated, by a government agency, to limit the environmental and economic damage. Clearly this involves trying to stem the source of the leak; for example, containing damaged vessels, capping blow-outs or repairing damaged well-heads. Of course, in the event of total loss of the vessel (sinking) this is not possible but in most cases it involves days or weeks of action. In addition, to stemming the leak at the source, the oil that has escaped becomes the focus of a clean-up.

It is generally the case that the fresher the oil (the less degradation/tarring, the less emulsification), the more amenable it is to clean-up. Recovering oil is easier and the ecological impacts are generally less while the oil is at sea compared to after it has washed ashore. Ideally one would recover the spilt oil to recycle/reuse. Two main technologies exist for recovering oil from the sea: the first uses the hydrophobic properties of the oil, the second uses the propensity of oil to float on top of water.

At sea, vessels equipped with floating booms seek to concentrate the oil and guide it into a collecting vessel (Figure 4.17a). Collecting vessels are equipped with either *oleophilic* oil mops or skimmers. The former consists of ropes or rollers covered in fibres made of oleophillic plastics, and

(a)

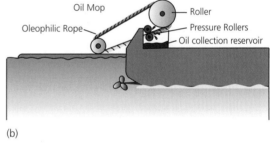

(b)

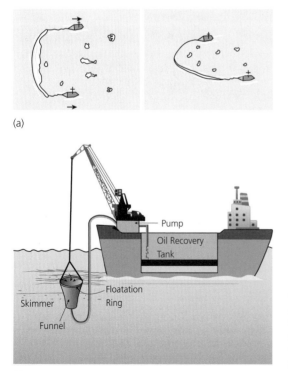

(c)

Figure 4.17 As oil floats on water it can be manipulated and to some extent recovered. (a) Vessels towing booms can contain and concentrate oil, moving it away from sensitive coasts and making it easier to recover. (b) An oil mop, consisting of oleophilic threads, is drawn through the spill and the oil is then recovered from the rope. (c) Oil skimmers are weighted so that the opening of the funnel is flush with the waters surface, oil therefore flows over the top and down into the funnel to be pumped away.

as these pass through the oil it binds to them, in preference to the water and can be lifted clear of the sea. On the vessel, the ropes or mops are then squeezed between rollers that force the oil off and into a collecting vessel (Figure 4.17b) before they are deployed into the slick again. Oil skimmers are devices that float on the sea surface and act like a funnel with the device weighted such that the rim of the funnel floats at the waters surface. Oil therefore flows over the rim and down into the funnel from where it can be pumped into the collecting vessel (Figure 4.17c).

Oil recovered from the sea will often be chemically altered and contain some seawater. Passing the oil through a normal refinery process will deal with these issues but at greater cost compared with the processing of fresh crude oil. To-date, even under ideal conditions oil recovery rates are generally less than 15% of the spilt oil. Specialist oil-recovery vessels and equipment are not universally available and are generally only docked near major oil-handling ports or oil fields. Their operation is also highly dependent on good sea conditions as strong wind and wave action makes booms and skimmers ineffective.

One widely applied response to offshore oil spills is the application of dispersants, either from vessels such as tugboats and/or from aircraft. Chemical dispersants are special formulations that break down the oleophilic bonds in the oil causing the slick to break up and the oil to form small droplets. These droplets become dispersed and mixed into the water column. The advantages of a dispersed slick are that it is no longer a hazard to marine birds and mammals, it will not strand and affect beach amenity values and the increased surface area-to-volume ratio of the oil should promote biological breakdown while maintaining access to oxygen and nutrients from the water column. In fact, some dispersants contain cocktails of bacterial starter cultures and nutrients to facilitate biodegradation. The disadvantage of dispersed oil is that it is more biologically available to be ingested by fish or drawn over gills of fish and filter feeders. If the environment is not sufficiently energetic and the oil droplets are not dispersed, it may coalesce back into a slick due to the physical mixing action of the water. Weathered slicks are far less amenable to dispersion than fresh oil due to photoxidation and the loss of

volatiles, so if dispersant spraying is to be used it needs to be done soon after the spill.

Oil that reaches the shore coats the intertidal zone, causing toxic and physical impacts on the organisms in the area. Oil on the shore continues to be subject to evaporation and photoxidation and so may remain on the shore during any subsequent tide. Alternatively, stranded oil may be removed by an incoming tide and moved to foul a different area of shore, resulting in a series of impacted sites rather than just one.

High-value resources—for example, those with significant economic, cultural or ecological value—may be protected by the placement of booms offshore, if conditions allow, to prevent the oil from reaching important sites. This requires oil response managers to identify high-value and essentially sacrificial shores. Different shore habitats vary in their sensitivity to oil spills (Table 4.17); for example,

Table 4.17 The environmental sensitivity index (ESI) to oil pollution for major shoreline habitat types; high numbers indicate greater sensitivity. Expanded by various authors from the original classification of Gunlach and Hayes (1980) (source Hissel 2013).

ESI Values	Coast type
1	Exposed rock cliff; exposed vertical seawall, made of wood, metal or concrete
2	Exposed rock platform or coast; exposed unstable escarpment behind a rock platform
3	Fine-to-average sand beach; unstable eroding escarpment
4	Coarse sand beach or heterogeneous sand beach
5	Very coarse sand beach
6A	Gravel or pebbles beach
6B	Sheltered beach of sand and rollers
7	Tidal flat exposed beach
8A	Sheltered rock coast or sheltered unstable escarpment
8B	Sheltered solid infrastructure
9	Sheltered flat beach
10A	Sea marshlands
10B	Mangrove swamps
10C	Freshwater marsh (herbaceous vegetation)
10D	Freshwater marsh (woody vegetation)

coral reefs and estuaries are the most sensitive habitats whereas wave exposed rocky shores are the least sensitive. In fact, under stormy conditions the high-energy nature of rocky shores means that they are essentially 'self-cleaning'.

Shore clean-up options involve removing as much of the oil as possible. This might be achieved by directly removing oiled substrates (i.e. oiled sand from a beach), removal of oiled vegetation (e.g. seaweeds, saltmarsh grass) by cutting the foliage but leaving the 'roots' to regenerate or by removing the oil itself. The latter may involve suction equipment (mounted on trucks or vessels), high-pressure washing, steam cleaning or the use of sorbent materials. Alternatively, the shore may be cleaned using chemical dispersants. The application of dispersants will break up the oil, causing it to become mobilised by the incoming tide. As with spraying dispersants at sea, if the system is energetic the dispersed oil will be carried away, but if the system is low energy the oil may be mobilised but will instead coalesce into a new slick that goes on to foul a new shore. On many types of shore, oil spill clean-up can actually do more ecological damage than the original spill, for example, from the toxicity of the dispersants and the physical damage from the clean-up operation. However, for economic reasons or to prevent the oil moving to other sites, clean-up may be a required response even if it does result in some ecological damage.

4.6 Wrecks, rigs and reefs

It is not uncommon for larger debris than that discussed in Section 4.2.1 to end up in the ocean, like sunken vessels. In 2013, approximately 10 500 shipwrecks were estimated to be present on the seafloor of the world's oceans. Shipwrecks end up on the seafloor during military operations, through their deliberate disposal in designated areas or as a result of accidents in dangerous shipping regions such as along the Seven Stones in Cornwall, UK, and the Skeleton coast off the Namib Desert (Table 4.18). In some regions, large 'ship graveyards' even exist (Table 4.18). These wrecks function as substrate for marine animals and can act as *artificial*

reefs. Sessile animals and algae rapidly attach to the surfaces of any objects in the sea because hard substrates are limited and larval supply is high and so man-made structures are quickly colonised. Fish and invertebrates begin to inhabit the many crevices and hollows within these artificial structures.

Offshore structures such as oil rigs can be decommissioned at sea and sunk both as a disposal option and to create artificial reefs. Several coastal states in the United States, for example, have a 'rigs-to-reefs' programme that decommissions oilrigs when their working life ends and submerges them for precisely this reason. The first such rigs were sunk in Florida in 1979. Disposing of rigs in this way saves the owners the great expense of removing them; in shallow waters it can cost US$10–15 million to decommission a rig. After decommissioning, ownership of

Table 4.18 Submerged ship graveyards around the world and the reasons they were formed. Data from: Richards 2013.

Location	Number	Reason
The Seven Stones Cornwall	200 wrecks	Major shipping route, wrecks on the Seven Stones reef
Landevennec, north-western France	Ten ships	Civilian and military vessels used as breakwaters, awaiting demolition, and for naval training
Coast of Nouadhibou, Mauritania, west Africa	> 300	Dumping of derelict ships
Skeleton coast	> 1000 intertidal and subtidal	Dangerous sea and navigational conditions; Benguela current meets Namib desert air forms dense fogs
Bikini Atoll, Pacific	Eight US and two Japanese vessels	Decommissioning of US Pacific fleet from Second World War
Mallows Bay Ghost fleet Maryland, US	230 including 100 wooden steamers	Disposal site for steamships built for First World War
Curtin Artificial Reef, Moreton Island, Australia	32 ships, plus cars	Reef creation
Ironbottom Sound, Solomon Islands	18 Japanese, 32 US, plus aircraft	Battle of Guadalcanal, 1942–45

Plate 1 (see also Figure 3.1a) The Roman latrines at Housesteads Fort on Hadrian's Wall in northern England. The excavated latrine building.
Source: <https://commons.wikimedia.org/wiki/File:Housesteads_latrines.jpg>. By Steven Fruitsmaak (Own work) [Public domain], via Wikimedia Commons. (see page 48)

Plate 2 (see also Figure 3.1b) The Roman latrines at Housesteads Fort on Hadrian's Wall in northern England: An artist's reconstruction. Water from a diverted stream flowed in at the upslope end and waste was carried away to the nearby stream. Cisterns of water were used for hand-washing whilst sponges on sticks (in the bowl in the middle background) were used in place of modern toilet paper.
Source: Original painting by Ronal Embleton, copyright Frank Graham. (see page 48)

Plate 3 (see also Figure 3.4) The foreshore at Lynemouth in Northumberland in 1997 when the colliery was in full production and minestone was deposited on the foreshore. The large cobbles are stained orange and the pool is rust-coloured; both are the result of the iron-rich material in the minestone. (Photo: C. Frid). (see page 61)

Plate 4 (see also Figure 4.2a) Shoreline litter. © Greg Martin, Surfers against sewage 2016. (see page 84)

Plate 5 (see also Figure 4.2c) Common terns (*Sterna hirundo*) on the shoreline at Seaforth Nature Reserve, Liverpool. © Paul Scott 2016 (sgpscott@flickr/Facebook). (see page 84)

Plate 6 (see also Figure 4.2b) Mesoplastics and microplastics on the strandline of a beach in Cornwall. © Tracey Williams (creator of the Lost at Sea facebook group: <https://www.facebook.com/groups/LostAtSeaGroup/>). (see page 84)

Plate 7 (see also Box 4.1 Figure 1a) Shipping containers lost at sea. The Svenborg Maersk losing containers in extreme weather in northern France in 2014. © Paul Townsend 2014 <https://www.flickr.com/photos/brizzlebornandbred/12852909293>. Available under Creative Commons Attribution-No Derivs 2.0 generic licence (CC BY-ND 2.0, <https://creativecommons.org/licenses/by-nd/2.0/legalcode>). (see page 90)

Plate 8 (see also Box 4.1 Figure 1b) Shipping containers lost at sea. The MV Rena grounded on a reef near Tauranga, New Zealand. © New Zealand Defence Force 2011 <https://www.flickr.com/photos/nzdefenceforce/6386334175/>. Available under Creative Commons Attribution 2 Generic licence <https://creativecommons.org/licenses/by/2.0/legalcode>. (see page 90)

Plate 9 (see also Figure 4.1c) Shipping containers lost at sea. Sunken container in the Monterey Bay National Marine sanctuary. Credit: © 2011 NOAA / MBARI. Description: After seven years on the deep-sea floor, this sunken shipping container had been colonized by a variety of deep-sea animals, including crabs and deep-sea snails. (see page 90)

Plate 10 (see also Figure 4.1d) Shipping containers lost at sea. Lego® octopus washed up on a shoreline in Cornwall that may be one of the 5 million pieces of Lego® lost from the shipping containers transported by the Tokio Express in 1997. © Tracey Williams (creator of the Lego Lost at Sea Facebook page: <https://www.facebook.com/LegoLostAtSea/>). (see page 91)

Plate 12 (see also Figure 4.9b) Turtle entangled in fishing nets. Credit: U.S. National Oceanic and Atmospheric Administration. (see page 98)

Plate 11 (see also Figure 4.9a) Examples of marine megafauna entangled in macroplastic litter. Caspian gull (*Larus cachinnans*) entangled in string at Ainsdale beach, Southport, United Kingdom © Paul Scott 2016 (sgpscott@flickr/Facebook). (see page 98)

Plates 14 and 15 (see also Figure 4.9d and e) Steller sea lions with plastic packing bands entangling theirs necks causing lacerations in (d). © Alaska Dept. of Fish and Game, the activity was conducted pursuant to an NMFS Permit. (see page 98)

Plate 13 (see also Figure 4.9c) Triggerfish entangled in fishing nets, at Acqua bella, Ortona, Abruzzo, Italy. © Claudio Stoppato 2013. © Claudio Stoppato. (see page 98)

Plate 16 (see also Figure 4.10) Remains of albatross chick with gut full of plastic pieces presumably ingested whilst still alive. ©Sparkle Motion/ Flickr 2014 <https://www.flickr.com/photos/54125007@N08/14876384246/>. Available under Creative Commons Attribution 2.0 Licence <https://creativecommons.org/licenses/by/2.0/legalcode>. (see page 100)

the rig is transferred to the state. These actions have caused some controversy with environmental groups because the industry is essentially being allowed to leave its trash in the oceans. However, the rigs have been shown to contribute marine habitat, stimulating local biodiversity.

The most significant rigs-to-reefs programme is the Louisiana Artificial Reef programme in the Gulf of Mexico where extensive oil exploration occurs with 3858 active rigs in 2006. The Louisiana programme receives charitable donations from the oil companies: typically half the savings from the rig decommissioning are donated to the state to fund the reef's upkeep. Approximately 10% of the oilrigs are sunk in this way and the largest artificial reef complex is present in the Gulf of Mexico. The submerged portions of most rigs will already have been colonised during their productive life. Decommissioning involves breaking the rigs into smaller parts. They are either cut and left *in situ*, cut and moved to approved areas of seabed or the rigs may be toppled using explosives and left *in situ*. Other countries with rig-to-reef programmes include Brunei, which has two designated reef sites, and Malaysia, where the Baram-8 platform was sunk. In some regions, the development of reefs from oil rigs has been less popular and controversy surrounding the environmental impacts has blocked their use; for example, the sinking of rigs is banned in the North-West Atlantic by the OSPAR Convention.

Some shipwrecks, and particularly military shipwrecks, may also pose a danger to shipping because of the unexploded munitions they hold. For example, the wrecked SS Richard Montgomery that sank during the Second World War resides in the mouth of the River Thames in London and still contains 1400 t of live explosives.

The sites of sunken ships or rigs can also become diving or fishing attractions and so support the tourist industry. Indeed, some ships are sunk specifically for this purpose. Australia has 21 artificial shipwreck reefs that were created for recreational purposes. Submerged artificial structures may also function as coastal defences; for example, breakwaters to dissipate wave energy offshore or trap moving sediments at eroded beaches.

Artificial reefs have been constructed from a variety of materials including construction debris, fly ash waste from power plants compressed into bricks, rubble or tyres. Many tyre reefs are now being removed because they do not function well as reefs and it is not certain whether or not they leach toxic heavy metals and peroxides. Similarly, many of the other large artificial reef structures have the potential to leach toxicants into the water column over time. However, the high numbers of animals inhabiting the reefs do not suggest acute toxicity (Section 2.1). Being permanent additions to the marine environment these structures can be considered an ongoing issue.

The reef communities that develop depend upon the structure, size and age of the artificial reef. For example, shipwrecks seem to attract larger fish. Artificial reefs all act to aggregate fish and so behave like *fish aggregation devices* (FADs) by providing shelter from predators. Artificial reefs have been shown to aggregate fish with densities being 20–50 times higher than in other proximal areas, but it is not clear whether the reefs actually stimulate fish populations or not. Although it has not been extensively studied, in some reefs the fish are attracted away from natural reefs and this can result in significant ecosystem change at the natural sites.

4.7 Radioactivity

4.7.1 The nature of radiation

Light, sound, heat and gravity waves are all examples of natural radiation. Strictly speaking, the term *radiation* refers to any form of energy that travels in the form of a wave and so will radiate out from a source. In terms of pollution, the term is usually used to refer to the hazardous forms of radiation, so strictly speaking the term *radioactivity* is used to refer to ionising radiation. Ionising radiation contains enough energy to break chemical bonds ionising atoms and molecules (greater than 10eV; eV= electron volts, 1eV is the energy gained by a single electron moving across an electric potential difference of one volt). Ionising radiation comes in six forms; α, β and neutron particles and γ-rays, X-rays and UV (ultraviolet) rays.

Ionising radioactivity is also a naturally occurring phenomenon, with certain isotopes being emitters of radioactive particles while various electromagnetic rays, including some emitted by the sun (UV, X-rays) that are also radioactive. Ionising radiation is biologically active because it can damage biological molecules. Exposure to very high levels of ionising radiation causes burns, hair loss, sloughing of skin, nausea/vomiting, ulcers/sores and confusion/weakness. These effects are only likely in the event of exposure to large amounts of radiation following nuclear war, catastrophic nuclear accidents, exposure in nuclear facilities or from certain medical procedures such as the radiotherapy used to treat some cancers. More common are the biological effects caused by prolonged or repeated exposure to radiation.

There is no safe lower threshold for exposure to ionising radiation but polluting radiation is that which occurs on top of a natural background level of exposure. For example, one of the effects of radiation exposure is damage to DNA molecules and the production of cancerous cells. Cancers arise through natural copying errors during cell replication, through errors promoted by a range of natural agents and through exposure to ionising radiation naturally (i.e. from the local geology such as granite rocks or from solar radiation). However, most copying errors are corrected or are not biologically detrimental. Only a proportion of exposures result in cancer. Anthropogenic sources simply add to the total exposure to cancer forming risk factors.

4.7.2 Sources of radiation

The majority of harmful UV and X-rays emitted by the sun are absorbed by the Earth's atmosphere and so are not biologically active. Atmospheric pollution, by chlorofluorocarbons (CFCs) and other artificial fluorocarbons, damaged the ozone layer and created the so-called ozone hole over Antarctica in the 1970s and 1980s. This led to a significant increase in the level of harmful UV reaching the Earth's surface. This may have impacted wildlife; for example, the mutation rates of plankton have been found to increase in laboratory studies of UV exposure. It also may have affected humans,

for example, through the increased incidence of skin cancers in Australians of Caucasian origin. The impact of the CFCs on the ozone layer was so substantial that their use as refrigerants, solvents and propellants (e.g. in fridges and aerosols) was rapidly phased out. This is an interesting example of the complexity of environmental interactions in nature. The harmful UV was not pollution, it was naturally occurring and while the CFCs were not causing direct harm to humans or wildlife, they caused indirect negative effects. Therefore, they meet the definition of a pollutant.

Nuclear reactors are used to produce energy and many scientific and medical applications also use ionising radiation sources. Nuclear reactors produce α, β, and neutron particles and γ-rays. Neutron radiation is the only form of radiation that can 'make' another substance radioactive. It is therefore an important tool in the development of materials used in many medical applications.

Nuclear radiation is often characterised by the *half-life* of the material. Nuclear emissions are caused by the shedding of energy by an unstable nucleus, and this energy is emitted as particles and/or high-energy photons (i.e. γ-rays). Over time, the energy content of the material decreases and eventually all the excess energy has been emitted and a stable substance remains. The rate of energy release is not linear, it follows a decaying exponential curve, and this rate is often characterised by the half-life: this is the time taken for the rate of emissions to reduce by half. For instance, most carbon naturally occurs as ^{12}C but some occurs as the heavy isotope ^{14}C. ^{14}C has a half-life of approximately 5730 years, meaning that after this time approximately half the ^{14}C in a sample will have decayed to the stable form of ^{14}N (through the emission of β particles). This is the basis of the carbon dating technique used in archaeology to age organic remains.

4.7.3 Nuclear waste

A nuclear reactor generating energy, whether a power station or the nuclear power unit of a submarine or warship, is shielded to prevent the radioactive particles/rays from spreading outside of the reactor. The reactor needs to be supplied with radioactive fuel (usually enriched uranium and/or plutonium) and the spent fuel remains radioactive. Thus, a nuclear reactor produces spent fuel throughout its operational life and when it reaches the end of its life, the reactor and the containment material/shielding are also radioactive and must be safely disposed of.

In addition, to nuclear power-generation plants, the nuclear fuel manufacturing plants and the spent fuel reprocessing plants also handle radioactive material. Both of these processes generate radioactive waste and during operation parts of these plants become radioactive. Medical and scientific businesses import radioactive material and, following its use, subsequently have radioactive material to discard. Military operations produce spent fuel, contaminated equipment and so on. Highly reactive material in the form of unused weapons that reach the end of their service life also need to be safely decommissioned.

Radioactive waste is characterised into low-level, intermediate-level and high-level based on its level of radioactivity. Low-level waste includes gloves and protective gowns worn by military, engineering, medical and scientific personnel when dealing with radioactive material, dilute solutions of radioactive reagents, biological material (i.e. tissue/organ samples) and shielding that has only become mildly radioactive. This material is diverse and bulky, making its treatment and disposal a challenge.

Intermediate-level waste is of high reactivity and so requires shielding but it does not usually need cooling as well. It generally arises from the nuclear power industry and military use and includes the casings from fuel rods and reactor shielding.

High-level waste, on the other hand, is highly radioactive and requires extremely specialised handling. It generally requires both cooling and shielding. However, the high level radioactivity means that it can usually be treated to separate out and concentrate the reactivity and in so doing manufacture fresh fuel or useful radionuclides.

4.7.3.1 Low-level radioactive waste

Low-level waste is usually encased in concrete, often in drums and is stored above ground or in shallow underground stores. There is considerable concern that the use of the limited underground storage space is rapidly being filled with low-level waste leaving less capacity for intermediate and high level waste.

4.7.3.2 Intermediate-level radioactive waste

Intermediate-level waste is initially encased in glass (vitrification), often after a degree of chemical treatment (i.e. ion exchange) to concentrate or stabilise the waste. Bulky items such as reactor components may simply be encased in concrete blocks. Intermediate waste with a short half-life can then be stored above ground or in shallow underground repositories. Material with a long half-life is usually placed into deep underground storage.

4.7.3.3 High-level radioactive waste

International agreements require that high-level radioactive waste remains the responsibility of the country that generates it. In many cases the waste may be shipped to a third party for treatment (i.e. separation and concentration of any remaining useful fuel). However, this residual high-level waste remains the property of and under the management of the originator. Some countries have nuclear reprocessing plants that will provide on-site storage of the final high-level waste underground for a fee, but these facilities can require the owner take their waste back at any time.

4.7.4 Marine waste disposal

Following the Second World War, the developing nuclear industry used the marine environment as a repository for radioactive wastes. The 1972 London Dumping Convention prohibited the disposal of high-level radioactive waste in the ocean. In 1993, however, Russia reported that the Soviet Union had continued deep-sea disposal of high level waste. Between 1954 and 1993, low and intermediate-level waste was subject to ocean disposal by the United States, Soviet Union, United Kingdom, France, Belgium, The Netherlands, Germany, Sweden, Italy, Japan, Switzerland and Korea. A moratorium on low-level waste dumping was

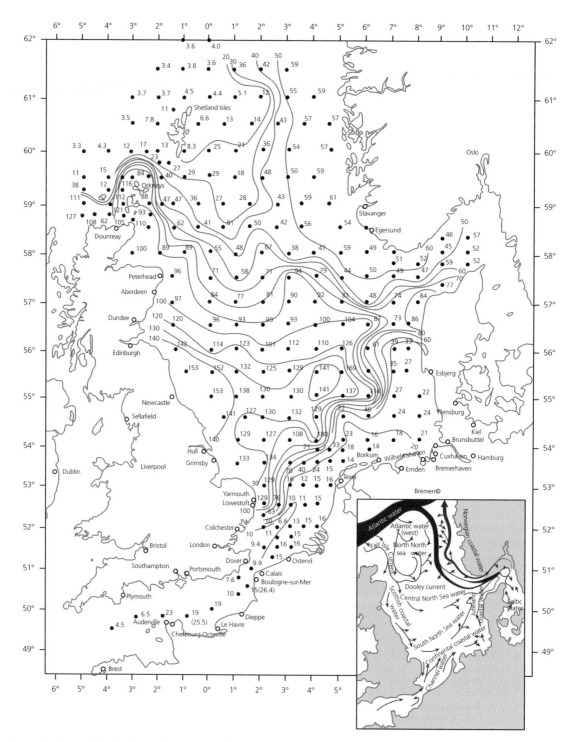

Figure 4.18 The distribution of the radioactive isotopes ^{137}Cs and [134]Cs (Bq/m³) in the surface waters of the North Sea in 1984. The distribution and the movement over time of the isotopes, allowed oceanographers to map the detailed circulation of the North Sea (inset) and revealed many previously unknown small-scale features. Data from Salomons et al. 1989, redrawn from Frid and Dobson 2013.

brought forward in 1983 but the total prohibition of radioactive waste disposal at sea only entered into force in February 1994. Monitoring at these disposal sites has shown elevated levels of radionuclides, suggesting that there has been leakage from the containers into the surrounding seawater. However, the levels of activity are low and so the environmental risks are considered low. Given the practical impossibility of recovering these containers of waste from the deep ocean after 30 to 50 years of corrosion, it has to be hoped that leakage rates remain low and the material becomes dispersed before it can come into contact with the human food chain.

While the direct dumping of nuclear waste into the marine environment is now prohibited, many coastal plants emit radioactive discharges via pipeline emissions. This includes very low levels of radioactivity in cooling water from coastal nuclear power stations, emissions from naval vessels and facilities and, most significantly, discharges from nuclear fuel manufacturing and reprocessing plants. These discharges are regulated as for other waste streams (Chapter 7) and, as a result, the quantities of radioactivity discharged are low and within what are considered safe limits. However, nuclear waste discharges are a hot topic and in many Western democracies considerable pressure has been exerted on regulators and operators to reduce these discharges. With a rapidly growing economy and major pollution problems arising from an energy sector dependent of burning coal, China is investing heavily in nuclear energy plants with an associated growth in emissions from fuel manufacturing and reprocessing plants and considerable quantities of highly radioactive material when the plants are decommissioned at the end of their operational lives.

The Sellafield (formerly known as Windscale) Nuclear Reprocessing Plant on the north-west coast of England discharged radioactive waste into the Irish Sea from 1952 until 2003. These discharges contained material in solution that was scavenged by muddy sediments and became concentrated within local estuaries. Here it posed a risk to local fishers and shellfish consumers.

Other radioactive isotopes remained in solution and were dispersed across the Irish Sea where they caused contamination in the Republic of Ireland and contributed to the political tensions with the United Kingdom. This radioactive material was transported around the Irish Sea and up the west coast of Scotland, into the wider Atlantic and around into the North Sea. The distribution of the isotopes allowed oceanographers to develop a more accurate understanding of the hydrography of the region (Figure 4.18). Following protracted pressures from environmental groups, discharges of radioactive waste to the Irish Sea ended in 2003, and over the following three years the three, now radioactive, pipelines were also removed.

In addition to the direct dumping of the radioactive waste and liquid discharges, the marine environment also receives nuclear material from the atmosphere in the form of radioactive fallout and washout. Historically, the main sources of atmospheric nuclear material was the above-ground testing of nuclear weapons; with the shift to underground testing this source ended. Since this time there have been a series of accidents at nuclear plants that have released material into the atmosphere including the fire at Windscale in north-west England (1957), the partial reactor meltdown at Three Mile Island in the United States (1979), fire and explosions at Chernobyl in the Soviet Union (1986) and the accident at Fukushima in Japan (2011). The Fukushima accident resulted from a tsunami triggered by an earthquake and, in addition to aerial releases, a large volume of contaminated cooling water was released into the sea. Estimates of the amount released vary between 5 and 27×10^{15} Bq mainly of ^{137}Cs and ^{134}Cs (Bq—a becqueral is the SI unit of radioactivity, one becquerel is defined as the *activity* of a quantity of radioactive material in which one *nucleus* decays per *second*). This material was mixed, diluted and dispersed over a vast area of the sea (Figure 4.19).

Other accidental deposits of nuclear material in the oceans include nine nuclear-powered submarines that have sunk, one nuclear-armed but diesel electric-powered submarine that sunk, and six nuclear weapons that were lost and never recovered.

(2.5 years after Fukushima)
920 days after release

10^{-12} 10^{-9} 10^{-6} 10^{-3} 1

surface concentration relative
to initial values off Japan

Fukushima-Daiichi
Nuclear Meltdown,
March 11, 2011

Model simulations on the long-term dispersal of Cs-137 released into the Pacific Ocean off Fukushima
Erik Behrens, Franziska Schwarzkopf, Joke Lübbecke, and Claus W. Böning   GEOMAR

Figure 4.19 Simulated trajectory of radioactivity in the Pacific Ocean released by the Fukushima nuclear accident in 2011. (Simulation by GEOMAR, Kiel. See also Behrens, et al.2012).

4.8 Biosecurity

The GESAMP definition of marine pollution (see Chapter 1) refers to the introduction of 'substances' and when the definition was derived it was based on the idea that substances were chemical in nature. However, living material can also be considered a substance, in which the introduction of living organisms might be thought of as 'biological pollution'. The ecological impacts associated with the entry of non-native species into ecosystems has been identified by the United Nations as one of the major environmental challenges of the twenty-first century. The prevention and regulation of such introductions is known as biosecurity.

Much of the focus on biosecurity relates to the invasion of natural systems by species that are diseases, pests or are the competitors of economically important species or those which interfere with human uses of the ecosystem. An example of the latter might include the algal species that form dense mats or nuisance blooms. These species are known as *non-native invasive species* (NIS).

Naturally, each species occurs within a particular range of environmental conditions (its habitat) and in certain parts of the world (its biogeographical region). These distributions are often the result of evolutionary adaptation and speciation. For example, while the Pacific and Atlantic Oceans are connected in the polar regions and through the deep ocean flows, the tropical and temperate surface waters have been isolated from each other for millions of years and so different species' assemblages have evolved independently in the different regions.

The distribution of species is not static: species move as a result of natural phenomena, such as range expansions/shifts associated with environmental changes and occasional long-distance dispersals (adults or larvae transported by a storm or on floating debris). However, the majority of such movements do not lead to the establishment of new populations because the existing communities are generally resistant to invasion from non-natives. Ecological resources are fully used and all the ecological 'niches' are filled. Thus, new entrants need to outcompete resident species and the invasion needs to be achieved with a sufficiently large *inoculum* to start a viable breeding population.

It is generally the case that degraded environments have limited species pools with potentially underutilised resources and so it is easier for invading species to establish. For instance, in estuaries which are naturally species poor due to the salinity regime. Ports and coastal waters often become polluted, resulting in the estuarine communities becoming degraded. In this way, other forms of marine pollution might enhance the opportunities for invasive species and biological pollution.

Analyses of historic invasions show that there are four main mechanisms of introduction: (i) deliberate introductions, (ii) accidental introductions alongside the deliberate movement of species, (iii) shipping-related relocations and (iv) *Lessepsian migrations*. Deliberate introductions have mainly been for fisheries and aquaculture reasons and include the introduction of Atlantic salmon to Australia and New Zealand. Such relocation have also moved disease agents and parasites around. For example, wild salmon populations in Norway are threatened by a liver fluke that was introduced with farmed salmon from Denmark, where the fluke is native. Other accidental introductions include the Chinese mitten crab to Europe in shipments of Pacific oysters. Most of these introductions occurred in the nineteenth and early twentieth

centuries and greater recognition of the issues has led to the implementation of tighter biosecurity measures and so a reduction in these sources.

Lessepsian migration is the term used to describe the migration of marine species through the Suez Canal, primarily due to the predominant flow, from the Red Sea into the Mediterranean Sea. Over 140 species, mainly fish, but including polychaetes, molluscs, bryozoans and crustaceans, have migrated into the Mediterranean and three species have migrated the other way, since the Suez Canal opened in 1869. These migrations have contributed to the eastern Mediterranean being one of the most heavily invaded regions of the world (Section 6.7).

Ships have been used for international trade since ancient times and have provided a vector for these biological 'hitchhikers'. However, the slow journey times and hence the exposure of coastal species to oceanic conditions meant that only low numbers of hitchhikers survived these early voyages. In more recent times the shorter transit times and the increasing use of seawater as ballast have increased the opportunities for vessel-mediated invasions (Figure 4.20). The Australian barnacle (*Elminius modestus*)

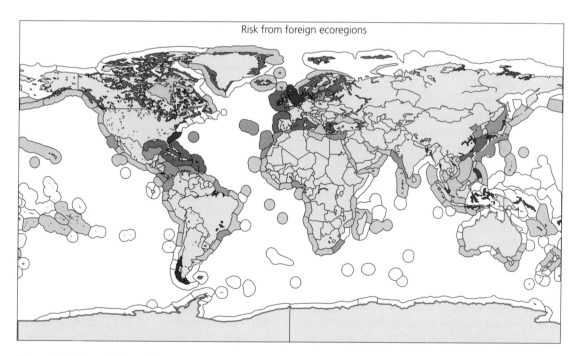

Figure 4.20 Regions with heavy shipping activities are most under pressure from invasive species. This map shows the ecological regions that are, due to the pattern of vessel movements, the highest risk areas for supplying invasive species to the north-eastern coast of the United States (Redrawn from Slosberg 2011). Low to high risk: shaded light grey to darkest grey (white indicates no risk due to no vessel movements).

now occurs in estuaries and coastal areas throughout the United Kingdom, having first being observed on the south coast just after the Second World War. The barnacle is believed to have arrived as fouling on the hulls of vessels arriving with troops and supplies from Australia during the war. The use of efficient anti-fouling treatments to increase vessel fuel efficiency and reduce the maintenance costs for vessels has the additional benefit of reducing the chances of the introduction of NIS via fouling (see Section 4.8).

With the introduction of steel ships it became possible to use seawater as ballast. When a vessel is empty or only partly loaded it floats higher in the water and is less stable. To overcome this buoyancy, the vessel will take on water, usually into special ballast tanks. This water is simply pumped in from the environment and so contains the plankton, including eggs and larvae and other living organisms that inhabit the coastal loading environment. As the vessel approaches the next port it pumps this water out, releasing any surviving organisms at the new location.

Given that the conditions in the port of loading were probably suboptimal and that conditions in the sealed ballast tanks would be challenging (dark, possibly anoxic, with contaminants present), the number of species released will be low. However, they are species that are hardy and well adapted to the harsh conditions within the ballast tanks and so are pre-adapted to the receiving environment at the port. The continual transport and introduction of these hardy, rapidly colonising species has given rise to an increasingly ubiquitous fauna in major shipping ports, something that is often referred to as a weedy fauna, drawing an analogy with the increasing global range of plant weeds found on waste ground in cities.

Concern surrounding the biosecurity issues associated with ships ballast water has prompted the IMO to introduce ballast water-treatment measures. These include the continual exchange of ballast water at sea: dumping coastal species mid ocean (where they will not survive) and then reloading with ballast water and subsequently dumping oceanic species in a polluted port (where they will not survive). However, there are many ballast treatment regimes being promoted that involve either the addition of toxicants (hypochlorite; Section 5.4.1) or nutrients (to promote eutrophication and so anoxia in the ballast tanks) or inducing mortality by heating, or any combinations of these treatments, but while they might

kill any invasive species it does mean the ballast discharge will contain toxic chemicals, thermal pollution, excessive nutrients and/or have a BOD. The regulation of biosecurity is an emerging issue that is lagging behind the development of the technology and its mandate.

4.9 Synthesis

- Plastics are now so ubiquitous in the environment that they are being incorporated into the sedimentary rock record and a distinctive new geological era the Anthropocene has been proposed. It is difficult to quantify the amounts of plastic in the ocean and it has taken time to determine the main sources and entry routes. Thus, it is still uncertain how much plastic is present and where it occurs. However, distribution by currents can help us to understand its sources and sinks. Of great concern is the recent observation that much of the plastic in our oceans is so small that it is not visible to the naked eye and even then we cannot account for all the plastic in these smaller size fractions. Further research on the distribution of ocean microplastics and its sources are needed in order for legislation and regulation to be developed that specifically targets microplastics.
- The damage caused by and the economic costs associated with marine plastics are high, clean-up is expensive and often ineffective overall and large losses of revenue occur across a number of industries. There are also losses of cultural ecosystem services and costs associated with the environmental damage caused by marine litter. Although, recycling and customer-focused schemes to reduce plastic use and recycle plastic products have had some success they are highly variable and their adoption is mostly slow. The scale of the plastic pollution problem requires concerted action at the source. Schemes that extend producer responsibility and the adoption of truly biodegradable plastics are the most promising long-term solutions. Petroleum plastic production is not sustainable because its production requires 8% of the oil produced globally and very little of it is reclaimed through recycling. Cost-effective production of biodegradable bioplastics (Section 4.2.5) could change the plastics industry and make it sustainable. Although some energy would still

be required to produce bioplastics it would require minimal or zero hydrocarbons as feedstock.

- The over-fertilisation of crops and the extensive livestock industry that support a growing world population have contributed considerably to nitrogen emissions (as N_2O and NH_3). These emissions result in the over-fertilisation of the oceans with nitrates. The discharges of sewage (Section 3.4) and detergents have added to the nitrogen and phosphorous burden. In 2011, according to the Food and Agriculture Organisation 10% of global land was used to produce crops and 23% was used for pasture. It is predicted that as global populations grow the amount of food required by these populations will increase by 70% to 100%. The additions of nutrients from agriculture are therefore expected to have more than doubled by 2050. The current levels of nutrient additions are proving difficult to manage and with population increases they will continue to be a major pollutant.

- Increasing globalisation has led to increased shipping and oil pollution and the increasing transport of plants and animals around the world. Petroleum hydrocarbons will continue to be an important marine contaminant for as long as society relies upon them as a primary energy source. As oil exploration goes to greater lengths to acquire oil perhaps the levels of contamination will even increase (e.g. Section 6.2). No amount of technological development will remove all of the risks posed by oil pollution. Clean-up activities are mostly effective and natural processes will ultimately ensure its breakdown. The build-up of PAHs in sediments and animals in coastlines around the world continues to be a legacy of our historic reliance on oil.

- Radioactivity, although not of upmost concern at present, continues to be a threat, and this could increase if the industry expands. Like plastics, radioactivity is another marine pollution challenge that has never been solved: its persistence is its greatest challenge.

- Many of these ongoing contaminants require coordinated action between regions and states to determine their sources, the quantities entering the sea and to reduce their impacts. The persistence of these contaminants (e.g. plastics, oil and radioactivity) is one of the reasons that they have not yet been 'solved'. Unlike those contaminants that we can consider to be essentially 'solved' (Chapter 3),

the sources of the contaminants discussed in this chapter are numerous and the pathways through which they enter the marine realm are more complex and so more difficult to monitor and regulate. In some cases, new legislation and regulations are needed (e.g. for microplastics, ammonia, biosecurity) and for others the existing regulations have not been effective (e.g. plastics and nutrients).

Resources

Artula Institute For Arts and Environmental Education (2016). Washed Ashore: Art to Save the Sea. http://www.washedashore.org/gallery.

Models for tracking drifting ocean plastic pollution: Adrift.org.au (2016). http://adrift.org.au/backward#what.

Lost nuclear weapons are referred to as 'broken-arrow' incidents and web searches on this term will provide details of them from both official and campaigning perspectives. https://en.wikipedia.org/wiki/List_of_military_nuclear_accidents.

Centre Environment Fisheries and Aquaculture Science (2016). HABS surveillance programmes and monitoring. http://www.cefas.co.uk/cefas-data-hub/food-safety/habs-surveillance-programmes-and-monitoring/.

European Commission (2016). FATE Maps—Nutrient pressure maps. http://fate-gis.jrc.ec.europa.eu/geohub/MapViewer.aspx?id=2.

Ghost Nets Australia (2017). http://www.ghostnets.com.au/about/.

Kommunenes Internasjionale Miljøorganisasjion (2016). Fishing for Litter. http://www.kimointernational.org/FishingforLitter.aspx.

Newquay Beachcombing (2016). Map of HP cartridges lost from a container spill from an unknown vessel in the Atlantic in 2014. https://www.google.com/maps/d/u/1/viewer?hl=en&authuser=1&mid=14KK9UerLS_SHi7r5l5gX8JMlMTc.

NOAA (2016). Harmful Algal Blooms. oceanservice.noaa.gov/hazards/hab/.Ocean Conservancy (2016). International Coastal Clean up. http://www.oceanconservancy.org/keep-the-coast-clear/organize-the-cleanup.html.

The Plastic Soup Foundation (2016). Beat the Microbead. https://www.beatthemicrobead.org/en/.

World Resources Institute (2016). Interactive Map of Eutrophication and hypoxia. http://www.wri.org/resource/interactive-map-eutrophication-hypoxia.

Bibliography

Algalita Marine Research Foundation (2016). http://www.algalita.org.

Allan, G.F., Maguire, G.B. and Hopkins, S.J. (1990). Acute and chronic toxicity of ammonia to juvenile *Metapenaeus macleayi* and *Penaeus monodon* and the influence of low dissolved-oxygen levels. *Aquaculture*, 91, 265–80.

Australian Academy of Science (2016). Nova. The Future of Plastics. http://www.nova.org.au/earth-environment/future-plastics.

Barnes, D.K.A., Galgani, F. Thompson, R.C. and Barlaz, M. (2009). Accumulation and fragmentation of plastic debris in global environments. *Philosophical Transactions of the Royal Society Series B*, 364, 1985–998.

Behrens, E., Schwarzkopf, F.U., Lübbecke, J.F. and Böning, C.W. (2012). Model simulations on the long-term dispersal of ^{137}Cs released into the Pacific Ocean off Fukushima. *Environmental Research Letters*, 7, 1–10.

Boardman, G.D., Starbuck, S.M., Hudgins, D.B., Li, X. and Kuhn, D.D. (2004). Toxicity of ammonia to three marine fish and three marine invertebrates. *Environmental Toxicology*, 19, 134–42.

Bowmer, T. and Kershaw, P.J. (2010). Proceedings of the GESAMP International Workshop on Microplastic Particles as a Vector in Transporting Persistent, Bioaccumulating and Toxic Substances in the Ocean, 28–30 June. GESAMP Report Study No. 82. GESAMP (2010 IMO/FAO/UNESCO-IOC/UNIDO/WMO/IAEA/UN/UNEP Joint Group of Experts on the Scientific Aspects of Marine Environmental Protection).

Browne, M.A., Crump, P., Niven, S. J., Teuen, E., Tonkin, A., Galloway, T., and Thompson, R. (2011). Accumulation of Microplastic on Shorelines Worldwide: Sources and Sinks. *Environmental Science and Technology*, 45, 9175–179.

Browne, M. A. (2015). Chapter 9. Sources and pathways of microplastics to habitats. In: M. Bergmann, M. Gutow and M. Klages (eds). *Marine Anthropogenic Litter*. Springer International Publishing AG, Switzerland, 229–44.

Camargo, J. A., Alonso, A. and Salamanca, A. (2005). Nitrate toxicity to aquatic animals: a review with new data for freshwater invertebrates. *Chemosphere*, 58, 1255–67.

Claessens, M., De Meester, S., Van Landuyt, L., De Clerck, K. and Janssen, C. R. (2011) Occurrence and distribution of microplastics in marine sediments along the Belgian coast. *Marine Pollution Bulletin*, 62, 2199–204.

Cózar, A., Echevarría, F., González-Gordillo, J.I., Irigoien, X., Úbeda, B., Hernández-León, S., Palma, A.T., Navarro, S., García-de-Lomas, J., Ruiz, A., Fernández-de-Puelles M.L. and Duarte, C.M. (2014). Plastic debris in the open ocean. *Proceedings of the National Academy of Sciences*, 111, 10239–244.

Davis, E.M. and Gloyna, E.F. (1967). *Biodegradability of non-ionic and anionic surfactant by blue-green algae. Environmental Health Engineering Research Laboratory*. Centre for Research in Water Resources, University of Texas, Austin, United States.

Eriksen, M., Lebreton, L.C.M., Carson, H.S., Thiel, M., Moore, C.J., Borerro, J.C., Galgani, F., Ryan, P.G. and Reisser, J. (2014). Plastic pollution in the world's oceans: more than 5 trillion plastic pieces weighing over 250,000 tons afloat at sea. *PLoS ONE*, 9, 1–15.

Eriksson, C., Burton, H., Fitch, S. Schulz, M. and van den Hoff, J. (2013). Daily accumulation rates of marine debris on sub-Antarctic island beaches. *Marine Pollution Bulletin*, 66, 199–208.

European Fertilisers Manufacturers Association (2009). Feeding the World and combatting climate change. Annual Report 2009. URL: www.fertilizerseurope.com

Frid, C. L. J. and Dobson, M. 2013. The ecology of aquatic management. Oxford University Press, Oxford.

Galgani, F. Hanke, G. and Maes, T. (2015). Global distribution, composition and abundance of marine litter. In: M. Bergmann, M. Gutow and M. Klages (eds). *Marine Anthropogenic Litter*. Springer International Publishing AG, Switzerland, 29–56.

Galgani, F., Leaute, J.P., Moguedet, P., Souplet, A., Verin, Y., Carpentier, A., Goraguer, H., Latrouite, D., Andral, B., Cadiou, Y., Mahe, J.C., Poulard, J.C. and Nerisson, P. (2000). Litter on the seafloor along European coasts. *Marine Pollution Bulletin*, 40, 516–27.

GESAMP (2007). Estimates of oil entering the marine environment from sea-based activities. Joint Group of Experts on the Scientific Aspects of Marine Environmental Protection. London: International Maritime Organisation.

Gregory, M. R. (1986). Plastic 'scrubbers' in hand cleansers: a further (and minor) source for marine pollution identified. *Marine Pollution Bulletin*, 32, 867–871.

Gundlach, E. R. and Hayes, M. O. (1980). Vulnerability of coastal environments to oil spill impacts. *Marine Technology Society Journal*, 12, 18–27.

Hammer, J., Kraak, M.H.S. and Parsons, J.R. (2012). Plastics in the marine environment: The dark side of a modern gift. In: D.M. Whitacre. *Reviews of Environmental Contamination and Toxicology 213*. New York, NY, Springer, 1–44.

Hissel, F. (2013). *Index of Vulnerability of Littorals to Oil Pollution*. http://www.coastalwiki.org/wiki/Index_of_vulnerability_of_littorals_to_oil_pollution.

Jambeck, J.R., Geyer, R., Wilcox, C. Siegler, T.R., Perryman, M. Andrady, A., Narayan, R. and Law, K.L. (2015). Plastic waste inputs from land into the ocean. *Science*, 347, 768–71.

Kuriyama, Y., Tokai, T., Tabata, K. and Kanehiro, H. (2003). Distribution and composition of litter on seabed of Tokyo Bay and its age analysis. *Nippon Suisan Gakkaishi*, 69, 770–81.

Lusher, A. (2015). Microplastics in the marine environment: distribution, interactions and effects. In: M. Bergmann, M. Gutow and M. Klages (eds). *Marine Anthropogenic Litter*. Springer International Publishing AG, Switzerland, 245–307.

Macfadyen, G., Campbell, H.F. and Rule, M.J. (2009). Abandoned lost or otherwise discarded fishing gear. UNEP Regional Seas Reports and Studies, No. 185. FAO Fisheries and Aqauculture Technical Paper, No. 523. FAO, Rome.

Mariani, L., De Pascale, D., Faraponova, O., Tornambe, A., Sarni, A., Giuliani, Ruggiero, G., Onorati, F. and Magalettis, E. (2006). The use of a test battery in marine ecotoxicology: The acute toxicity of sodium dodecyl sulphate. *Environmental Toxicology*, 21, 373–9.

Marine Conservation Society (2016). Beachwatch: The UK's Biggest Beach Clean Up and Survey. https://www.mcsuk.org/beachwatch/.

Miller, D.C., Poucher, S., Cardin, J.A. and Hansen, D. (1990). The acute and chronic toxicity of ammonia to a marine fish and a mysid. *Environmental Contamination and Toxicology*, 19, 40–8.

Molnar, J.L., Rebecca L., Gamboa, C.R. and Spalding, M.D. (2008). Assessing the global threat of invasive species to marine biodiversity. *Frontiers in Ecology and Environment*, 6, 485–92.

Morris, R.J. (1980). Floating plastic debris in the Mediterranean. *Marine Pollution Bulletin*, 11, 125.

Mouat, J., Lopez-Lozano, R. and Bateson, H. (2010). Economic impacts of marine litter. *Kommunenes Internasjionale Miljøorganisation* International Secretariat Report, Shetland, United Kingdom.

NASA Earth Observatory (2009). Oil Seeps in the Gulf of Mexico. https://earthobservatory.nasa.gov/IOTD/view.php?id=36873.

National Oceanographic and Atmospheric Administration (2016). NOAA Marine Debris Programme. https://marinedebris.noaa.gov

Ocean Conservancy (2015). 30th Anniversary International Coastal Clean up Annual Report. http://www.oceanconservacny.org.

Persson, A., Molin, G., Andersson, N. and Sjoholm, J. (1990). Biosurfactant yields and nutrient consumption of *Pseudomonas fluorescens* 378 studied in a microcomputer controlled multifermentation system. *Biotechnology and Bioengineering*, 36, 252–5.

Pham, C. K., Ramirez-Llodra, E., Alt, C. H. S., Amaro, T., Bergmann, M., Canals, M., Company, J. B., Davies, J., Duineveld, G., Galgani1, F., Howell, K.L., Huvenne, V. A. I., Isidro, E., Jones, D.O.B., Lastras, G., Morato, T., Gomes-Pereira, J. N., Purser, A., Stewart, H., Tojeira, I., Tubau, X, Van Rooij, D., and Tyler, P. A. (2014). Marine litter distribution and density in European seas, from the shelves to deep basins. *PLoS ONE*,9, e95839.

Plastics Europe (2015). *Plastics—The Facts 2014/2015*. Brussels: Association of Plastics Manufacturers. http://www.plasticseurope.org.

Rabalais, N.N., Diaz, R.J., Levin, L.A., Tuener, R.E., Gilbert, D. and Zhange, J. (2010). Dynamics and distribution of natural and human-caused hypoxia. *Biogeosciences*, 7, 585–619.

Richards, N. and Seeb, S.K. (2013). *The Archaeology of Watercraft Abandonment*. New York, NY: Springer Science and Business Media.

Ryan, P.G., Moore, C.J., van Franeker, J.A. and Moloney, C.L. (2009). Monitoring the abundance of plastic debris in the marine environment. *Philosophical Transactions of the Royal Society Series B*, 364, 1999–2012.

Salomons, W., B. L. Bayne, E. K. Duursma and Förstner, U. (1988). Pollution of the North Sea: an assessment.Springer-Verlag, Berlin.

Slosberg, M. (2011). Risk Assessment for the Spread of Invasive Marine Species from Vessels Arriving in Northeastern US. Michigan Institute of Technology, Cambridge MA.

Statista (2016). *Facts on Cleaning Products Industry in the U.S.* http://www.statista.com.

Surfers Against Sewage (2016). Marine Litter. https://www.sas.org.uk.

Swedmark, M., Broanten, B., Emanuelsson, E. and Grammo, A. (1971). Biological effects of surface active agents on marine animals. *Marine Biology*, 9, 183–201.

Taylor, J.R., DeVogelaere, A.P., Burton, E.J., Frey, O., Lundsten, L., Kuhnz, L.A., Whaling, P.J., Lovera, C., Buck, K.R. and Barry J.P. (2014) Deep-sea faunal communities associated with a lost intermodal shipping container in the Monterey Bay National Marine Sanctuary, CA. *Marine Pollution Bulletin*, 83, 92–106.

Thompson, R.C., Olsen, Y., Mitchell, R.P., Davis, A., Rowland, S.J., John, A.W.G., McGonigle, D. and Russell, A.E. (2004). Lost at sea: Where is all the plastic? *Science*, 304, 838–838.

United Nations Environment Programme (2007). *Marine Litter in the South Asian Seas Region*. South Asia Cooperative Environment Programme, Colombo, Sri Lanka.

United Nations Environment Programme (2015). *Biodegradable Plastics and Marine Litter. Misconceptions, Concerns and Impacts on Marine Environments*. United Nations Environment Programme, Nairobi.

US Environmental Protection Agency (1989). *Ambient Water Quality Criteria for Ammonia (Saltwater) 1989*. EPA report 440/5–88–004. Washington, DC: Office of Water Regulation and Standards Criteria and Standards Division.

Votier, S. C., Archibald, K., Morgan, G. and Morgan, L. (2011). The use of plastic debris as nesting material by a colonial seabird and associated entanglement mortality. *Marine Pollution Bulletin*, 62, 168–72.

Wade, E. and Hittori, A. (1990). *Nitrogen in the Sea: Forms, Abundance and Rate Processes*. Boca Raton, FL: CRC Press.

Wang, W., Wang, H., Yu, C. and Jiang, Z. (2015). Acute toxicity of ammonia and nitrite to different ages of Pacific cod (*Gadus macrocephalus*) larvae. *Chemical Speciation and Bioavailability*, 27, 147–55.

Wright, S.L., Thompson, R.C. and Galloway, T.S. (2013). The physical impacts of microplastics on marine organisms: A review. *Environmental Pollution*, 178, 483–92.

CHAPTER 5

Emerging problems

5.1 Introduction

Over the last century, humankind has made major technological advances and chemical production has increased substantially. Today (12 October 2016), 121.67×10^6 unique organic and inorganic chemical compounds are registered with the Chemical Abstracts Service (CAS) registry (American Chemicals Society 2016), and each day 15 000 new chemicals are added. The US Environment Protection Agency (EPA) listed 86 000 compounds as being manufactured and processed by or imported into the United States in 2016 (according to the Toxic Substances Control Act inventory, 2016). We use hundreds of thousands chemicals in our daily lives for the purposes of cleaning, promoting health and treating disease, for industrial applications and in agriculture. Consider the developments in modern medicine, for example: all modern medical compounds were developed in the last 200 years, starting with the discovery of anaesthetics by Humphrey Davy in 1799. Today, more than 100 000 tonnes of pharmaceutical products are sold every year, with 35 unique new molecules being produced each year (Section 5.2). The development of nanotechnology has led to the production of compounds one-millionth of a millimetre in diameter for commercial, medical and industrial applications (Section 5.5). Advances in the scale and complexity of offshore operations—for example, to extract fossil fuels (Section 5.4), minerals or generate renewable energy— and increasing use of the sea for transportation mean that we are changing the marine environment in numerous ways that were not previously have been anticipated (e.g. the introduction into the sea

of light, sound, electromagnetic frequencies (EMF) and brine; Sections 5.6–5.7). Our knowledge of the quantities of new compounds being released into the environment is very limited, as is our ability to measure their environmental concentrations (this is particularly the case for nanomaterials and pharmaceutical products). The legislation and regulation surrounding their use is lagging far behind their rates of production and release. Some of these new technologies are so complex and so small that we do not know what effect they will have once they enter the environment in terms of their physicochemical behaviour or their effects upon the biota. The large number of different synthetic compounds, and/ or those that are natural but polluting, now present in the environment brings with it the further challenge of needing to understand how complex mixtures of compounds behave (Sections 2.2.2 and 2.2.3) particularly those that may have synergistic effects on wildlife and, through the food chain, on human health.

This chapter is entitled 'emerging' problems because although some of the compounds discussed herein were first introduced to the environment several years ago, here we focus on issues that have only recently been recognised; for example, because of increased sensitivity of analytical technologies, the development of new compounds or as a result of the scale of the problem expanding (i.e. the extent of offshore infrastructure). Furthermore, all of these will represent major challenges to ecotoxicologists and regulators for many years to come because at present the environmental fate and toxicology of most of these contaminants are poorly known, and there is little legislation or regulation.

Marine Pollution. Christopher L. J. Frid & Bryony A. Caswell.
© Christopher L. J. Frid & Bryony A. Caswell 2017. Published 2017 by Oxford University Press.
DOI 10.1093/oso/9780198726289.001.0001

5.2 Pharmaceuticals, personal care products and veterinary products

5.2.1 Use of pharmaceutical compounds

The large-scale production of medicines began in 1827 in Darmstadt in Germany when Merck began manufacturing and selling alkaloids (one of which is morphine). The development of the pharmaceutical industry was driven by war; first the American Civil War (1861–65), followed by the First World War (1914–18), and then the Second World War (1939–45), due to increasing demand and disruptions to international trade in medicinal compounds. The industry underwent major advancements after the isolation of insulin by Frederick Banting in 1921, closely followed by the identification of penicillin by Alexander Fleming in 1928. From this period onwards, collaborations between companies and governments saw the rapid expansion of pharmaceutical research and in the 90-odd years since Banting's discovery, around 4000 compounds have been developed for use in medicines across the world today.

Pharmaceuticals are a persistent feature of our modern daily lives: we use them to treat burns, scrapes and cuts, a headache or a cold, to control allergies and indigestion, as contraceptives and so on (Table 5.1). Pharmaceuticals are also contained within other, perhaps less obvious, products such as impregnated tissues, wipes, wound dressings, creams and moisturisers. These products are collectively referred to as Pharmaceuticals and Personal Care Products (PPCPs). Prescription medicines that are used to treat diseases such as cancer, arthritis, asthma, mental health, diabetes, autoimmune conditions, ulcers or to regulate blood pressure, decrease inflammation and vaccinate against infection are all used in substantial volumes (Table 5.1). The production of pharmaceuticals is big business and in 2015 generated $US 954 116 \times 10^6$. Pharmaceuticals represented approximately 3% of total global merchandise export trade in 2014 according to the World Trade Organisation (2014). Several hundred-thousand tonnes of pharmaceutical substances are sold across the globe. These figures exclude the personal care products (PCPs) used in our daily lives.

Table 5.1 Pharmaceutical product sales (as an indicator of production) of the top ten therapeutic classes for 2015; total production in 2015 was $US954 116 \times 10^6$. Data from: IMS Health (2015).

Therapeutic class	US$ million
Oncology	78 939
Antidiabetics	71 471
Pain	56 191
Autoimmune	41 928
Antihypersensitives	41 393
Respiratory agents	40 037
Antibacterials	38 361
Mental health	34 870
Viral hepatitis	32 027
Dermatologics	29 484
Anticoagulants	27 230

The global production of pharmaceuticals, based on economic export statistics, indicate that the biggest producers are Germany, followed by Switzerland, Belgium and the United States, each of which sold more than US$44 000 \times 10^6$ worth in 2013 (Table 5.2). Data on the consumption and use of pharmaceuticals are limited and vary geographically (Table 5.2) and annually because of the differences in disease incidence, legislation and economics. However, market predictions suggested the largest sales in 2016 were for oncology and anti-diabetic medicines, respiratory agents and autoimmune drugs (see Table 5.1 for the top ten in terms of sales volumes). In addition to being used in medicines for human consumption, pharmaceuticals are widely applied in intensive farming practices. For example, they are applied to fruit crops or used in animal husbandry for veterinary reasons and growth promotion (Section 5.3). The PPCP market is expanding all the time as technology becomes more advanced and demand grows. On average 35 *new molecular entities* were and are predicted to be produced each year between 2013 and 2017. These products include novel small compounds, complex molecules purified from natural materials or combination products where at least one component is being launched for the first time.

Table 5.2 Top ten producers and consumers of pharmaceutical products on the basis of the economic value of their exports and imports (World Trade Organisation 2014); ordered by rank expenditure.

Top producers		Top consumers	
Country	Exports (US$ million)	Country	Imports (US$ million)
Germany	75 624.6	USA	67 345.8
Switzerland	62 382.9	Germany	46 479.9
Belgium	48 965.5	Belgium	40 872.3
USA	44 387.0	France	29 794.0
France	37 788.0	UK	28 937.6
UK	33 141.1	Italy	24 737.3
Ireland	28 202.3	Switzerland	23 874.1
The Netherlands	25 098.7	Japan	21 830.9
Italy	24 924.3	The Netherlands	17 415.8
Spain	13 867.7	China	16 195.8

5.2.2 Routes into marine environments and marine organisms

PPCPs can enter the environment via a number of routes (Figure 5.1) such as, and perhaps most obviously, point-source discharges from pharmaceutical production plants. The emissions from the manufacturing process are assumed to be minimal in Europe and North America due to industrial regulations of production (adhering to good manufacturing practice) and the high economic value of the product. However, it is not clear that this assumption is valid because the issue has received scant independent scientific investigation. One reason for this is because the monitoring data are not made publicly available by industry. The production process is complicated, consisting of several separate steps often undertaken at multiple locations with differing regulations and compliance limits. There is, however, some evidence to suggest emissions from pharmaceutical plants in the developed world can be significant. For example, high levels of the anti-depressant venlafaxin have been found in the River Rhein downstream of a production plant in Switzerland.

The importance of inputs of PPCPs from production plants into the marine environment is of greater magnitude in Asia and it will become increasingly important as the Asian pharmaceutical sector grows over the next few decades. India and China have the fastest growing pharmaceutical markets representing between 33% and 39% of total market growth, in terms of spending, by 2015. A waste treatment plant near Hyderabad, India that received effluent from 90 different industrial plants that produce active pharmaceutical ingredients (APIs) found 23 different pharmaceutical compounds, 11 of which occurred in concentrations of between 100 µg l^{-1} and 31 mg l^{-1} (6/11 were antibiotics) (Larsson et al. 2007). The highest levels were for the antibiotic ciprofloxacin and the amount released in one day is estimated to be equivalent to the total amount consumed by a population of 9 million people in 5 days. Based on these levels it is hardly surprising that we are finding more and more antibiotic-resistant bacterial strains.

Other examples of releases of large quantities of APIs include 2.7 g l^{-1} of salicylic acid in discharges from an aspirin production plant in Bangalore, and 51 ng l^{-1} of ethinylestradiol, an oestrogen used in contraceptives that causes the feminisation of fish and reduces egg fertilisation rates (Section 5.3), from a plant in China (see Bisarya and Patil 1993 and Larsson et al. 2007 for further details).

Discharges of PPCPs can also enter the ocean via the waste-water from hospitals or other healthcare providers and private households (Figure 5.1). These compounds may enter the environment by direct disposal within waste-water, for those with topical applications, or in human wastes. Pharmaceutical compounds may be broken down in the body (metabolised) or they may be transformed once they enter the environment. Within our bodies, medicines are metabolised by enzymes or the bacteria that inhabit our skin or gut, or it may occur abiotically by hydrolysis in the gut. The extent of metabolism in the human body varies, depending on the physicochemical properties of a compound and how it is administered. Some compounds are almost completely metabolised prior to excretion, whereas others (e.g. X-ray contrast media) may complete the journey through our bodies having undergone no changes at all. However, complete metabolism is uncommon. The toxicology of the metabolites (the products of metabolism) may be

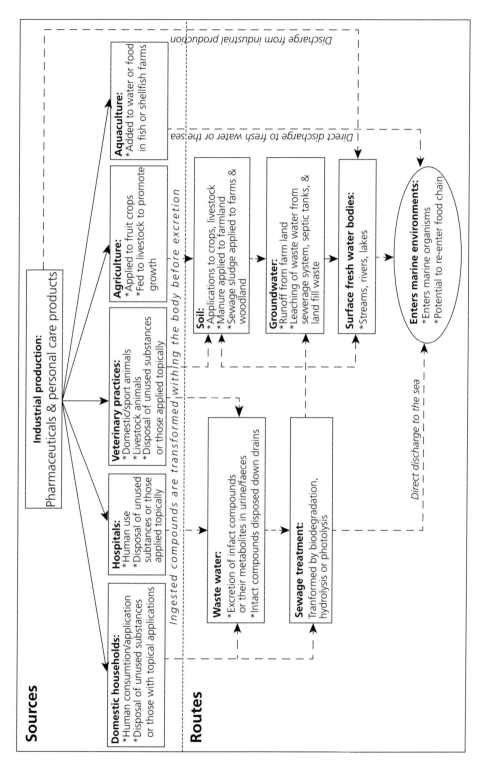

Figure 5.1 Sources and routes of Pharmaceutical and Personal Care Products (PPCP) into the marine environment. Arrows indicate movement of PPCPs through the environment and the waste-water industry.

very different from the parent compound; for example, their solubility may have changed or they may have lower biological activity. In the environment PPCPs are transformed by biotic (e.g. bacteria or fungi) and abiotic processes (e.g. photo-oxidation or hydrolysis) (Section 2.2). They may also be transformed during water-treatment processes such as photolysis, chlorination or ozonolysis (Section 3.2.2). As well as changing the solubility of the parent compound, the changes in chemical properties can increase toxicity; for instance, the so-called prodrugs (medicines administered in an inactive form) that are not fully metabolised before leaving the body, or through processes such as sulfation that increase mutagenicity.

In the United States and the European Union, it is currently legal to disposed of unused drugs in household waste (and ultimately landfill) and down the drain. Surveys show that in Kuwait, 11%, in Germany, 25%, in Austria, 33% and in the United States, 50% of unwanted drugs are disposed of by the public down the drain. However, this may be the least desirable option because they may then enter the environment directly (via sewage) or through landfill leachates. Take-back programmes have been created in a number of countries (ten US states, Australia, Canada, Sweden and the EU) where unused drugs can be returned to pharmacists for reuse or disposal by incineration. The disposal pathway is thought to be relatively small compared with entry to the marine environment through the human body, however, like many other aspects of *ecopharmacology* (the study of the fate and affects of drugs in the environment) there are little data available to support this proposition. The most environmentally favourable disposal option appears to be incineration.

5.2.3 Compounds used in veterinary practice

Pharmaceutical compounds are widely used in animal husbandry (e.g. growth promotion) and veterinary practices (Table 5.3) and this represent a major route into the environment (Figure 5.1). The most widely used pharmaceuticals in veterinary medicine and animal husbandry are antimicrobials, pharmaceuticals used to kill or inhibit the growth of microbes. The use of antimicrobials in veterinary practices and animal husbandry has rivalled usage

Table 5.3 Total UK sales of antibiotics* in 2011 for veterinary applications for both food-producing and non-food producing animals. Data from: Veterinary Medicines Directorate (2012).

Group	Tonnes of active ingredient
Food animals	
Cattle only	12
Pig only	62
Poultry only	23
Sheep only	< 1
Fish only	2
Pig and poultry combined	162
Multi-species products	29
Non-food animals	
Dog only	11
Horse only	21
Other	3

*Excludes antiprotozoals (283), antifungals (8.7) and coccidiostats (antiparasitics) (277) sold in the United Kingdom in 2011. Data exclude antibiotic growth promoters which are no longer sold in the European Union since their ban in 2006.

within human medicine. For example, in 1996, 10 200 t of antibiotics were used in the EU and of this total, 50% were given to animals. These compounds are either given to the animals in their feed or by direct topical applications.

Pharmaceuticals are also used in arable farming where they are directly applied to crops particularly fruit crops (antimicrobials such as antibiotics or antifungals). Products that are directly applied to crops and/or animals become washed off in the rain and those that are ingested by animals are subsequently excreted after passage through the body. These compounds enter the soil and pass into water bodies that eventually flow into the sea. Topical applications of antimicrobials for animals can lead to the release of these compounds into the dust in their stables. A study of a pig-fattening farm in Germany found antimicrobials present in quantities from mg to kg in dust from pig houses. Exposure to this dust may have negative health effects for farm workers. Pharmaceuticals are also used in aquaculture to treat disease by directly adding them to the water. The compounds that are routinely topically applied to herd animals or those used in aquaculture have the potential to enter the marine environment in large quantities. The greatest volumes of veterinary pharmaceuticals in the environment are probably antibiotics and antiparasitics (Table 5.3).

Because animal manure and sludge from human sewage (Section 3.2) is applied to food crops, drug residues can be introduced into agricultural soil and crops. Studies show that between 40 mg l^{-1} and 66 mg l^{-1} of the antibiotics tetracycline and sulfamethazine occur in liquid manure from animals given the recommended doses of antibiotics. The manure is applied to crops and so may be introduced into the human food chain. The risk of exposure via this route depends on how long the slurry (animal manure) is stored before it's applied because some degradation may occur in the interim. A number of studies have shown that a range of veterinary compounds (chlortetracycline, florfenicol, trimethroprim, levamisole, diazinon, enrofloxacin and sulfadiazine) have been taken up by winter wheat, cabbage, green onions, corn, lettuce and carrots, but uptake does seem to vary between plant species and depending on the ambient conditions.

The leaching of veterinary pharmaceuticals into groundwater is another route through which they can enter the human food chain, this time via drinking water. Studies of antimicrobials in groundwater have shown that for some commonly used compounds (e.g. tetracyclines, macrolides and sulphonamides), concentrations are at or below the limit of detection (see Hamscher et al. 2000 and Blackwell et al. 2007 for details). However, this varies for each compound, the concentrations used, how it is applied, its chemical nature and the environmental conditions.

5.2.4 The fate and affects on marine ecosystems and human health

Most PPCPs, their metabolites and transformation products, whether from point or non-point sources, ultimately end up in aquatic environments where they represent a potential threat to human and ecosystem health. The concentrations of PPCPs present in the environment, their source, route of entry (Section 5.2.2) and ecological impacts have undergone little scientific investigation. Thus, the science of ecopharmacology and the associated legislation is lagging behind that for most other pollutants

(Chapter 7). Over the next few decades we can expect it to be a major growth area of research, as global populations increase so will the demand, use and disposal of pharmaceuticals.

The APIs within PPCPs have properties that make their effects and interactions complex; for example, their activity and function differs with changing pH, they can form zwitterions, and their methylation and degradation products may have enhanced biological activity. Also, these compounds commonly occur in formulations with other APIs, adjuvants or dyes. In addition to this complexity, they also interact with other APIs and pollutants within human bodies, other organisms and the environment. As for most other pollutants (Section 2.2.2), there is a knowledge gap surrounding the effects of mixtures and their interactions with environmental pollutants, despite this being the most realistic exposure scenario. We also know little of the chronic effects of pharmaceuticals on organism health. Although PPCPs are not all classified as persistent wastes (Section 1.1), they are continually discharged in high concentrations and so are ubiquitous.

Toxicity data are available for less than 1% of the APIs used in the world today (Tables 5.4 and 5.5). Most data on their toxicity are for freshwater taxa and there is little information on the chronic effects on marine fish and invertebrates. A summary of the current knowledge on API toxicity in pharmaceutical and veterinary products is presented in Tables 5.4–5.6. The toxicity data suggest that the amounts of APIs usually found in the environment are too low to inhibit growth and reproduction in classic OECD (Organisation for Economic Co-operation and Development) laboratory experiments (e.g. using algae and *Daphnia*; Section 2.4.3). Some pharmaceuticals occur in the environment at concentrations that are 1000 to 10 million times lower than those *known* to kill or effect 50% (the LC50 and EC50, respectively; see Section 2.4.1) of most studied aquatic organisms. The NOECs (no observed effect concentrations) are up to one-thousandfold lower. This is, of course, not the case for all pharmaceutical compounds in the environment; for example, the analgesic diclofenac occurs at 0.01 µgl^{-1} to 10 µgl^{-1} in sewage treatment plants and the lowest-observed effect concentration (LOEC) for fish species is around 1 µg l^{-1}. The beta-blocker propanolol occurs in sewage treatment plants at 0.01–1 µg l^{-1}

Table 5.4 Acute toxicity (24 h to 96 h) of a range of pharmaceutical compounds to aquatic phytoplankton, zooplankton, benthic invertebrates and fish; hyphen indicates no data available. NSAID = Non-Steroidal Anti-Inflammatories. Data from: Kummerer 2008.

Pharmaceutical compound	Acute toxicity range (mg l^{-1})			
	Phytoplankton	Benthos	Zooplankton	Fish
Contraceptives				
Ethinylestradiol	0.05–1.7	–	3.2–12	–
Analgesics and NSAID				
Diclofenac	14.5–75	–	22–71	–
Ibuprofen	7–332	–	9–110	153–202
Paracetamol	110–138	–	29–5310	372–854
Acetyl salicyclic acid	–	–	137–1485	–
Salicyclic acid	–	–	104–1960	–
Naproxen	–	–	33–171	548–723
β-blockers				
Propanolol	0.7–252	20–35	0.8–416	26–39
Metoprolol	–	–	8.6–202	–
Antidepressants				
Fluoxetine	–	0.3–1	0.67–2.15	–
Antiepileptics				
Diazepam	–	–	4.2–10 321	16–23
Carbamazepine	31–104	–	13.4–104	–
Blood lipid lowering agents				
Clofibrate	12–40	–	28–38.4	–
Clofibric acid	40–333	–	71–202	79–548
Bezafibrate	–	–	181–226	–
Anti-ulcerants				
Cimetidine	–	–	490–684	–
Ranitidine	–	–	519–723	–
Antidiabetics				
Metformin	–	63–137	162–239	–
Cytostatic and cancer therapy				
Methotrexate	45–267	–	–	–
Stimulant				
Caffeine	–	–	162–5024	145–764

and the NOEC for propanolol in phytoplankton and fish is 1 µg l^{-1}; the serotonin uptake inhibitor fluoxetine occurs at concentrations of up to 0.0001 mg l^{-1} in sewage treatment plants and phytoplankton have a NOEC for fluoxetine of 0.001 mg l^{-1}. Although the measured environmental concentrations of many APIs are low, their effects on organisms in the environment are largely unknown. The conventional toxicity test methodologies are not sensitive to APIs and are not of sufficient duration to establish the chronic effects of APIs on reproduction, neurobehaviour and immunology.

As discussed in Chapter 2, it is not sufficient to study the toxicity of a compound in isolation, although this is the methodology most toxicity testing employs. A study of the effects of mixtures of antibiotics on marine diatoms illustrates this problem. The antibiotic median inhibition concentrations (IC50s) for the marine diatoms *Cylindrotheca closterium* and *Navicula ramosissima* were higher than those found for freshwater diatom species (Table 5.5), and both additive and synergistic toxicity effects were found. In addition to mixture effects, pharmaceuticals may also have indirect effects on other trophic levels through bioaccumulation (Section 2.2.3) in the food chain. Residues of the anti-inflammatory drug diclofenac, thought to have entered the food chain via veterinary practices, have been associated with population decline and renal failure in the oriental white-backed vulture, *Gyps bengalensis* (Table 5.5). The full extent of the potential for bioaccumulation of APIs is not yet known.

Most acute toxicity testing with aquatic organisms has focused on small planktonic freshwater invertebrates such as *Daphnia* and *Ceriodaphnia*, also known as water fleas (Table 5.4). The acute toxicity of pharmaceuticals to aquatic organisms ranges from 0.05 mg l^{-1} to 10 321 mg l^{-1} and the phytoplankton appear to be the most sensitive group, followed by the benthos and fish, with the zooplankton

generally being the least sensitive to most of the pharmaceutical compounds tested so far. In terms of therapeutic action, those with the highest acute toxicity are the beta blockers and anti-depressants, such as fluoxetine, some anti-epileptics, followed by blood lipid agents, anti-ulcerants and anti-diabetics, with the analgesics and nonsteroidal anti-inflammatory drugs (NSAIDS) being the least toxic (Table 5.4).

Many APIs exhibit chronic toxicity which, although not immediately apparent, has substantial effects beyond the individual organism and can cause the collapse of whole populations and species via its disruption of reproductive processes (Section 5.3). In terms of the chronic toxicity of pharmaceutical products to aquatic organisms, the contraceptive ethinylestradiol is the most toxic (at concentrations of ng l^{-1}); followed by selective serotonin uptake inhibitors which evoke a chronic effect at concentrations between ng l^{-1} and μg l^{-1}; and, neuroactive compounds, blood lipid-lowering compounds, beta-blockers and NSAIDs are toxic in concentrations of 0.5 μg l^{-1} to 1.8 mg l^{-1}. The pharmaceuticals for cancer treatment exhibit the lowest chronic toxicity to aquatic organisms (49 μg l^{-1} to 19 mg l^{-1}; Table 5.5). Similar to acute toxicity tests, the majority of chronic toxicity tests are

Table 5.5 The chronic toxicity (e.g. more than 96 h) of selected pharmaceuticals and personal care products (PPCP) to aquatic organisms. LOEC = lowest observed effect concentration; NOEC = no observed effect concentration; EC = effect concentration. Data from: Oaks et al. 2004, Sorensen 1998, Taggart et al. 2007, Kummerer 2008, Hagenbach and Pinckney 2012 and Cortez et al. 2012.

PPCP	Test species	Chronic toxic effect
Contraceptives		
Ethinylestradiol	*Danio rerio* (zebrafish) *Pimephales promelas* (Fathead minnow)	Full life-cycle exposure to 3 ng l^{-1} causes elevated vitellogenin and gonadal feminisation; 4 ng l^{-1} caused vitellogenin induction 0.3 ng l^{-1} caused decreased egg fertilisation and sex ratio change.
Cancer treatments		
Tamoxifen	*Acartia tonsa* (copepod)	EC50 49 μg l^{-1}
	Paracentrotus lividus and *Spherechinus granularis* (sea urchins)	0.0037–3.7 mg l^{-1} impairs embryo development
Flutamide	*Pimephales promelas*	0.9 mg l^{-1} affects male sex characteristics 0.5 mg l^{-1} reduced fecundity 1–19 mg l^{-1} affects male courtship behaviour
Fadrozole	*Pimephales promelas*	0.05 and 0.96 mg l^{-1} for 21 d inhibited ovarian and testes growth
Anti-diuretics		
Furosemide	*Ceriodaphnia dubia* (water flea) *Brachionus calcyiflorus* (marine rotifer)	LOEC 0.3 and 1.25 mg l^{-1} inhibits population growth. Photoproduct of furosemide has LOEC of 0.02–0.31 mg l^{-1}

(continued)

Table 5.5 (*Continued*)

PPCP	Test species	Chronic toxic effect
Selective Serotonin Reuptake Inhibitors		
Sertraline	*Ceriodaphnia dubia*	LOEC 45 µg l^{-1} and NOEC 9 µg l^{-1}
Fluoxetine	*Ceriodaphnia dubia*	56 µg l^{-1} Increases fecundity
Fluoxetine and ibuprofen	*Gammarus pulex*	10–100 ng l^{-1} decreases behavioural activity
Fluvoxamine	*Dresseina polymorpha* (zebra mussel)	0.032 µg l^{-1} induces spawning
Neuroactive compounds		
Carbamazepine (antiepileptic)	*Ceriodaphnia dubia*	NOEC (7 d) 25 µg l^{-1}
	Brachionus calcyiflorus	NOEC (2 d) 377 µg l^{-1}
	Danio rerio early life stages	NOEC (10 d) 25 µg l^{-1}
Diazepam	*Hydra vulgaris* (freshwater hydroid)	10 µg l^{-1} inhibits polyp regeneration
Blood lipid lowering agents		
Clofibrate	*Daphnia*	NOEC reproduction 10 µg l^{-1}
Clofibric acid	*Ceriodaphnia dubia*	NOEC (7 d) reproduction 640 µg l^{-1}
	Brachionus calcyiflorus	NOEC (2 d) reproduction 246 µg l^{-1}
	Danio rerio early life stages	NOEC (10 d) 70 mg l^{-1}
β- blockers		
Propanolol	*Ceriodaphnia dubia*	NOEC 125 µg l^{-1}
		LOEC 250 µg l^{-1} affects reproduction
	Hyalella azteca	100 µg l^{-1} for 27 d affects reproduction
	Oryzias latipes (medaka)	0.5 and 1 µg l^{-1} for 4 weeks reduced egg release
Non-steroidal Anti-Inflammatory Drugs		
Acetyl salicyclic acid	*Daphnia magna and D. longispina*	1.8 mg l^{-1} causes reproductive effects
Diclofenac	*Onchorhynchus mykiss* (rainbow trout)	LOEC 5µg l^{-1} causes kidney and gill abnormalities
	Salmo trutta (brown trout)	0.5–50 µg l^{-1} causes gill and kidney inflammation, decreased haematocrit (volume of red blood cells in blood)
	Gyps bengalensis (white backed vultures)	> 0.007 mg kg^{-1} caused renal failure and up to 1 mg kg^{-1} in prey caused population decline
Ibuprofen	*Lepomis machrochirus* (bluegill sunfish)	NOEC (4 d) 10mg l^{-1}; LC50 (4 d) 173 mgl^{-1}
	Mysidopsis bahia (mysid)	NOEC (4 d) 30 mg l^{-1}
	Skeletonema costatum (marine diatom)	EC50 (4 d) 7.1 mg l^{-1}
Antibacterials		
Triclosan	*Perna perna*	IC50 0.490 mg l^{-1} (egg fertilisation) and IC50 0.135 mg l^{-1} (embryo development)
Antibiotics		
Ciproflaxin,	*Cylindrotheca closterium and Navicula ramosissima*	IC50 (growth) 55.43 mg l^{-1} and 72.12 mg l^{-1}
Lincomycin	*Cylindrotheca closterium and Navicula ramosissima*	IC50 (growth) 14.16 mg l^{-1} and 11.08 mg l^{-1}
Tylosin	*Cylindrotheca closterium and Navicula ramosissima*	IC50 (growth) 0.27 mg l^{-1} and 0.99 mg l^{-1}

undertaken with freshwater organisms (Table 5.5) in particular species of *Daphnia* and *Ceriodaphnia*. The focus of toxicity testing on such a small number of biologically simple test species limits both the quality of toxicity assessments and our ability to predict and regulate the impacts of pollutants in the environment (Section 2.4.3).

Some studies of the chronic effects of PPCPs on higher marine invertebrates and vertebrates exist although more are needed. For example, there is considerable interest in triclosan toxicity due to its prevalence in various personal care products (toothpaste, moisturiser, shampoo, soap, clothing, toys and many other household items) and the environment, in organisms and human breast milk, its ability to bioaccumulate and its toxic by-products that are produced during methylation and photodegradation (e.g. dioxins and furans). The effects of triclosan on egg fertilisation and embryo development of the mussel *Perna perna* were inhibited at 0.490 mg l^{-1} and 0.135 mg l^{-1}, respectively (Table 5.5). Although these IC50s are above the environmentally relevant levels (ng l^{-1}) of triclosan reported in some studies, physiological stress has been detected at just 12 ng l^{-1}.

Decades of use of large quantities (100 000–200 000 t yr^{-1}) of antibiotic compounds in medical and veterinary practices (Table 5.3) is receiving increasing scientific attention regarding their fate and ecotoxicology although it is another area of ecopharmacology of which we know relatively little. However, more is known than for many other groups of APIs. The antibiotic compounds have a complex chemistry with differing functionality depending on ionic state and environmental pH. Antibiotics are the most widely used compounds in veterinary practices (Table 5.3, Section 5.2.2), although recent legislative changes have seen a decline in their use for what is called 'growth promotion'.

Algae are particularly sensitive to antibiotics that inhibit their growth by disrupting their metabolism, chloroplast replication, protein translation and transcription. The blue-green algae are the most sensitive taxa to a wide range of antibiotics (between 0.16 mg l^{-1} and 19 mg l^{-1}). For example, *Microsytis aeruginosa* growth is inhibited at 0.1 mg l^{-1} and the freshwater green algae *Selenastrum* and *Chlorella* are inhibited at 1–2 mg l^{-1} (Table 5.6). Higher invertebrates and

fish seem to be less sensitive to antibiotics (Table 5.6). Studies of small aquatic crustaceans (*Daphnia* and *Artemia*) showed chronic effects on the larvae and eggs by inhibiting hatching and physiological development (e.g. phototaxis, the development of pigmentation). The effects of antimicrobials on fish occur only at much higher concentrations,

Table 5.6 Veterinary treatments of known toxicity to non-target aquatic organisms, species tested and toxic effects. Abbreviations MIC = Minimum Inhibitory Concentration; LC50 = median lethal concentration; NEL = no effect level (equivalent to NOEC) (see Chapter 2 for a detailed account of the meaning of these toxicity indicators). Data from: Sarmah et al. 2006 and Sorensen et al. 1998.

Veterinary treatment	Test species	Toxic effects
Antibiotics		
Bacitracin	*Daphnia magna*	LC50(2 d) 30 mg l^{-1}
	Artemia salina	LC50(2 d) 21.8 mg l^{-1}
Flumequine	*Artemia salina*	LC50(3 d) 96 mg l^{-1}
	Aeromonas salmonicida	MIC (3 d) 16 µg l^{-1}
Furazolidone (fish farm use)	Sediment bacteria	Antibiotic resistance
	Chlorella pyrenoidosa (algae)	EC50 (2 d) 1.3 mg l^{-1}
	Daphnia magna	LC50 (2 d) > 30 mg l^{-1}
	Salmo gairdner	LC50 (2 d) > 30 mg l^{-1}
	Lebistes reticulatus (guppy)	LC50 (2 d) 25 mg l^{-1}
Nitrofurazone	*Selenastrum capricornutum*	EC50 (96 h) 1.45 mg l^{-1}
	Daphnia magna	LC50 (48 h) 28.67 mg l^{-1}
Streptomycin	*Vibrio harveyi* (blue-green algae)	LC50 (0.25 h) 19 mg l^{-1}
Sulphadimethoxine	*Artemia salina*	LC50 (2 d) 0.9 g l^{-1}
		LC50 (3 d) 0.5 g l^{-1}
		LC50 (4 d) 19 g l^{-1}
Chloramphenicol	*Vibrio harveyi* (blue-green algae)	EC50 (0.25 h) 0.16 mg l^{-1} biolumicens test
Antiparasitics		
Ivermectin	*Salmo gairdneri* (rainbow trout)	LC50 (4 d) 3 mg l^{-1}
		NOEC (4 d) 0.9 mg l^{-1}
	Lepomis macrochines (bluegill sunfish)	LC50 (4 d) 4.8 mg l^{-1}
		NEL > 9.1 mg l^{-1}
	Chlorella pyrenoidosa (algae)	
Bromocyclen	*Daphnia magna*	LC50 (48 h) 0.7 mg l^{-1}
		NOEC 0.1 mg l^{-1}

or no toxic effects are seen (Table 5.6). A number of antiparasitic compounds (e.g. ivermectin) used in veterinary treatments are toxic to some fish and crustacean in concentrations of mg l^{-1}.

The great interest in the environmental fate and effects of antibiotics stems from the concerns that *over*use of PPCPs in the environment that has induced biological resistance in pathogens. There is now a growing body of evidence which shows that several pathogenic bacteria (e.g. *Staphylococcus aureus* and *Enterococcus* sp., that cause a range of tissue infections; *Neisseria gonorrhoeae*, which causes gonorrhoea; and *Mycobacterium tuberculosis*, which causes tuberculosis) have become resistant to widely used antibiotic compounds such as methicillin, vancomycin and penicillin. Although the development of resistance is not yet fully understood it represents a significant threat to human health as pathogenic bacterial infections will become increasingly difficult to treat.

5.2.5 Legislation surrounding the release of pharmaceutical and veterinary products into the environment

Compulsory risk assessments for all new compounds were introduced in the 1980s within the European Union, but pharmaceutical drugs were excluded from this legislation. However, pharmaceutical products are highly likely to affect living organisms due to their biological activity and because they, or their metabolites, are water soluble, they are very mobile. Most pharmaceutical products are not readily biodegradable. The available evidence now shows that APIs are widely distributed throughout the biosphere and they represent a serious threat to human and ecosystem health (Section 5.2.4). The European Commission (EC) established directives from 1998 (2001/83/EC Community code relating to medicinal products for human use and its amendment 2004/27/EC, 2001/82/EC Community code relating to veterinary medicinal products and its amendment 2004/28/EC, and Regulation EC/726/2004 Community procedures for the authorisation and supervision of medicinal products for human and veterinary use and establishing a European Medicines Agency, the latter covers

medicines containing genetically modified organisms), which stated that for all *newly* licensed compounds, an environmental risk assessment (ERA) should be conducted. This should be an 'evaluation of the potential environmental risks posed by the medicinal product', and 'specific arrangements to limit it shall be envisaged'. These directives mostly apply to APIs through the consumption pathway (by animals or humans) and not the production pathways. Furthermore, the directives state that marketing authorisation for human pharmaceutical products *cannot* be refused on the basis of the ERA. The PPCP production stage is governed by the Directive 96/61/EC which is concerned with Integrated Pollution Prevention and Control which applies to industrial sources.

In contrast with the legislation for human medicines, authorisation for new veterinary medicines can be refused if the environmental risk is high. The directives for both types of pharmaceutical product (PP) also specify that any major changes to, or extensions of, existing authorisation that could increase the environmental exposure should be assessed. While new drugs or major changes require an ERA for PPs authorised prior to the adoption of EU legislation there is no requirement for PPs for human or veterinary use to be assessed. These 'old medicines' are considerable in number and use, but there is no system for their evaluation or regulation. Some compounds considered of minor environmental risk are exempt from the legislation (e.g. amino acids, lipids, peptides, proteins, vitamins, electrolytes and carbohydrates), as are vaccines and herbal medicines. Compounds containing radioactive materials are covered by additional legislation (96/29/Euratom and 97/43/Euratom).

The EC provides guidelines that outline the scope and procedures required for assessing the environmental impacts of PPs. For human medicines, this guidance is based on the 2006 guidelines of the European Agency for the Evaluation of Medicinal Products (EMEA) and for veterinary medicines on the Veterinary International Conference on Harmonization (VICH). The EMEA guidance is based on the Predicted Environmental Concentrations (PEC) in surface waters with an action limit of 0.01 µg l^{-1}; this is used for guidance to decide whether a full

ERA is to be completed. Pharmaceuticals present in concentrations below this limit are not subject to second phase testing on their environmental fate and biological effects. Further information on these testing procedures may be found in the Committee for Medicinal Products for Human Use report (2006). The legislation surrounding PPs of course varies between countries. In the United States, environmental pharmaceuticals have been regulated since 1977 by the FDA under the auspices of the National Environmental Policy act 1969. Similar to the European Union, the FDA requires an ERA be conducted for all new PPs and guidance documents similar to the EMEA are available.

Assessment of PPs using the EMEA guidelines only consider the aquatic component and not terrestrial or groundwater components. Second phase testing includes assessments for groundwater and terrestrial systems for compounds bound to sewage sludge that are subsequently applied to agricultural land. For veterinary medicines, the terrestrial and aquatic exposure pathways are considered if the PP is applied to terrestrial or aquatic animals. The veterinary guidelines use environmental concentrations of $1 \, \mu g \, l^{-1}$ for aquatic systems and a PEC of 100 $\mu g \, kg^{-1}$ for soil.

Pharmaceutical products have not yet been listed in the Drinking Water Directive (DWD 98/83/EC), the Water Framework Directive (WFD, 2000/60/EC), the Groundwater Directive (2006/118/EC) or Urban Waste-water Directive (91/271/EEC), despite their persistent nature and the threats to human health. Nor are they monitored in European waters. Pharmaceutical products are not listed in the Sewage Sludge Directive (86/278/EEC) either, although this is another potential route into the human diet through the application of sludge to crops from where it subsequently enters groundwaters (and so potentially drinking water supplies). Some PP candidates for the WFD priority pollutant list (toxic, persistent compounds that have the potential to bioaccumulate) have been proposed and rejected. Pharmaceuticals are not included under the Regulation, Evaluation, Authorisation and Restriction of Chemicals (REACH) Regulations (EC No. 1907/2006) because it exempts 'substances used in medicinal products' that are included in the human and veterinary use directives (2001/83/EC and 2001/82/EC). Endocrine-disrupting compounds (EDCs) (Section 5.3) are considered under the European Endocrine Strategy which considers the short, mid- and long-term effects of EDCs and is charged with identifying and evaluating suspected EDCs in the aquatic environment under the WFD. More data on the environmental occurrences of APIs, their ecotoxicological effects, particularly the low-level chronic and cumulative effects, are required to provide the evidence to support their inclusion within the directives and regulations.

5.3 Endocrine-disrupting compounds

One of the environmental effects that pharmaceutical compounds can have on animals, vertebrates and invertebrates, is to disrupt their endocrine systems. Compounds with this activity are called *endocrine-disrupting compounds* or *endocrine disruptors*. They can be natural or synthetic (Table 5.7) and their key attribute is that they mimic the chemistry of natural hormones that can negatively affect the normal functioning of the endocrine system in wildlife and humans. We have already touched upon a number of EDCs in earlier chapters and a summary of the most common EDCs is provided in Table 5.7. The US EPA defines EDC's as 'exogenous agents that interfere with the synthesis, secretion, transport, binding, action or elimination of natural hormones in the body that are responsible for the maintenance of homeostasis, reproduction, development and/or behaviour'. Three main forms of disruption occur: (1) impairment or mimicking of oestrogen production (oestrogenic EDCs), (2) impairment or mimicking of testosterone production (androgenic EDCs) and (3) interference with thyroid function (thyroidal EDCs). These processes may involve direct interaction between the EDCs and the hormone receptors or they may interact with the proteins responsible for the delivery of hormones to their site(s) of action. EDCs can affect the three hormone systems individually or simultaneously. Organism sensitivity to EDCs and their effects vary throughout ontogeny with earlier life stages being the most susceptible, especially during tissue development, and the effects can be latent, for example, only manifesting later in life (Table 5.7).

Table 5.7 Established effects of a range of EDCs on animals. These are a small fraction of the 800 chemicals used commercially that are thought to interfere with endocrine system functioning. Data from: Bergman et al. 2012.

Endocrine disruptor	Origin	Effects
Diethylstilbestrol (DES)*	Synthetic oestrogen used for hormone replacement therapy, prevention of miscarriage, cancer treatment, growth hormone in beef and poultry	Sterility, decreased hormone levels, impairs reproductive development and ovarian function, causes various cancers, cysts, lesions, and uterine abnormalities
Phytoestrogens	Natural compounds produced by plants	Male and female reproductive development, size of reproductive organs, erectile dysfunction, advanced onset of puberty
Ethinylestradiol	Contraceptive	*Gammarus pulex* feminisation and fathead minnow mortality
Oxybenzone	Sunscreen	Suspected EDC
Triclosan	Antimicrobial in many personal care products, kitchen utensils, clothing, toys, bedding	Little data but seems to interfere with testosterone/oestrogen production and reproductive success
Bisphenol A (BPA)*	Found in plastics (including those for food consumption), water pipes, PVC, paper products	Mammary and prostate cancer, diabetes, obesity, decreased sperm count, early puberty and neurological problems
Phthalate esters*	Soft toys, flooring, medical equipment, cosmetics and air fresheners	Male reproductive development, size of reproductive organs, lesions, reduced male and female fertility, uterine abnormalities, and reduced sperm production
Perfluorooctanic acid (PFOA)	Surfactant (in Teflon, Gore-tex)	Affects thyroid hormone levels, fertility, delays puberty, and is carcinogenic
Alkyl phenol ethoxylates* (e.g. nonylphenol ethoxylate)	Surfactant	Reproductive development, size of reproductive organs, miscarriage, and sperm production
Dichlorodiphenyltrichloroethane (DDT)*	Insecticide (especially for disease vectors)	Development of female reproductive organs, egg shell thinning, and suspected to impact fertility
Linuron	Herbicide	Male reproductive development, nipple retention, size of reproductive organs
Dicarboximide fungicides (Vinclozolin, Procymidone)	Fungicides used to control rot, blight and molds on fruit/veg crops (including vineyards)	Androgen receptor antagonist. Development of male reproductive organs, size of reproductive organs, sperm production, sperm abnormalities
Tributyltin (TBT)*	Antifouling paints on ships and offshore structures	Invertebrate and vertebrate development, inhibits germ cell production in reproductive organs, miscarriage, and masculinisation
Dioxins	By products of industrial processes: including bleaching paper pulp, pesticide production, waste incineration	Reproductive development in males and females, delayed puberty, decreased sperm production, size of reproductive organs, decreased fertility, and endometriosis
Polychlorinated biphenyls (PCBs)*	Industrial coolants and lubricants	Sperm production and motility, male and female reproductive development, size of reproductive organs, and endometriosis
Polycyclic aromatic hydrocarbons (PAHs): Benzoapyerne, Pyrene, Anthracene	Found naturally in hydrocarbons (fossil fuels and tar) and from combustion of organic matter	Tumour forming
Lead*	Personal care products, fuel combustion, lead-based paints, lead pipes	Development of male and female reproductive organs, including sperm counts
Cadmium	Mining, ferrous metal production, waste incineration, fuel combustion, fertilisers, natural sources	Sperm production and motility, development of sexual organs, various cancers, and endometriosis
Manganese	Additive in unleaded petrol, batteries, ceramics	Affects testosterone and estradiol levels, sperm production

*banned for some applications in some countries. Endometriosis: growth of endometrial tissue outside the uterus (may interfere with reproduction).

EDCs have effects at low concentrations that are environmentally relevant. This is presumably due to hormones, which have high affinity for their receptors, also acting at very low concentrations. The effects of EDCs may be additive to the extent that in isolation no effect is observed but in combination it is apparent. The dose–response relationship (Section 2.4) of a hormone or EDC and the manifestation of a biological effect is non-linear and thus can vary with concentration. For example, at low concentrations the effects may be very large but at high concentrations may be very small. This response may be additive, with the EDC contributing to the effects of the natural hormone. The dose–response relationship may also vary with the number of receptors present in the body (the more receptors, the greater the impact). Thus, predicting the effects of the EDCs is complex and differs from those for other toxicants.

The existence of EDCs is not a new problem (Sections 2.4.4, 3.3 and 3.4); we have known of their existence for more than 70 years. Approximately 800 chemicals in commercial use are known or suspected to have the ability to interfere with endocrine system functioning (Table 5.7). Although their toxicity is complex and there is still much that is not known. It is now well established that they can have serious implications, at the concentrations in which they are found in the environment, for the reproductive success of many species and they are a threat to human populations. We use large quantities of EDCs and reproductive disorders are on the increase. Exposure routes are numerous, including through food-storage containers, clothes, children's toys, personal care products and medicines, in addition to exposure to the EDCs that are released into our environment by industrial processes that subsequently find their way into our food and water. Laboratory testing of EDCs on animals does not always indicate whether there will be an effect in humans. A recent report commissioned by the World Health Organisation (Bergman et al. 2012) on the state of the science with respect to EDCs concluded that the risks from them may be substantially underestimated because major knowledge gaps still exist. For example, suitable laboratory methods for detecting the effects of EDCs on humans and wildlife are lacking; relatively few of the 800 known, or

suspected, EDCs have been studied in-depth, the relative importance of environmental exposure routes are unknown, and the links between exposure and the development of endocrine diseases are poorly understood.

Some of the substances that are known EDCs have been banned for particular applications in specific countries or throughout the world. For example, dichlorodiphenyltrichloroethane (DDT) was banned for global agricultural use in 2004 and it is now only used in the developing world for disease-vector control (Section 3.4), the use of tributyltin (TBT) was banned in 1980 for use on ships less than 25 m in length, and in 2008 all TBT use was banned by the International Maritime Organisation (Section 3.3). Polychlorinated biphenyl (PCB) manufacture was banned in 1979, and diethylstilbestrol use was banned in the 1970s and 1980s for various applications (although production continued until 1997). Restrictions on the use and production of these EDCs have in many cases proven effective in reducing the incidence of endocrine disruption in humans and wildlife. However, for many of these compounds, restrictions on their use has not resolved the issue because many are persistent organic wastes that remain present in the environment, organisms and our own bodies: they are permanent additions to the environment.

5.4 Other industrial compounds

5.4.1 The by-products of disinfection

The use of chlorine compounds to disinfect drinking water is a widespread practice that began in 1905 in response to an epidemic of typhoid fever in England. Chlorine is a strong oxidising agent that destroys pathogens and oxidises odour-forming compounds. Chlorine-based disinfectants react with organic compounds (e.g. fulvic and humic acids) to produce toxic halogenated disinfection by-products (DBPs) such as trihalomethanes (e.g. chloroform, bromoform, dibromochloromethane) and haloacetic acids. Although produced in relatively small concentrations, the trihalomethanes are known carcinogens that represent a threat to human health and so their levels in treated drinking waters are monitored. In the United States, in addition to the treatment of drinking water, chlorination is also used in some sewage treatment processes. Disinfection products are also used to prevent the growth of biofilms within power plant cooling systems and desalination plants (Section 5.7), in aquaculture and to disinfect ships ballast water (Section 4.8).

Three compounds are commonly used to disinfect drinking water: chlorine, hypochlorite and chloramine of which so-called free chlorine (Cl−) has the longest history of use. Chloroform, one of the main by-products of disinfection, volatilises readily from soil and water to form phosphogene, dichloromethane, formyl chloride and other compounds in the atmosphere, and it is toxic to humans through both air and water. The biodegradation of disinfection products in water is slow and bioaccumulation within animal tissue is minimal. The ingestion of chloroform has been shown to have carcinogenic impacts on mammals, and some of the haloacetic acids have also been shown to promote tumour formation in mammals. The World Health Organisation recommends a 'tolerable daily intake' of less than 10 µg of chloroform per kg of body weight each day. In the United States, safe drinking water standards are 70 µg l^{-1}, in Europe they are less than 200 µg l^{-1} and in Australia less than 250 µg l^{-1} of chloroform. The nature of the DBPs depends on the disinfectant used, how much was used, the nature of organic matter in the receiving environment, the temperature and pH of the receiving waters. The toxicity of chloroform to aquatic taxa has been shown to produce behavioural and physiological changes and mortality in invertebrates (annelids, molluscs, phytoplankton and zooplankton, and crustaceans; Table 5.8) and fish. It has also been found to interfere with reproduction and development. In the European Union, chloroform has an Environmental Quality Standard (EQS) of 12 µg l^{-1} in estuarine and marine waters. The World Health Organisation (Bergman et al. 2012) estimated the Predicted No Effect Concentrations (PNECs) of chloroform for aquatic life to be 72 µg l^{-1} and the PECs 0.2–0.5 µg l^{-1}.

The use of chloramine (NH$_2$Cl) to disinfect drinking water is increasing because it is more stable against environmental photodegradation. Also, in

Table 5.8 Toxicity of chloroform to a range of aquatic taxa. Data from: Kegley et al. 2016 (and specific references provided therein).

Taxa	Concentration (mg l^{-1})	Endpoint
Algae		
Skeletonema costatum	437	Biomass; EC50 (5 d)
Chlorella vulgaris	3.2	Photosynthesis; EC50 (3 h)
Chlamydomonas angulosa	3.2	Photosynthesis; EC50 (3 h)
Chlamydomonas rheinhardtii	13.3	Biomass; EC50 (72 h)
Microcystis aueroginosa	185	LOEC
Crustacea		
Penaeus duorarum	134	LC50 (24 h)
	81.5	LC50 (48 h)
	81.5	LC50 (96 h)
Streptocephalus proboscideus	771	LC50 (24 h)
Artemia salina	30.4–37.0	Mobility; EC50 (24 h)
Gammarus minus (early life stages)	245	LC50 (96 h)
Daphnia magna	6.3	Reproduction (NOEC (21 d))
Mollusca		
Dreissena polymorpha	33.0	LOEC (10 mins)
Fish		
Danio rero	6.1	Behavioural changes; NOEC (14 d)
Lepomis macrochirus	17.1	LC50 (12 h)
	17.1	LC50 (24 h)
	16.3	LC50 (48 h)
	16.2	LC50 (96 h)
Micropterus salmoides	45.4–55.8	LC50 (12 h)
	45.4–55.8	LC50 (24 h)
	45.4–55.8	LC50 (48 h)
	45.4–55.8	LC50 (96 h)
Oncorhynkiss mykiss	0.02	Enzyme activity; NOEC (4 d)
Eggs	1.2–2.0	LC50 (23 d)
	18.4–37.1	LC50 (12–96 h)
Oryzias latipes	108	Tumour production (10 d)
Poecillia reticulata	300	LC50 (96 h)

the United States, the increasing use of chloramine by utility companies is an attempt to comply with existing regulations on DBPs. The increasing use of chloramine, instead of, or in addition to chlorine and hypochlorite, is leading to the emergence of a number of new nitrogenous (n-DBPs) and iodinated DBPs (i-DBPs) such as halonitromethanes, haloacetonitriles, haloamides, iodoacetic acid, iodotrihalomethanes and nitrosamines.

Unfortunately, these new DBPs are between 1 (n-DBPs) and 250 times (i-DBPs) more toxic than the by-products of chlorine and hypochlorite. One example of an n-DBP is the nitrosamine N-nitrosodimethylamine (NDMA), a by-product of chloramine disinfection that is listed on the US EPA's draft contaminant candidate list for 2015 because it has been shown to be carcinogenic to rodents when inhaled or ingested; it is classed as a suspected human carcinogen. The EPA (2012) estimated that NDMA concentrations of 0.07 µg l^{-1} represent a 1 in 10 000 excess lifetime cancer risk to humans. The i-DBPs are toxic to animal cells (cytotoxic) and are 250 times more toxic than the DBPs from traditional drinking water disinfection. However, despite this higher toxicity the regulation of these emerging DBPs has not yet been established. It is thought that not all DBPs from chloramine have yet been identified, with the methodology and standards still being developed, but it is likely that many will have carcinogenic properties or cause developmental defects.

The World Health Organisation concludes that the risk of death from microbial pathogens is more than 100 to 1000 times greater than the risk of cancer from DBPs. With the DBPs from chloramination being more toxic than those produced by the more traditional methods, it seems preferable to revert to these older methods as they produce less toxic by-products than chloramination.

5.4.2 Perchlorates

Perchlorates (ClO$_4^-$) are the salts of perchloric acid and are strong oxidising agents that occur naturally in some nitrogenous mineral deposits and are produced on an industrial scale by electrolysing sodium chloride solutions. Four perchlorates are manufactured and used in industry: perchloric acid, ammonium perchlorate, potassium

perchlorate and sodium perchlorate. The largest applications of perchlorates are as propellants in rocket fuels, and they are also added to fireworks, flares, some munitions and matches. Perchlorates can also be introduced to the environment through disinfection processes (Section 5.4.2), bleaches and organic fertilisers applied to land, and so can enter the marine environment through waste-water and land run-off. Perchlorates are highly water-soluble and very mobile in soils, and when released into the atmosphere (in munitions or fireworks/flares) eventually fall-out or are washed out in rainwater and enter aquatic environments.

Perchlorates are very stable and so persist in the environment for long periods of time (more than 10 years) and may eventually be broken down by microbes. They have been found in animal tissues (aquatic insects, mammals, fish and amphibians) and plants demonstrating that they bioaccumulate. They may also be produced as the by-products of the disinfection process (Section 5.4.2). It is unclear to what extent perchlorates are distributed throughout the oceans due to limited data, however it seems likely that they are present in estuaries at high concentrations. Perchlorates also occur in high concentrations with in sewage; for example, 0.56–380 µg l^{-1} have been found in sewage treatment plants in China and 0.1–260 µg l^{-1} in sewage treatment plants in New York. Perchlorates have been described as an emerging contaminant because they are highly persistent and bioaccumulating, they have the potential to disrupt endocrine systems (and so are a risk to human health), and our knowledge of their prevalence and environmental impacts is limited.

Perchlorates were initially thought to be a predominantly localised issue in the vicinity of rocket manufacturing or testing facilities and perchlorate, munitions or pyrotechnic manufacturers. However, improvements in analytical methodology in 1997 increased their detection limit by two orders of magnitude. Subsequently, perchlorates were found throughout the environment. In 2011, the United States EPA listed perchlorates on the dangerous contaminants list for drinking water after they were detected above the minimum reporting level within 160 public water systems across 26 states (affecting more than 2 million people) in 2005. The mean perchlorate concentrations in

Table 5.9 Ecotoxicology of sodium (indicated by 'Na') and ammonia perchlorate (indicated by 'NH$_3$') to aquatic taxa (data only available for freshwater taxa). Data from: the European Chemicals Agency (2016).

Taxa	Concentration	Endpoint
Algae		
Selenastrum capricornutum	615 mg l^{-1} Na	IC25 (96 h)
Pseudokirchneriella subcapitata	12.5 mg l^{-1} Na	Growth; NOEC (48–96 h)
	> 435.7 mg l^{-1} Na	Growth; EC50 (72 h)
	> 435.7 mg l^{-1} Na	Growth; EC10 (72 h)
	> 505 mg l^{-1} NH$_3$	Growth; EC50 (72 h)
	182 mg l^{-1} NH$_3$	Growth; EC10 (72 h)
Invertebrates		
Daphnia magna	> 100 mg l^{-1} Na	EC50 (24 h)
	490 mg l^{-1} Na	EC50 (48 h)
	> 341 mg l^{-1} NH$_3$	Mobility; EC50 (48 h)
Ceriodaphnia dubia	10 mg l^{-1} Na	Reproduction; NOEC (7 d)
	25 mg l^{-1} Na	Mortality; NOEC (6 d)
	33 mg l^{-1} Na	LOEC (7 d)
	10 mg l^{-1} NH$_3$	Reproduction; NOEC (6 d)
	25 mg l^{-1} NH$_3$	Mortality; NOEC (6 d)
Fish		
Pimephale promelas	1655 mg l^{-1} Na	EC50 (96 h)
	155 mg l^{-1} Na	NOEC (7 d)
	280 mg l^{-1} Na	LOEC (7 d)
	9.6 mg l^{-1} NH$_3$	NOEC (7 d)
	24 mg l^{-1} NH$_3$	LOEC (7 d)
Oncorhynchus mykiss	500 mg l^{-1} NH$_3$	LC100 (96 h)
Embryos	> 200 mg l^{-1} NH$_3$	LC50 (96 h)
	200 mg l^{-1} NH$_3$	LOEC (96 h)
	100 mg l^{-1} NH$_3$	NOEC (96 h)
	> 400 mg l^{-1} NH$_3$	LC50 (48–96 h)
Danio raro	10 mg l^{-1} NH$_3$	Behaviour and body condition; NOEC (12 weeks)
	0.01 mg l^{-1} NH$_3$	Thyroid endpoints; LOEC (12 weeks)
	> 1000 mg l^{-1} Na	LC50 (96 h)
	11.48 mg l^{-1} Na	Behaviour and body condition; NOEC (12 weeks)
	0.01 mg l^{-1} Na	Thyroid endpoints; LOEC (12 weeks)
Amphibians		
Xenopus laevis	59–75 µg l^{-1} NH$_3$	Morphology; NOEC (70 d)
	5–18 µg l^{-1} NH$_3$	Morphology; LOEC (70 d)
	250–500 µg l^{-1} NH$_3$	Development; NOEC (21d)
	< 59 µg l^{-1} Na	Morphology; NOEC (70 d)
	5–18 µg l^{-1} Na	Morphology; LOEC (70 d)
	62.5–500 µg l^{-1} Na	Development; NOEC (21 d)
Xenopus tropicalis	> 1500 µg l^{-1} NH$_3$/Na	Metamorphosis and size; NOEC (40 weeks)
	170 µg l^{-1} NH$_3$/Na	Thyroid endpoints; NOEC (40 weeks)
	56 µg l^{-1} NH$_3$/Na	Vitellogenin production; NOEC (40 weeks)

these water systems were approximately double the minimum reporting level of 4 µg l^{-1}. Perchlorates interfere with thyroid function in humans and had previously been used as a treatment for hyperthyroidism in the 1950s. In adult humans, perchlorates affect iodine uptake and thus thyroid function and its regulation of metabolic processes through the production of hormones. In children, the interruption of thyroid function can affect normal growth and development. The maximum acceptable oral dose of perchlorates in the United States is 25 µg kg^{-1} of body weight, and the World Health Organisation's (2011) provisional maximum daily intake is less than 10 µg kg^{-1}.

Humans are also exposed to perchlorates through their food and this could either be due to them being a constituent of some fertilisers, their presence in soils and water supplies, or because they bioaccumulate in plant tissues. The EC concluded that more research is needed to identify the possible exposure routes of humans. Perchlorates have been found in high concentrations in cow's milk, up to 9 µgl^{-1} in China and 6.4 µgl^{-1} in Texas and at 10.5 µg l^{-1} in human breast milk. The presence of perchlorate in milk presumably occurs via consumption of plant tissues that have accumulated perchlorates. High levels of perchlorate, up to 350 µg kg^{-1} (and ranging from 4–110 µg kg^{-1}), have been detected in fruits and vegetables grown in Europe. Consequently in 2015 the EC recommended that perchlorates be monitored in foodstuffs.

With the concerns surrounding perchlorate pollution being relatively recent most existing ecotoxicological data are obtained from tests with freshwater taxa (Table 5.9) and the PNECs are determined from this data. The PNEC for freshwater taxa is 1 mg l^{-1} and for marine taxa is 0.1 mg l^{-1} for sodium perchlorate, and the PNEC of ammonium perchlorate for freshwater taxa is 1.2 mg l^{-1} and for marine 0.12 mg l^{-1}. Chronic exposure to perchlorates produces sub-lethal effects such as disrupting frog metamorphosis, normal thyroid functioning, impairing the mobility of *Daphnia*, the development of fish embryos and reproductive processes. Algal growth is inhibited by concentrations of between 12.5 and 500 mg l^{-1}. The toxicity of perchlorates is much higher for aquatic animals and amphibians experience changes in development and metamorphosis at µg l^{-1} concentrations. The toxicity of ammonium perchlorates to fish is generally ten-fold higher than for the sodium perchlorates (Table 5.9). Fish thyroid function is inhibited by as little as 0.01 mg l^{-1} of either perchlorate after 12 weeks exposure and amphibian morphological development is inhibited by 5 µg l^{-1} of ammonia perchlorate.

5.4.3 Pollution from hydraulic fracturing industries

High-volume hydraulic fracturing, *fracking*, is a technique for the extraction of fossil fuels, in the form of oil or gas, from buried shale and coal seams. Traditional oil wells drill boreholes into underground reservoirs containing oil and/or gas (Section 4.5). These reservoirs are relatively rare and there is a high likelihood that the oil/gas in such reservoirs will be exhausted in the not too distant future, and during this time the price will increase substantially. Oil and gas also occur within the matrix of rocks such as shale and tar sands; the oil and gas filling the spaces between the mineral particles in the rock. Extracting oil/gas from such deposits was previously too costly to be profitable, but with advances in technology and the rising oil/gas price the exploitation of these deposits is now an attractive prospect for the hydrocarbon industry. The fracking industry has developed quickly in the United States and, based on its apparent success, many other countries are now trying to exploit subsurface deposits in the same way.

The basic process of fracking involves drilling a borehole, both vertically and horizontally, into subsurface, organic-rich deposits and then pumping down a fracking fluid under high pressure that penetrates the fissures in the rock causing them to expand/shatter; that is, fracturing the rock matrix (Figure 5.2). Fracking creates voids and channels in the rock into which the oil and gas flows, and the fracking fluid/oil/gas mixture can then be pumped out. The fracking fluids comprise large volumes of water combined with added sand and chemicals that are injected into the well at high pressure. Like the 'mud' used to lubricate, cool and pressurise the drill heads in traditional oil exploration (Section 4.5.2), the fracking fluid is a complex solution and the composition varies between sites based, in part, on the geology of the rocks being drilled.

In theory, water under pressure could fracture the oil-bearing rocks but fracking fluids contain added chemicals for several purposes: to reduce the growth of limescale, to control bacterial growth, as gelling agents to transport the sand, acids to prevent blocking, clays to stabilise, detergents to help

Table 5.10 Toxic compounds used in offshore fracking in the USA (US House of Representatives Committee on Energy and Commerce 2011, Broderick et al. 2011). Regulation under EU and/or US law: W indicates regulated in water and A in air.

Compound	Effects	Regulation
Methanol	Affects swimming behaviour; hazardous air pollutant	W, A
Ethylene glycol	Toxic by ingestion, Hazardous air pollutant	W
Diesel fuel	Carcinogen; hazardous air pollutant	W, A
Naphthalene	Carcinogen; hazardous air pollutant, bioaccumulates	W, A
Xylene	Hazardous air pollutant, toxic by skin contact and ingestion, affects reproduction	W, A
Hydrochloric acid	Hazardous air pollutant	A
Toluene	Carcinogen, hazardous air pollutant	W, A
Ethyl Benzene	Carcinogen, hazardous air pollutant	W, A
Diethanolamine	Hazardous air pollutant	W, A
Formaldehyde	Carcinogen, mutagen, highly toxic to aquatic life; hazardous air pollutant	W, A
Sulphuric acid	Carcinogen	A
Thiourea	Carcinogen	
Benzyl chloride	Carcinogen; hazardous air pollutant	A
Dimethyl formamide	Hazardous air pollutant	A
Phenol	Hazardous air pollutant	A
Benzene	Carcinogen; causes leukaemia, hazardous air pollutant	W, A
Di(2-ethylhexyl) phthalate	Carcinogen, hazardous air pollutant	W, A
Acrylamide	Carcinogen	A, W
Hydrofluoric acid	Hazardous air pollutant	A
Acetaldehyde	Carcinogen, hazardous air pollutant	A
Acetophenone	Hazardous air pollutant	A
Phthalic anhydride	Hazardous air pollutant	A
Ethylene oxide	Carcinogen	A
Pb	Carcinogen, hazardous air pollutant	W, A

Compound	Effects	Regulation
Propylene oxide	Carcinogen	A
p-Xylene	Hazardous air pollutant	A
Nonyl phenol ethoxylate	Irritant, breaks down into EDCs, persistent	A, W
Methylisothiazolinone	Genotoxic and neurotoxic	
Cu	Highly toxic to aquatic organisms	W
Boron compounds	Growth and development	W
Crystalline silica quartz (X-cide)	Benthic biodiversity	
Glyoxal	Genotoxic, affects reproduction; negatively affects kidneys and pancreas	
Isotridecanol ethoxylated	Very toxic and bioaccumulates	
C12–14 ethoxylated alcohols	Bioaccumulates	
Trisodium nitrilotriacetate	Skin irritant, carcinogen	W
Glutaraldehyde	Irritant to skin, eyes, lungs	A, W
Dazomet	Toxic to aquatic organisms and mammals	W
Sodium tetraborate	Respiratory and skin irritant, reproductive effects	W, A

penetration, friction reducers to facilitate pumping and 'breakers' to facilitate backflow (the return of the product to the surface). Some wells also use chemical tracers to evaluate well performance and efficiency. In a 2011 review of the industry, the US Committee on Energy and Commerce found that US oil and gas producers used more than 2500 different fracking products containing 750 compounds and that around 3000 were used in a five-year period. These chemicals include relatively harmless compounds, such as citric acid, and many that are toxic to humans and wildlife, including copper, lead, benzene, naphthalene, diesel, methanol and formaldehyde (Table 5.10). The list of chemicals obtained from the US fracking companies included 29 regulated under the US Drinking Water Act. Some of the chemicals used in fracking are highly toxic, persistent and bioaccumulative (Section 2.2.3). These compounds include carcinogens, mutagens and endocrine disruptors (Section 5.2.4). The frack-ing water that returns to the surface contains a wide variety of these chemicals plus the target hydrocar-bons. It also contains the compounds that are mo-bilised from the rock as part of the fracking process including lead, aluminium, arsenic, and naturally radioactive substances.

The waste products produced by fracking thus seem to be more dangerous than those from con-ventional oil and gas drilling (Section 4.5). The threats to the environment and human health com-bined with strong international interest in exploit-ing fracking technology, that represent a quick fix to fossil fuel depletion, have placed fracking at the top of the environmental and public agenda.

The high toxicity of some fracking chemicals and the large volumes of waste-water produced (e.g. 8 585 t produced at Preese Hall in Lancashire during six operations) mean that many tonnes of fracking chemicals may be released from each op-eration. The Californian government has granted oil companies permission to dispose of more than 34 million tonnes of fracking waste-water per year (from 19 wells). Pollutants enter the environment via the same pathways as in conventional oil and gas production (e.g. spills, blow outs, leakage, flaring; Section 4.4) and, like traditional oil ex-ploration, the cumulative quantities entering the environment from ongoing operations will be con-siderable. Entry into the environment may be di-rect into seawater for offshore operations, into the atmosphere as a result of flaring and evaporation, as leachate into ground and surface waters, and into soil from onshore fracking operations or via the disposal of production waste waters. Fracking uses larger quantities of water compared with tra-ditional oil exploration. For this reason and because of the lower hydrocarbon yields requiring substan-tially more wells that are operated for longer, and so of the two fracking operations are considered to represent the greater pollution risk. The large volumes of contaminated production water must be stored and/or transported for disposal, and so they may be a source of air pollution because vola-tile compounds evaporate from the open storage lagoons. Another common practice is to re-inject this production water into the seafloor or under-ground once back on land which has the potential

to contaminate groundwater supplies or leach back out into seawater. The wells used for fracking are thought to be more prone to failure than traditional oil wells; for example, in the Gulf of Mexico one-third of offshore fracking-well casings experienced damage after five years of operation, increasing to half after 20 years. Wells containing contaminated fracking fluids remain a pollutant risk after pro-duction has ceased. In some cases the fluids may be separated at the surface and be reused after stabili-sation (Figure 5.2).

Thus far in the United States, there has been widespread air and water pollution associated with the fracking industry with rivers and streams being polluted (including drinking water reserves). Leg-islation and regulations surrounding the industry have been under review but have so far failed to protect the environment from the negative effects: growth of the industry has again exceeded the de-velopment of regulation. There is a risk that similar situations will occur in other countries seeking to develop the industry. For example, in Europe, al-though the REACH regulatory framework could be applied to fracking it is presently inadequate because, as already described, the process of frack-ing differs from traditional hydrocarbon drilling practices. Fracking uses large quantities of highly toxic compounds whose environmental fate is un-known because environmental impact assessments have not yet been completed. In 2014, the European Commission made recommendations regarding the environmental and human health threats from fracking although these are not binding. The im-pacts of fracking pollutants on human health have not yet been established but these risks are in need of a full assessment because many of the chemicals used are known to be highly toxic to wildlife and humans (Table 5.10).

5.5 Nanomaterials and nanoecotoxicology

5.5.1 What are nanomaterials?

Advancements in microscopy in the 1980s led to the discovery in 1985 of fullerenes: carbon molecules arranged in diverse structures, or allotropes, such as hollow spheres or tubes. Since then the rapidly advancing field of nanotechnology (manipulation

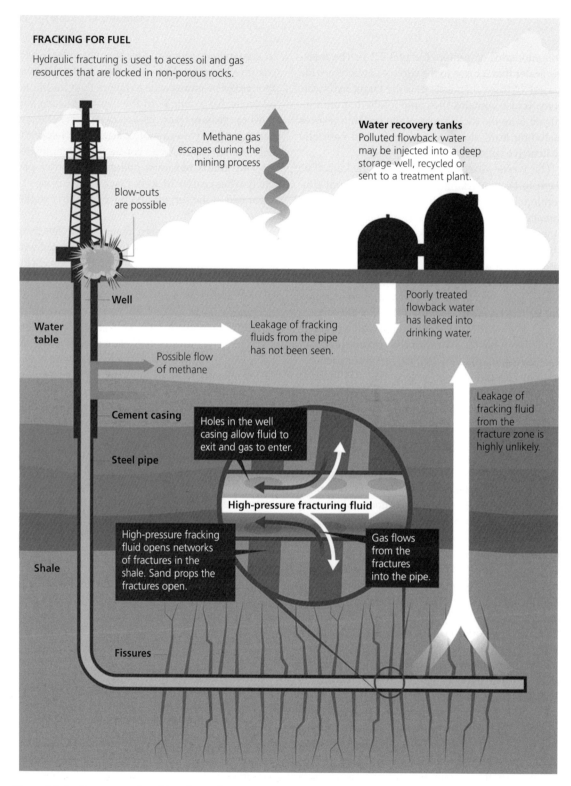

FRACKING FOR FUEL

Hydraulic fracturing is used to access oil and gas resources that are locked in non-porous rocks.

Methane gas escapes during the mining process

Blow-outs are possible

Water recovery tanks
Polluted flowback water may be injected into a deep storage well, recycled or sent to a treatment plant.

Well

Water table

Leakage of fracking fluids from the pipe has not been seen.

Poorly treated flowback water has leaked into drinking water.

Possible flow of methane

Cement casing

Holes in the well casing allow fluid to exit and gas to enter.

Leakage of fracking fluid from the fracture zone is highly unlikely.

Steel pipe

High-pressure fracturing fluid

High-pressure fracking fluid opens networks of fractures in the shale. Sand props the fractures open.

Gas flows from the fractures into the pipe.

Shale

Fissures

Figure 5.2 Fracking process and possible pathways of contamination. High-pressure water (often with a sand and a variety of chemicals added) is injected into gas-/oil-bearing rocks to fracture them and so release the valuable hydrocarbons. Reprinted by permission from Macmillan Publishers Ltd: [Nature] (Howarth, R.W., Ingraffea and Engelherder, T. 2011. Natural gas: Should fracking stop? Nature, 477, 217–275), copyright (2011). (http://www.nature.com).

of matter with at least one dimension between 1 nm and 100 nm; Figure 5.3) means an increasing number of nanomaterials (NMs) and the commercial products utilising them have become available. Nanoparticles (NPs) are now present in the environment. A nanoparticle is defined, by the American Society for Testing Materials (ASTM) and the International Standards Organisation (ISO) as a particle with lengths of between 1 nm and 100 nm in two or three dimensions. Particles in this size range have properties that are unique to their size because they behave differently to larger sized particles. The ASTM states that they may 'exhibit a size related intensive property'. The 1 nm lower limit is used to exclude individual or small groups of atoms. The European Scientific Committee on Emerging and Newly Identified Health Risks (SCENHIR) uses a more complex definition that subdivides NPs into 1–100 nm, 100–500 nm and more than 500 nm in size.

The unique properties of NPs have led to their increasing use in technology, medicine and commercial products (Table 5.11). At atomic or subatomic scales quantum effects manifest and the behaviour of matter and energy changes, meaning that NPs behave differently from larger particles of the same composition. NPs differ from their larger counterparts in transparency, chemical reactivity, electrical and thermal conductivity, colour, crystal structure, chemical reactivity and magnetism. Additionally, changes in the surface-area-to-volume ratio of NPs may change their electrical, optical and mechanical properties.

More than 1600 commercial products containing NPs are now available to consumers. The main

producers are the United States (46%), Europe (27%) and East Asia (17%). Most commercially available products that contain NPs are from the health and fitness, household, food and beverage and the automotive industries (in order of quantities produced). Estimates of the global annual production of NPs are between 7800 t y^{-1} and 38 000 t y^{-1} nano-titanium dioxide, 2.8–20 t y^{-1} nano-silver, 35–700 t y^{-1} nano-cerium dioxide, 55–1101 t y^{-1} carbon nanotubes and 2–80 t y^{-1} of fullerenes. The European Commission (2012) estimated the global market for NMs was 11 $\times 10^6$ t with a value of €20 $\times 10^9$.

NPs are used in a wide array of commercial products (Table 5.11). Zinc and titanium dioxide are used in cosmetics and PCPs, including titanium in toothpaste, moisturisers and shampoos, zinc and titanium NPs in sunscreen for their light-reflecting properties. NPs of silver are used in deodorants, PCPs, cleaning products, clothing and other household textiles for silver's antibacterial properties. Similarly, the food and beverage industry are employing NPs (e. g. zinc, silver, copper, silicon) for their antibacterial and antifungal activity in food packaging and as health supplements. Silica NPs confer water- or stain-repelling properties to clothing and other surfaces. They are used to strengthen materials used in vehicle construction (e.g. carbon nanotubes and silicon dioxide) and those with conductive properties that are used in a variety of electronic products (e.g. gold, silver, copper). They are also used in air purification technology (e.g. titanium, silver, gold, carbon), fuel cells, anti-caking agents, as catalysts, in lubricants and sealants. NPs are also increasingly being used for specialist medical applications (e.g. gold for targeted drug delivery, optical bio-imaging, laser phototherapy, genomics, immunoassays and within biosensors). NPs are also being developed to manipulate intracellular processes for the purposes of tissue engineering. Polystyrene NPs are also used within laboratory bioassays.

5.5.2 Occurrence of nanomaterials in the environment and their toxicity

Nanoparticles can occur naturally and estimates suggest that millions of tonnes are produced naturally each year. Carbon nanotubes, carbon fullerenes and silicon dioxide NPs have been found in

Table 5.11 Composition and number of commercially available nanoparticle products in 2013 representing 46% of all nanoparticle products (total = 1628). Data from: The Consumer Products Inventory (2015).

Nanomaterial	Products
Carbon	87
Silver	383
Titanium	179
Zinc	36
Silicon and silica	52
Gold	19

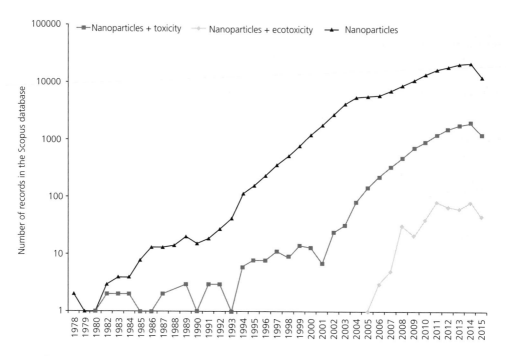

Figure 5.3 Number of records of peer-reviewed scientific publications in the Elsevier Science Scopus database (on 14 August 2015), spanning 1978–2015, containing the following term combinations in the article title, abstract, or keywords: 'nanoparticles', 'nanoparticles AND ecotoxicity' and 'nanoparticles AND toxicity'.

10 000-year-old ice cores, and NPs of iron and silicon have been found in sedimentary rocks from the Cretaceous–Tertiary boundary (around 75 million years ago). These natural NPs are formed by weathering, soil formation and vulcanism (see Handy et al. 2008 for more details). Some of these natural NPs have been shown to be toxic, for example, to mammals through atmospheric exposure. At present, the scale of industrial NP production seems to be less than the natural occurrences of NPs. However, industrial production is likely to increase, and manufactured NPs can differ in form (e.g. manufactured NPs are stabilised by the addition of organic matter and surfactants) and environmental persistence from those that occur naturally and so could differ in toxicity. Our knowledge of NP physicochemistry, environmental occurrence and fate, and their toxic effects are presently very limited.

Because NPs behave differently from their larger-sized counterparts, the environmental fate and ecotoxicology cannot be predicted from the known behaviour of the larger equivalents. Research on the physicochemical behaviour of NPs is thus needed

before their ecotoxicology can be better understood. Similarly, the methods for determination of the NP abundance within organisms and the environment are still being developed (e.g. test methodology, chemical characterisation and definition of laboratory reference materials). The environmental occurrence and ecotoxicology of NPs cannot be accurately assessed until these methods are developed further.

Owing to their small size, NPs may be taken up into cells and their uptake has been demonstrated in the human gut, the gut of *Daphnia* fish gill cells (Figure 5.4) and within the cells of mussels. Because NPs can accumulate within cells, they have the potential to biomagnify up food chains and represent a threat to human health (Section 2.3.3). NPs may also increase cell membrane permeability and so can actually enhance the uptake of NPs and other compounds. The likely distribution of NPs within the body is not known but will depend upon the solution chemistry of the different bodily fluids: development of methodologies for the determination of NP concentrations are needed before their distribution within the body can be accurately determined.

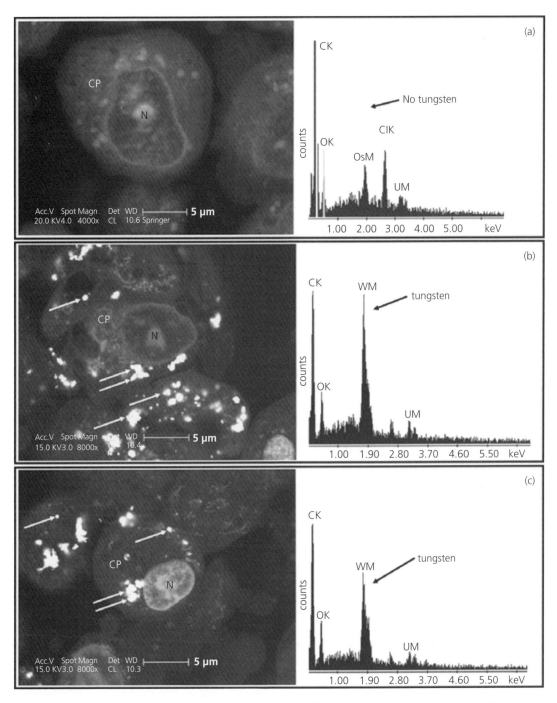

Figure 5.4 Evidence for the uptake of tungsten nanoparticles in rainbow trout gill cells. Back scatter electron detector images of (a) control cells with no tungsten NPs added (magnification × 4000), (b) 30 μg l⁻¹ tungsten NPs (magnification × 8000) and (c) 33 μg l⁻¹ tungsten NPs. Corresponding energy-dispersing x-ray spectroscopy data (right hand side) for the different elemental profiles. NPs are present in both the cytoplasm (CP) and as aggregates outside cells; none are present in the nuclei (N). Scale bar = 5 μm. Reprinted from *Aquatic Toxicology*, 93, Kühnel, D., Busch, W., Meißner, T., Springer, A., Potthoff, A., Richter, V., Gelinsky, M., Scholz, S., and Schirmer, K., Agglomeration of tungsten carbide nanoparticles in exposure medium does not prevent uptake and toxicity toward a rainbow trout gill cell line, 91–9, Copyright Elsevier (2009), with permission from Elsevier.

Recent reviews of the effect concentrations for NPs found that approximately 80 values have been derived to-date with a focus on titanium dioxide (31%), fullerenes (18%), zinc oxide (17%), silver (13%), single-walled carbon nanotubes and copper oxide (9% each) (see Kahru and Dubourguier 2010 for details). One-third of these tests were conducted on crustaceans, one-third on bacteria, 15% on algae and 15% on fish. Toxicity tests show that NPs are toxic to a broad range of taxa; for example, bacteria, algae, invertebrates, fish and mammals. Most ecotoxicological data are based on a narrow range of freshwater model species typically employed in regulatory tests (Section 2.4.3), such as the water flea, *Daphnia magna*, and fathead minnow, *Pimephales promelas*. The respiratory structures of mammals and fish are thought to be very sensitive to NPs including titanium oxide and single-walled carbon nanotubes. Chronic toxicity studies have been conducted with fathead minnow, Japanese medaka, rainbow trout, and zebrafish, with invertebrate tests focusing on *Daphnia* spp. and *Hyalella azteca* using mobility, feeding, moulting and fat metabolism as indicators of chronic toxicity. The impact of gold NPs on the infaunal marine bivalve *Scrobicularia plana* after 16 days exposure to 100 µgl^{-1} of gold NPs (of 5–40 nm size) included the impairment of burrowing behaviour, induced production of metallothioneins and biomarkers for oxidative stress (e.g. catalase, superoxide dismutase and glutathione-*S*-transferase; Section 2.4.4), and gold NPs were found to accumulate within tissues. A range of tests have also been completed with model bacterial species such as *Escherichia coli* and *Bacillus subtilis*, and for these species fullerenes have been found to slow growth at 0.5–3.0 mg l^{-1} and toxicity varied depending on whether the tests were run in a liquid suspension or within soil.

Although data are limited, it seems at present as though the most sensitive taxa to the NPs tested so far are algae and crustaceans. The acute toxicity of NPs is less than 0.1 mg l^{-1} for silver and zinc oxide, 0.1–1 mg l^{-1} for fullerene and copper oxide, 1–10 mg l^{-1} for single-walled carbon nanotubes and multi-walled carbon nanotubes, and 10–100 mg l^{-1} for titanium oxide. Chronic toxicity seems to occur in the µg l^{-1} range. It is thought that NPs may

aggregate and adsorb onto the bodies of organisms before being taken up into cells via a variety of possible mechanisms that have not yet been studied in detail. Several environmental factors may influence toxicity through their effects on NP aggregation chemistry; for example, pH, ionic strength, the presence of organic matter and water hardness (or alkalinity). The shape, size and surface area of NPs and their surface charge can also affect their aggregation chemistry and subsequent ecotoxicology. The high surface area and surface charge mean that NPs are able to act as contaminant carriers thus potentially increasing an organismse exposure to other toxicants. For example, the polycyclic aromatic hydrocarbon (PAH) phenanthrene adsorbs to NPs of fullerene, increasing its uptake, and the toxicity of this mixture can be ten times higher than phenanthrene alone.

5.5.3 The regulation of nanomaterials

The development of regulations for the introduction of NPs into the environment is fraught with problems concerning the methodology for effectively determining their fate within the environment and organisms, the shortage of data on NP production, the lack of knowledge on their natural occurrence, and ignorance of their ecotoxicology. Even the regulations for classifying wastes that are discharged into the environment, as hazardous or non-hazardous, do not accommodate the specific properties of NPs (which are very different from their bulk form; see Section 5.5.1) and so their hazardous properties may not be completely considered. There are presently no requirements for the labelling of commercial products that contain NPs, no guidance for the disposal of these products, and so they will most likely end up being disposed of with municipal wastes. It is almost impossible for legislators to manage the risks associated with nanomaterials when it is at present unquantifiable.

In 2004 the UK's Royal Society and Royal Academy of Engineering reviewed the risks associated with nanotechnology. The report included a series of recommendations: (i) that NMs be treated as new chemicals because of their novel properties, (ii) that industry and research treat NMs as hazardous until known otherwise, (iii) that release into the environment be avoided if possible, and (iv) that products containing NPs undergo new regulatory testing before commercial release. A regulatory review by the European Commission (2012) has since produced a strategy and code of conduct, for responsible nanosciences and nanotechnologies research, and has conducted a number of reviews of the policies applying to NMs. The European Commission concluded that NMs should be regulated within the REACH framework (Chapter 7) and under the regulations for the Classification Labelling and Packaging (CLP) of substances and mixtures. It was also noted that modifications to the existing annexes be made specifically regarding NMs. REACH legislation applies to the manufacture, marketing and use of substances alone or in preparations whereby substances manufactured in exceeding a tonne be registered, and information should be provided on the supply chain. However, the mass threshold used to trigger regulation such as this are not appropriate for NPs, which are of very low mass and unlikely to exceed this threshold; NMs which fulfil the criteria of being hazardous under EC regulations must be classified and labelled. The development of EU regulations to cover the production and disposal of NMs is still in its formative stages due to the many knowledge gaps. The Scientific Committee on Emerging and Newly Identified Health Risks provides expert opinions to the European Commission on the effects of novel compounds such as NMs.

The United States FDA recognises the unique properties of NMs, the need to identify sources, environmental mobility, and the risks to human health, animals and the environment. The FDA claims that NMs will be regulated under the existing Toxic Substances Control Act that regulates the manufacturing but, as seen from this discussion of the existing EU legislation, it is not clear that the toxic substances control act will provide adequate regulation for these unique materials.

5.6 Light and electromagnetic radiation

The term *electromagnetic (EM) radiation* covers, literally, the whole spectrum of wave energies from high-energy gamma rays, X-rays, through the ultraviolet, visible light, infrared, microwaves, to low-energy radio waves. Physicists distinguish between the higher energy types of electromagnetic radiation as *ionising radiation*, where the energy is high enough to cause material to ionise and break chemical bonds. Ionising forms of radiation include gamma rays produced by some types of radioactive decay, X-rays and the higher energy part of the ultra-violet portion of the spectrum. The effects of ionising radiation as pollutant are discussed in Section 4.6. In this section, we consider the non-ionising parts of the EM spectrum.

The lower energy (from 10 eV to 3 eV; eV stands for electronvolt and is a unit of energy, equivalent to that gained by moving a single electron across a potential difference of 1 volt) is often referred to as soft UV or simply UV light, having wavelengths between 100 nm and 380 nm. While non-ionising, this radiation is energetic enough to cause damage to biological material, in humans we most commonly experience this as sunburn. However, in the sea the water quickly absorbs the UV radiation and so any effects are limited to the organisms present in surface waters or to those that spend some time exposed above the water's surface. Although the level of UV impinging on the planet has not altered, the atmospheric composition has changed due to human activities, allowing more UV to reach the sea surface. The most dramatic example of which was the seasonal ozone hole over Antarctica identified in the 1980s and the thinning of the ozone layer over the whole planet (Section 4.7).

A number of whale species undertake large-scale migrations, feeding in the productive areas around Antarctica but migrating to tropical or sub-tropical areas to breed. Off the coast of California, recent surveys have shown that the levels of skin damage, from sunburn, have increased in recent years. The proportion of blue whales with visible damage increased by 56% between 2007 and 2009. The whales seem to respond to this by developing a 'tan' and individual blue whales have developed more pigmentation and so become darker in colour as the season progresses.

Laboratory studies have also shown that microscopic phytoplankton suffer cell damage and genetic mutations when exposed to UV at levels experienced by organisms under the ozone hole. This has raised concerns about damage to the organisms that form the base of the productive Antarctic food web. These laboratory studies only use an approximation of the situation in nature. In the actual ocean, the algae would be continually circulating by the waves and currents so that they would not remain in the region of high UV for prolonged periods.

Visible light is not energetic enough to cause cell damage but it has been identified as a marine pollutant of growing concern. Artificial lighting has been revolutionary for human societies, extending our active day, helping to reduce crime and increasing road safety, for example. However, our cities and roadways are clearly visible from space on the night-time side of the planet, mapped out by a web of lights (Figure 5.5a). The effect of artificial light on birds and nocturnal animals in terrestrial systems

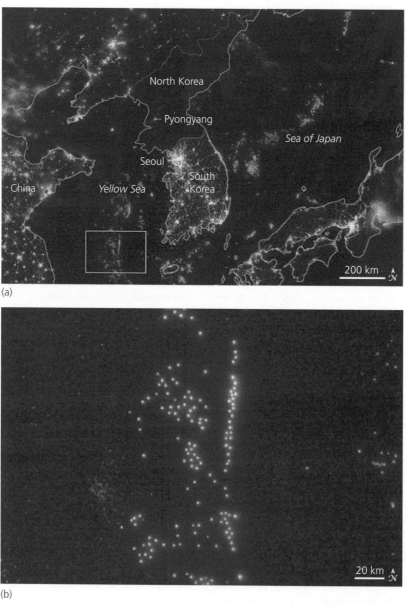

(a)

(b)

Figure 5.5 (a) Satellite image showing the Yellow Sea and the Sea of Japan taken on 24 September 2012. The lights of Tokyo in Japan and Seoul in South Korea and other cities are clearly visible but the most striking thing is that the lights of the Korean and Japanese squid-fishing vessels (enclosed by the inset box) in the Sea of Japan are of a similar magnitude to the light emitted by these two mega-cities. The enlargement (b) of the lights in the Yellow Sea show individual vessels of the Chinese fleet aligned along the boundary between the jurisdictions of China and Korea. Credit: National Aeronautic and Space Administration.

has long been recognised as one of the impacts of urbanisation on the natural world, but compared to the felling of forests and the concreting over of natural habitats, the effects of light are often seen as minor. In the marine environment neither lights nor concreting over of habitats is as widespread, but in certain circumstances light pollution is ecologically significant.

Coastlines in many parts of the world are highly urbanised and the United Nations predicts that the number of people living at the coast will grow faster than the general rate of population increase in the next 50 years. This will mean even greater urbanisation of the coastal strip.

Light pollution from coastal roads and hotels has been shown to affect the breeding of turtles. They have excellent navigational abilities and generally return to breed at the site where they were hatched. However, as long-lived organisms, they return to breed for the first time many years (it varies between species) after they hatched. As the turtles approach the breeding beach, high levels of light will discourage the females from coming ashore to lay their eggs. This causes them to delay laying or seek alternative, less suitable, locations.

A few weeks after laying, the young hatch on the night of a full moon. They dig themselves out of the nest and naturally head towards the brightest horizon: in nature this would be the sea reflecting the full moon. However, if there are road lights or illuminated buildings at the top of the shore the young turtles head up the beach instead of down to the sea. These hatchlings are either predated or they die from overheating/exhaustion when the sun comes up.

A more obscure form of light pollution is from fishing vessels using lights to attract squid. Most ships at sea have lights at night but they are widely dispersed across the ocean and the effect of the lights is negligible. However, in some parts of the world fishing boats use lights hung over the sea at night to attract their targets. In the South China Sea and the Sea of Japan, Korean, Japanese and Chinese fishing vessels use powerful lights shone on to the sea surface at night to attract squid to the lures. The number of vessels and the light emitted from the powerful halogen lights means that these fleets are visible from space,

appearing as bright as some of the major cities of the region (Figure 5.5b). The changes in the underwater light regime have been sufficient to alter the vertical migration pattern of zooplankton in the area. Many species of zooplankton normally spend the daylight period at depth, avoiding visual predators, but they migrate into surface waters at night to feed on the phytoplankton. Where the fleet is fishing, the high light levels suppress this migration and the zooplankton remain in the food-poor, deep waters all day and all night. The implications of this on the nutrition and productivity of zooplankton and consequently on higher trophic levels are not yet clear.

5.7 Noise and vibration

When underwater explorer and inventor of the aqualung, Jacques Cousteau, published his autobiography he called it (and the accompanying movie) the *Silent World*. However, anyone who has ever snorkelled or dived on a coral reef will have immediately noticed the clicking, crunching and scraping sounds that provide a continual accompaniment to their explorations. Water transmits vibrations/ sounds more rapidly than air and the density of the medium means that low-frequency sounds can be transmitted over extremely long distances. This difference in the speed of sound in water makes it very difficult for human divers to locate the source of sounds or to interpret them correctly. However, some species of whale take advantage of this long-range transmission and the natural sound channels, set up by layers of water of a particular density, to communicate over extremely long (ocean-scale) distances. Toothed whales emit sounds to source their prey and many whales seem to use sound to navigate.

Fish and many invertebrates also use the propagation of vibrations as a sensory mechanism. In the case of fish, they use their lateral lines to sense vibrations, the approach of a predator, and, potentially, to communicate with other members of a shoal when swimming.

The noise and vibrations arising from human activities therefore have the potential to interfere with these natural uses of vibration/sound by marine organisms. For example, juvenile European eels have been shown to exhibit physiological signs of stress (e.g. increased ventilation rate) when played recordings of ships moving in a harbour compared to those individuals who were played sounds from the harbour when the ships were all stationary. Individuals exposed to the sounds of shipping were also less likely to respond to an ambush predator and were more rapidly caught by pursuit predators than in the control treatments. These effects were apparent in both laboratory experiments and field trials suggesting that anthropogenic noise was affecting both the physiology and behaviour of these eels in ways that could reduce their survival and fitness.

5.7.1 Shipping and construction

Shockwaves, a type of vibration, from underwater explosions are a directly damaging class of noise pollution. The shockwave spreads out from the centre of the explosion and any organism with gas voids in its body or vibration sensory organs will suffer damage. This therefore includes all air-breathing species: marine mammals, birds, reptiles, fish with gas bladders and invertebrates such as siphonophores with gas-filled floats and some species of cephalopod and crustacea and other invertebrates that have vibration receptors such as statocysts. Underwater explosions are generally

due to military activity and construction work, but many are also due to being used as sound sources for seismic surveys.

These surveys are a series of techniques used by geologists to examine the subsurface geological structure. They are important for the mapping of geological structures, in understanding tectonics (and hence earthquake and volcanic-eruption risk analysis and prediction), and in explorations for hydrocarbon reserves. A low-frequency sound is produced and a series of microphones listen for the echoes reflected back by sediments/rocks of different densities, in a manner analogous to sonar imaging of the seafloor (Figure 5.6).

Far and away the most pervasive source of noise/vibration in the sea is shipping noise. Until 150 years ago, ships moved almost silently across the sea surface driven by the wind, and then in the space of a few decades shipping became powered, driven by steam and then diesel engines. The engines drive propellers or, in some fast vessels, water powers jet drives, so there is both the noise of the engines themselves and also the noise and cavitation (generation of bubbles) of the water from the propulsion device. In the 1940s, ships started to emit pulses of sound, first as a military technology to detect enemy submarines, then subsequently by virtually every seagoing vessel to examine the depth of the water and so avoid running aground. There are no records

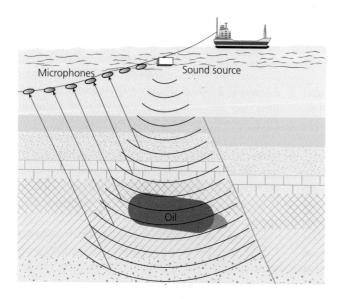

Figure 5.6 Seismic surveys use low-frequency sound waves to image the subsurface geology. A sound source may be placed on the seabed and microphones positioned at varying distances from it, or an array may be used that includes the sound source, and the detectors may be towed behind a survey vessel. Differences in rock density and discontinuities affect the speed of propagation of the sound and the nature of its reflection.

of how vibration-sensitive species behaved before shipping became so noisy and so it is impossible to assess whether there has been any change in their behaviour. It is quite likely that higher vertebrates, marine mammals, reptiles, and birds have become habituated to the sounds from shipping in the same way that terrestrial birds, mammals and reptiles are habituated to the noise from motorways carving through their habitat. In 2004, the International Whaling Commission's Scientific Committee (International Whaling Commission 2014) concluded that low-frequency (less than 1000 Hz) ambient noise levels had increased by two orders of magnitude over the last 60 years. This could be 3–5 decibels (dBs) per decade, thereby significantly reducing the potential for long-range communication between large baleen whales.

5.7.2 Offshore renewable energy

The latter part of the twentieth century saw a rapid expansion in the use of wind turbines to generate low carbon electricity. Initially these were sited on land but the advantages of turbines sited in the sea were quickly recognised. The wind over the sea is stronger, less turbulent and more predictable than that over land and larger turbines can be built offshore than would be permitted onshore where their presence would impinge on the natural visual appeal of the landscape. As the wind turns the turbines, some of the energy is converted into sound, some of which is transmitted down the supporting column(s) and emitted as vibrations underwater. When offshore windfarms were first proposed opponents claimed that they would interfere with bird flight paths and generate noise underwater that would drive fish and marine mammals away. To-date, no studies have demonstrated either of these impacts at operational windfarm sites.

Accompanying the continuing pressure to generate low-carbon energy, interest is also growing, in parts of the world with strong tidal currents, in deploying tidal energy devices. These technologies are at the trial stage at present and a variety of designs are being investigated, but many of the designs involve turbine blades rotating underwater and these will emit vibration/noise. Again, the opponents of developments claim these will have impacts on fish and marine mammals (and possibly reptiles and birds), while others suggest that the rotation speeds will be so slow compared to ships' propellers that the vibrations emitted, if any, will be quickly dampened and the biota will acclimatise. Studies to test these assertions continue. However, given the global need for low-carbon energy, it is difficult to envisage a situation where noise/vibration concerns would halt development of offshore wind or tidal turbine farms. They may exclude developments from particularly sensitive areas, or particular turbine designs may be required to mitigate underwater noise.

5.7.3 Military activities

An increasing number of mass strandings of cetaceans around the world have been linked to military exercises occurring nearby. The evidence is circumstantial but the number of cases reported is growing. The argument is that modern military *active*-sonar systems (i.e. those that emit pulses of sound rather than just listening for enemy vessels) use much higher power than earlier systems and, perhaps more importantly, are now utilising the frequency ranges used by marine mammals. The laws of physics dictate that frequencies that are most efficient for the military are also those that cetaceans have evolved to use.

One of the challenges of understanding the noise pollution produced by military activities is that they are usually exempt from the controls and reporting requirements applied to civilian activities. The US Navy, for example, recognises the impact of its use of mid-frequency active sonar and in recent years has reduced its use in training exercises (it would not, of course, agree to restrict its use in a combat situation). In spite of this, the US Navy expects its activities each year to change the behaviour of up to 1.78 million marine mammals. This includes interfering with their migration routes, frequency of surfacing and behaviours such as nursing, breeding, feeding, or sheltering. It estimates that up to 26 animals might suffer direct mortality, while a further 336 could sustain injuries such as temporary or permanent hearing loss or a decompression sickness like condition known as *bubble lesions*.

While the Scientific Committee of the International Whaling Commission recognised in 2004 that not all mass strandings were associated with military activity, they did raise concerns about the increasing number of stranding events since the introduction of mid-frequency military sonar in the 1960s. A tabulation of strandings from all the oceans, and both the northern and southern hemispheres where military activity was known to have occurred. This data also highlighted evidence showing that sound emitted from the air guns used in seismic surveys impacts the behaviour of some cetaceans.

Following the stranding of approximately 100 melon-headed whales in Madagascar in 2008, an independent international scientific team concluded that the stranding was most probably the result of acoustic stimuli from a multi-beam echo sounder system that was being used nearby for seabed mapping by a contractor for an international oil company. This means that mass strandings of whales have now been directly attributed, with reasonable certainty, to military activity, seismic exploration and seabed-habitat mapping.

5.8 Brine

Brine is the term used to describe solutions with a high salt content, typically more than 5% (i.e. 5 g salt in 100 g water, giving a salinity in excess of 50 compared to typical ocean water of 35). Brines are produced by some offshore oil/gas operations (when salts are flushed from the geological structures containing oil), by some industrial plants, from some mines and desalination plants.

Given the projected human population growth in tropical coastal regions, concerns about changing weather patterns and the demand for freshwater for agriculture (to feed the growing human population), there is expected to be considerable growth over the next decade in the number of desalination plants operating in the Middle East, Mediterranean, Australia, the Indian sub-continent, the United States, and parts of China. In addition to the elevated salinity, brine from desalination plants is often heated (Section 3.7). Brine discharges would be regulated, as would any other point-source discharges from an industrial plant.

Putting some more salty water into an ocean full of seawater might initially not seem like a problem, but brine is so dense that it sinks as a mass to the seafloor. The brine will flow away from the discharge point as a density current, flowing down the slope of the seabed and accumulating in any hollows or depressions. As it flows, the brine mixes with the surrounding seawater and gradually loses its distinctly saline nature. The faster it flows, the more turbulence, and hence mixing, is created. In areas of high wave action or strong tidal currents, the additional turbulent mixing further speeds up the dispersion and assimilation of the excess salt from the brine.

As the brine, which is usually low in dissolved oxygen, flows over the seabed it will exclude fish and other mobile animals that usually avoid low-oxygen conditions. The combination of osmotic stress and hypoxia can cause the mortality of plankton and benthos that are not able to actively move away from the brine. A hot-brine discharge will exacerbate these effects by raising the organism's metabolic rates and having a lower capacity to hold dissolved oxygen. Any dead organisms then decompose, causing further hypoxia and potentially liberating toxic compounds such as hydrogen sulphide.

In high-energy areas, waves and tidal currents will disperse the brine and the affected area will be small. This can be reduced further using diffuser heads or similar mixing devices on the brine discharge pipes to promote mixing/dilution of the brine and adding more oxygen (Figure 5.7). While a hot effluent may potentially have greater biological effects, a warm effluent is less dense than a cold one and so tends to rise into the water column and so enters the more turbulent mixed near surface layers.

5.9 Synthesis

- The Pharmaceutical and Personal Care Products (PPCPs) that have become a part of our daily lives are big business, their production is increasing, and more novel compounds appear every year. PPCPs are ubiquitous, many are toxic and the discharges are nearly continuous, and so they represent a considerable threat to aquatic organisms and human health.
- There are very little data on the quantities of PPCPs discharged into the environment partly due

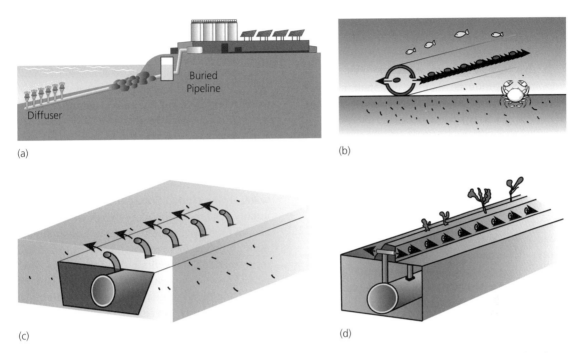

Figure 5.7 (a) Typical layout of a sub-sea brine discharge pipe with a diffuser head and possible options for the multi-port diffusers (b) on the seabed with port holes, (c) buried in a trench with a short riser, and (d) a deep tunnel with long risers.

to the lack of public disclosure by the industry and partly because there is no routine monitoring; for example, of PPCPs in sewage. This quantification is complicated by the need for advancements in analytical methodology. The application of animal and human sewage to crops represents a threat to human health because these crops can take up the PPCPs and veterinary compounds from the sewage. There is evidence that some bioaccumulate in marine animals and that some are EDCs. Other consequences of these environmental discharges include the development of antibiotic resistance in natural bacterial populations, attributed to the excessive use of antibiotics in medical and veterinary practice, and the ultimate consequences of these changes for human health are not yet apparent.

• In addition to lacking data on the inputs and quantities of PPCPs present in the environment, toxicity data are available for less than 1% of the APIs used. Most of the existing data are for acute exposures of freshwater algae and zooplankton,

and there is little information on the chronic effects on marine animals or humans. Many PPCPs are probably present in the environment at concentrations that will exhibit chronic affects. Improvements in understanding the ecotoxicity of PPCPs also requires the development of methodologies to detect chronic effects such as reproductive, neurological and immunological impacts.

• The large quantities of PPCPs introduced into the environment, coupled with their potency, means that they pose considerable risks to ecosystems and humans. However, the current legislation, what little exists, has many flaws and is not fit for purpose. For example, in Europe, where there exists probably some of the most advanced environmental legislation in the world, there are a number of problems, such as:

1) Old medicines (pre-1998) are not regulated, although they include those used in the greatest quantities;

2) For new human medicines, authorisation cannot be refused on environmental grounds (although it can be for veterinary products);

3) This legislation only applies to the marketing of products; that is, not their production, which is covered by different legislation which is not specific to PPCPs.

- It is understandable that the production of important medicines should be a priority, but the existing European Commission (EC) legislation neglects the risks to human health from environmental exposure to PPCPs, their impacts on the environment and the plants and animals on which we depend for important ecosystem services. In Europe, PPCPs are not incorporated into any of the key environmental legislation (e.g. the Water Framework, Drinking Water, Ground Water, Urban Wastewater or Sewage Sludge directives). They are not regulated under REACH and nor are they monitored in European waters.

- Although they are not new, endocrine-disrupting compounds (EDCs) are a considerable risk to the long-term viability of human and animal populations (Section 5.3). Legislation and regulations have proven effective in reducing their impact in the past, and the affected species have recovered, despite this having taken many years (Section 3.4). The risks from endocrine disruptors may only just be starting to emerge: there are more than 800 suspected EDCs and in many cases they have not been sufficiently tested and our knowledge of the risks and their modes of action are deficient. Similar to that concerning PPCPs, the development of a toxicological understanding requires methodological developments for detection in the environment and assessing their toxic effects. Our experiences with a number of banned EDCs, some of the most toxic to date (e.g. DDT, TBT and PCBs), that remain as a legacy in the environment and animals shows that EDCs should remain a top priority.

- The by-products from newly emerging methods of drinking water disinfection (DBPs), using chloramine, are carcinogenic and are up to 250 times more toxic than the residues from the previously used technology (Section 5.4.1). These compounds are a problem in the United States where disinfection is used as part of sewage and drinking water treatments. The ecotoxicology and regulation of the new DBPs is similar to the other emerging pollutants, it is lagging behind that of their production/use. Disinfection is likely to be used in larger quantities as demand for drinking water increases with a growing world population, as will the acquisition of clean drinking water through desalination (Section 5.8). Recent improvements in analytical precision led to perchlorates being found in drinking water in the United States at double the water-quality guidance limit; they exceeded the World Health Organisation daily limit in human breast milk and were 35 times higher in fruit than the recommended daily limit (Section 5.4.2). They persist in the environment for (more than 10 years) and bioaccumulate in plants and animals. They affect early development in animals and are EDCs. As with other emerging contaminants, their occurrence in marine ecosystems is poorly documented.

- Continued demand for fossil fuels has meant increasing interest in alternative sources of hydrocarbons such as fracking (Section 5.4.3). During the process of fracking, large quantities of wastewater are produced that are contaminated with hydrocarbons and heavy metals and also the chemicals used to facilitate hydrocarbon extraction. Many different chemicals are used including 38 that are regulated under US and/or EC drinking-water guidelines. These are highly toxic compounds that persistent in the environment and bioaccumulate. Due to the amounts of fracking fluids produced and the higher incidence of well failures, the pollution risks are at present thought to be greater than for traditional hydrocarbon exploitation. Effective regulation has not yet been established for fracking. It is still being developed in the European Union but differences from traditional practices may mean existing legislation (e.g. REACH; Chapter 7) is unsuitable.

- Technological advancements have seen a diversity of nanoscale compounds (approximately 1600 at the time of writing) being released onto the market, with 11×10^6 t in 2014 being used in PCPs, textiles, food packaging and for medical and industrial applications (Section 5.5). Differences of scale mean NPs have unique properties and unpredictable behaviour. Of all the emerging contaminants, these are the least well-known or understood, even in terms of much of their basic chemistry. Regulators in Europe and the United States have recognised that much of the existing legislation does not or cannot be applied to NMs and it needs redefining. Many challenges exist, beginning with the lack of knowledge on their natural occurrences, and this cannot be determined until suitable detection methods are developed. Knowledge and understanding of NMs' basic chemical properties and behaviour mean interpretations of ecotoxicological assessments are unreliable. There is a shortage of toxicological data, but we do know they are taken up into animal cells and are toxic to a range of taxa. Many of the qualities used to traditionally classify toxicity do not apply to NMs or they are unknown because of their unique quantum properties. Both the United States FDA and European Commission claim that NMs will be regulated using existing legislation, but this does not seem to be adequate (Section 5.5). The Royal Society and Royal Academy of Engineering in the United Kingdom (2004) recommended that the use and release of NMs be restricted, using a precautionary principle until more information is available.

- A number of diffuse contaminants are emerging including light, noise and brine (Sections 5.6–5.8). These contaminants can have a significant effect on the reproduction, navigation and stimuli for migratory behaviour in a range of taxa (from zooplankton to mammals). At its most extreme, noise pollution can cause mass mortality of cetaceans (e.g. during underwater explosions or military activities) or chronic conditions such as hearing loss or lesions. The amounts of these diffuse pollutants and their ecological effects are still being investigated. The regulation of noise is fragmented; engine noise and the emissions of electronic sounders are regulated under the specifications of the vessels, offshore operations (seismic, drilling and energy devices) are generally controlled via an environmental assessment and permitting/ planning approval process.

Resources

Further information on novel compounds

American Chemicals Society (2016). The Chemical Abstracts Service registry. http://www.cas.org.

Environmental Protection Agency USA (2015). Pharmaceutical and Personal care Products (PPCPs). http://www.epa.gov/ppcp/.

Environmental Protection Agency USA (2016). The Toxic Substances Control Act inventory. https://www.epa.gov/tsca-inventory.

Consumer Products Inventory. (2015). The Project on Emerging Nanotechnologies http://www.nanotechproject.org/cpi/.

Scientific Committee on Emerging and Newly Identified Health Risks (2017). http://ec.europa.eu/health/scientific_committees/emerging/index_en.htm.

US Environmental Protection Agency (2015). Control of Nanoscale Materials under the Toxic Substances Control Act. http://www.epa.gov/oppt/nano/.

Extraction of oil and gas by fracking

European Commission (2015). Environmental Aspects of Unconventional Fossil Fuel Extraction. http://ec.europa.eu/environment/integration/energy/uff_studies_en.htm.

European Commission (2017). Research in nanosciences and technologies. http://ec.europa.eu/research/industrial_technologies/nanoscience-and-technologies_en.html

Light pollution and turtle nesting issues

Sea Turtle Conservancy (2016) Information about Sea Turtles: Threats from Artificial Lighting. http://www.conserveturtles.org/seaturtleinformation.php?page=lighting.

Military technology and marine mammals

Natural Resources Defence Council (2016). US Navy implicated in new mass stranding of whales.http://switchboard.nrdc.org/blogs/mjasny/us_navy_implicated_in_new_mass.html.

Bibliography

AMEC Environment and Infrastructure UK Limited (2014). Technical support for assessing the need for risk management framework for unconventional gas extraction. Final Report to European Commission DG Environment. European Commission, Brussels.

Anderson, A. (2005). *Making Medicines: A Brief History of Pharmacy and Pharmaceuticals*. London: Pharmaceutical Press.

Bergman, A., Heindel, J. J., Jobling, S., Kidd, K. A., and Zoeller, T. (2012). State of the Science of Endocrine Disrupting Chemicals 2012. Geneva: United Nations Environment Programme and World Health Organisation Press.

Bisarya, S.C. and Patil, D.M. (1993). Determination of salicylic acid and phenol (ppm level) in effluent from aspirin plant. *Research and Industry*, 38, 170–2.

Blackwell, P.A., Kay, P. and Boxall, A.B.A. (2007). The dissipation and transport of veterinary antibiotics in a sandy loam soil. *Chemosphere*, 67, 292–9.

Boxall A.B.A., Fogg L.A., Kay P., Blackwell P.A., Pemberton E.J. and Croxford A. (2003). Veterinary medicines in the environment. *Reviews in Environmental Contamination and Toxicology*, 180, 1–91.

Broderick, J., Anderson, K., Wood, R., Gilbert, P., Sharmina, M., Footitt, A., Glynn, S. and Nicholls, F. (2011). Shale Gas: An Updated Assessment of Environmental and Shale Gas Impacts. Report for the Co-operative. University of Manchester: Tyndall Centre for Climate Change Research.

Committee for Medicinal Products for Human Use (CHMP) (2006). Guideline on the Environmental Risk Assessment of Medicinal Products for Human Use. London: European Medicines Agency.

Cortez, F.S., Pereira, C.D.S., Santos, A.R., Cesar, A., Choueri, R.B., de Assis Martini, G. and Bohrer-Morel, M.B. (2012). Biological effects of environmentally relevant concentrations of the pharmaceutical Triclosan in the marine mussel *Perna perna* (Linnaeus, 1758). *Environmental Pollution*, 168, 145–50.

European Commission (2012). Communication from the commission to the European parliament, the council and the European and economic social committee. Second Regulatory Review on Nanomaterials. European Commission, Brussels.

European Union (2016). European Chemicals Agency. https://www.echa.europa.eu/en/web/guest/information-on-chemicals.Grote, M., Schwake-Anduschus, C., Michel, R., Stevens, H., Heyser, W., Langenkämper, G., Betsche, T. and Freitag, M. (2007). Incorporation of veterinary antibiotics into crops from manured soil. *Landbauforschung Volkenrode*, 57, 25–32.

Hagenbach, I.M. and Pinckney, J.L. (2012). Toxic effect of the combined antibiotics ciprofloxacin, lincomycin, and tylosin on two species of marine diatoms. *Water Research*, 46, 5028–36.

Halling-Sørensen, B., Nors Nielsen, S., Lanzky, P.F., Ingerslev, F., Holten Lützhøft, H.C. and Jørgensen, S.E. (1998). Occurrence, fate and effects of pharmaceutical substances in the environment –A review. *Chemosphere*, 36, 357–93.

Hamscher, G., Sczesny, S., Abu-Qare, A., Hoper, H. and Nau, H. (2000). Substances with pharmacological effects including hormonally active substances in the environment: identification of tetracyclines in soil fertilized with animal slurry. *Deut Tierarztl Wocj*, 107, 332–4.

Handy, R.D., Owen, R. and Vasalmi-Jones, E. (2008). The ecotoxicology of nanoparticles and nanomaterials: current status, knowledge gaps, challenges, and future needs. *Ecotoxicology*, 17, 315–25.

Hendren, C.O., Mesnard, X., Dröge, J. and Wiesner, M.R. (2011). Estimating production data for five engineered nanomaterials as a basis for exposure assessment. *Environmental Science and Technology*, 45, 2562–569.

IMS Health Incorporated (2015). Top Line Market Data. Global Prescription Sales Information. http://www.imshealth.com/en/about-us/news/top-line-market-data.

International Whaling Commission (2014). Scientific Committee (IWC-SC) Report Annex K: Report of the Standing Working Group on Environmental Concerns Submitted at the IWC56 meeting, July 2004. International Whaling Commission, Cambridge, UK. http://www.acousticecology.org/docs/IWC56-noisesymposium.doc

Kahru A. and Dubourguier, H.-C. (2010). From ecotoxicology to nanoecotoxicology. *Toxicology*, 269, 105–119.

Kegley, S. E., Hill, B. R., Orme, S. and Choi, A.H. (2016). PAN Pesticide database. Pesticide Action Network, North America. http://www.pesticideinfo.org/Search_Ecotoxicity.jsp.

Kühnel, D., Busch, W., Meibner, T., Springer, A., Potthoff, A., Richter, V., Gelinsky, M., Scholz, S. and Schirmer, K. (2009). Agglomeration of tungsten carbide nanoparticles in exposure medium does not prevent uptake and toxicity toward a rainbow trout gill cell line. *Aquatic Toxicology*, 93, 91–9.

Kummerer, K. (2008). *Pharmaceuticals in the Environment: Sources, Fate, Effects and Risks*. 3rd edn, Berlin: Springer-Verlag.

Larsson D.G., de Pedro C. and Paxeus N. (2007). Effluent from drug manufactures contains extremely high levels of pharmaceuticals. *Journal of Hazardous Materials*, 148, 751–5.

Martinez-Levasseur, L.M., Birch-Machin, M.A., Bowman, A., Gendron, D., Weatherhead, E., Knell, R.J. and Acevedo-Whitehouse, K. (2013). Whales use distinct strategies to counteract solar ultraviolet radiation. *Nature Scientific Reports*, 3, 1–6.

Oaks, J.L., Gilbert, M. Virani, M.Z., Watson, R.T., Meteyer, C.U., Rideout, B.A., Shivaprasad, H.L., Ahmed, S., Chaudry,M.J.,Arshad,M.,Mahmood,S.,Ali,A.andKhan,A.A. (2004). Diclofenac residues as the cause of vulture population decline in Pakistan. *Nature*, 427, 630–3.

Parsons, E.C.M., Dolman, S.J., Wright, A.J., Rose, N.A. and Burns, W.C.G. (2008). Navy sonar and cetaceans: Just how much does the gun need to smoke before we act? *Marine Pollution Bulletin*, 56, 1248–257.

The Royal Society and Academy of Engineering (2004). Nanoscience and nanotechnologies: opportunities and uncertainties. The Royal Society, London, UK.

Sarmah, A.K., Meyer, M.T. and Boxall, A.B.A. (2006). Veterinary drugs. A global perspective on the use, sales, exposure pathways, occurrence, fate and effects of veterinary antibiotics (VAs) in the environment. *Chemosphere*, 65, 725–59.

Simpson, S.D., Purser, J. and Radford, A.N. (2014). Anthropogenic noise compromises antipredator behaviour in European eels. *Global Change Biology*. doi: 10.1111/gcb.12685.

Taggart, M.A., Cuthbert, R., Das, D., Sashikumar, C., Pain, D.J., Green, R.E., Feltrer, Y. Schulz, S., Cunningham, A.A. and Mehrag, A.A. (2007). Diclofenac disposition in Indian cow and goat with reference to Gyps vulture population declines. *Environmental Pollution*, 147, 60–5.

United States Environmental Protection Agency. (2012). Drinking Water Standards and Health Advisories. Washington, DC: Office of Water.

United States House of Representatives Committee on Energy and Commerce (2011). Chemicals used in hydraulic fracturing. http://www.conservation.ca.gov/dog/general_information/Documents/Hydraulic%20Fracturing%20Report%204%2018%2011.pdf.

Veterinary Medicines Directorate (2012). Sales of Antimicrobial Products Authorised for Use as Veterinary Medicines in the UK in 2011. The Veterinary Medicines Directorate, Surrey.

World Health Organisation (2011). Evaluation of certain contaminants in food: seventy-second report of the Joint FAO/WHO Expert Committee on Food Additives. WHO technical report series; no. 959. World Health Organisation Press, Geneva, Switzerland.

World Trade Organisation (2014). International trade and market access data: interactive tool. https://www.wto.org/english/res_e/statis_e/statis_bis_e.htm?solution=WTO&path=/Dashboards/MAPS&file=Map.wcdf&bookmarkState={%22impl%22:%22client%22,%22params%22:{%22langParam%22:%22en%22}}

CHAPTER 6

The state of seven seas

6.1 Introduction

The previous chapters have shown that while ultimately there is a biological basis to the impacts of human activities on marine ecosystems, the actual patterns of change, the extent and nature of the pollution is the result of a complex series of factors. These include the social and economic history of an area, the nature of the various impacting activities, the regulatory approach employed, public perceptions and expectations. In this chapter we look at seven sea regions that vary in their history of development, exploitation and regulation.

For each of the seven chosen case-study areas, we briefly document the nature of the regional sea, the impacts and pressures it has received and how these have (or have not) been managed and the present perception of their health. However, given the diverse nature of these regions, the format of the sections differ as we seek to provide an account that does justice to the history, issues and challenges faced by each region.

6.2 The North Sea

The Greater North Sea is one of the busiest maritime areas in the world. Offshore activities related to the exploitation of oil and gas reserves, offshore wind farms and maritime traffic are all extensive. Two of the world's largest ports are situated on North Sea coasts while the coastal zone is used intensively for recreation. The area has also supported extensive fisheries for millennia.

The North Sea is surrounded by the highly populous and developed countries of northern Europe, and human activities in the region have a long history. There is documentary archaeological evidence of trade across the North Sea dating back to the Stone Age. Artefacts from this period show extensive fishing by coastal populations. By Roman times (2000 years ago), there were well-developed trade routes and coastal markets that had blossomed into towns that were exploiting natural resources, building and modifying the coast (the Romans carried out land reclamation to build at the coast and extend fertile agricultural land) and discharging their waste into the sea. Tax records from the Middle Ages show that over-exploitation was already an issue for some fisheries while records from the United Kingdom show the King being petitioned to restrict 'destructive' fishing gears as early as the twelfth century. By the eighteenth century, the Industrial Revolution was underway and new waste streams were flowing into the North Sea. Following the Second World War, the drive to be self-sufficient in food production led to the widespread use of artificial fertilisers and pesticides with consequential leaching of these into waterways and eventually the North Sea. Thus, the North Sea is possibly the sea area with the longest history of intense human impacts.

The North Sea occupies the continental shelf of north-west Europe. It opens into the Atlantic Ocean to the north and via the Channel, to the south-west, and connects with the Baltic Sea to the east. The Greater North Sea (including its estuaries and fjords) has an area of approximately 750 000 km^2 and a volume of around 94 000 km^3, with depths not exceeding 700 m. The seabed is mainly composed of mud, sandy mud, sand and gravel. The variety of marine landscapes provide an important diversity of habitats.

Marine Pollution. Christopher L. J. Frid & Bryony A. Caswell.
© Christopher L. J. Frid & Bryony A. Caswell 2017. Published 2017 by Oxford University Press.
DOI 10.1093/oso/9780198726289.001.0001

Surrounded by densely populated and highly in-dustrialised countries (Belgium, Denmark, France, Germany, The Netherlands, Norway and the United Kingdom), the North Sea has received high inputs of industrial wastes. Other major activities in the North Sea include fishing, the extraction of sand and gravel and offshore activities related to the exploitation of oil and gas reserves including the laying of pipelines and the development of off-shore renewable energy systems, principally wind farms but with increasing development of tidal and wave systems. The North Sea is one of the most fre-quently traversed sea areas of the world with two of the world's largest ports situated on its coasts (Rot-terdam and Hamburg), and the coastal zone of the Greater North Sea is used intensively for agricul-ture and recreation.

The North Sea ecosystems are rich and complex. Approximately 230 species of fish are known to in-habit the area. Some 10 million sea birds are present at most times of the year. Marine mammal species occur over large parts of the North Sea including long-finned pilot whales, harbour porpoises, com-mon dolphins, white-sided dolphins, Risso's dol-phins, killer whales, grey seals, harbour seals and hooded seals.

6.2.1 Hydrography of the North Sea

Thinking of the North Sea (Figure 6.1) as an ap-proximately rectangular basin (surface area 575 000 km^2), the southern part is relatively shallow (less than 50 m depth) shelving to deeper water (c. 200 m depth) in the northern areas, with a deep (600 m) trough, the Norwegian Rhinne, on its north-eastern margin. Most of the water mass within the basin is of Atlantic origin, primarily flowing in from the north, with other inputs through the Dover Straits. Low-salinity water enters from the Baltic Sea through the Skaggerak. The prevailing south-westerly winds cause the bulk of the water mass to circulate in an anticlockwise gyre in the North Sea, with the main outflow being through the Norwe-gian Rhinne, a trough carved by glaciers in the last Ice Age (Figure 6.1).

Offshore parts of the northern North Sea undergo seasonal stratification while in the southern bights (see Figure 6.1), owing to their shallow (less than 50 m) depth and strong tidal flows, no stratification occurs. The southern portion of the North Sea is further influenced by large freshwater inputs from the major European rivers, which means this area is hydrographically distinct from the other areas. Nu-trient distributions and concentrations within the North Sea are influenced by the anticlockwise gyre with most of the natural nutrient inputs into the North Sea originating from the Atlantic. However, over the last 40 years high-nutrient discharges into the southern North Sea via the major continental rivers have been observed. These increased nutri-ent inputs from rivers passing through some of the most industrialised areas in the world are thought to be responsible for exceptional phytoplankton blooms, particularly along the Dutch coast and in the German Bight.

The naturally nutrient-rich waters drive a high productivity that has supported one of the world's richest fishing grounds. The North Sea contains valuable oil resources in the north, the southern re-gion is an important gas production area and the waters off the eastern coast of the United Kingdom are sources of aggregates used in the construction industry. Recent years have seen the rapid devel-opment of extensive arrays of offshore wind energy turbines. Other areas have been used for the dis-posal of sewage sludge, dredge spoil and fly ash.

6.2.2 Ministerial conferences: Quality status reports

The North Sea, being surrounded by the countries of western Europe, has been used as a source of food, for transport and as a sink for wastes since prehistoric times. However, increasing understand-ing of human impacts on coastal areas and the in-crease in environmental awareness in Europe has seen attention focused on the North Sea. Initially, this was in response to fisheries issues (e.g. the establishment of the International Council for the Exploration of the Sea (ICES) in 1902) and subse-quently extended to water quality (e.g. the Oslo Convention, Paris Commission and Ministerial Conferences on the North Sea) and biodiversity protection. Like many sea areas the North Sea is bordered by many nations and with the adoption of the Exclusive Economic Zone provisions of the

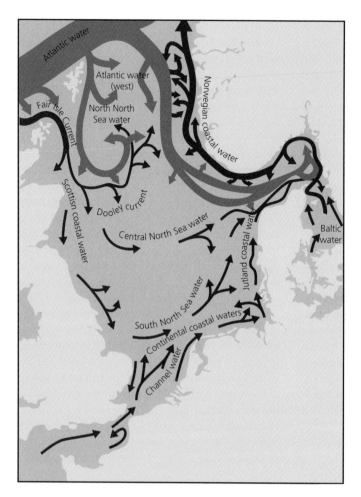

Figure 6.1 Surface circulation in the North Sea. Much of the detail of these currents have been derived by tracing the path of artificial radionuclides released by the French and UK nuclear waste reprocessing plants at Cap Le Harve and Sellafield. (Combined from various sources).

UNCLOS (United Nations Convention on the Law of the Sea) the whole of the North Sea lay within the jurisdiction of a nation. While many of these nations were members of the European Union and so bound, to some extent, by common approaches to fisheries and waste management, not all countries were, and the inflow from the Baltic was a significant source of some substances of concern.

By the 1980s, environmental non-government organisations (eNGOs; e.g., Greenpeace, Friends of the Earth and the World Wildlife Fund) were raising public awareness of what they characterised as a looming environmental catastrophe in the North Sea. Some even claimed that the North Sea would be 'dead' within ten years. Following a 1988 intergovernmental conference on the protection of the North Sea, Andy Booth, the leader of Greenpeace's North Sea clean-up campaign, claimed that by not acting, the participating countries had actually signed the sea's death warrant.

The growing public concern in North Sea states that the large inputs of various harmful substances via rivers, direct discharges and dumping of waste at sea could cause irreversible damage to the North Sea ecosystems prompted political action. The growing influence of 'green' parties in some countries also produced public and state dissatisfaction with the slow progress made by the established international organisations in developing the means for protecting the marine environment. These pressures resulted in the first International Conference on the Protection of the North Sea, in Bremen in 1984.

Five further North Sea Conferences were held (London in 1987, The Hague in 1990, Esbjerg in

1995, Bergen in 2002 and Gothenburg in 2006). The conferences united governments from Belgium, Denmark, France, Germany, The Netherlands, Norway, Sweden, Switzerland and the United Kingdom in an effort to protect the North Sea ecosystems. The North Sea Conferences (NSCs) provided a high-profile forum for developing a political framework and agreement on the actions that were needed to protect the North Sea. They resulted in politically based (rather than legally binding) commitments, but the high-profile nature of the meetings resulted in many of these commitments being translated into national regulations or legally binding international conventions.

Two of the most significant outcomes from the conferences were the adoption of the precautionary principle at the conference in London in 1987, and in 2002 in Bergen when the commitment was made to develop a conceptual framework for an ecosystem approach and to implement this approach for the management of fisheries and environments. At the conference in Gothenburg in 2006, the ministers declared that many of the issues that had been discussed over the years were now being treated in other forums and, while committing to following up on these commitments, they had no plans for any further meetings.

From a practical point of view, one of the key legacies of the NSCs was the production of periodic Quality Status Reports (QSR). Initially produced by international panels of scientific experts using data shared by the various nations, the QSRs served as a common starting point for the political teams to identify the key issues and prioritise actions. The first QSR was produced to support the London (1987) NSC. However, their value as periodic synoptic overviews of the state of the marine environment was quickly realised and they were expanded from 2000 to cover not just the North Sea but the whole of the North East Atlantic (the region covered by the OSPAR conventions). Similar documents are now also produced in other parts of the world as part of the marine management and reporting cycle. At the time of writing, the most recent North Sea QSR was produced in 2010, with an intermediate assessment scheduled for 2017 and the next full QSR to be produced in 2021.

In the 2010 QSR for the North Sea region it was noted that since 2000:

- **The status of some fish stocks had improved.** The development of long-term management plans for key stocks and substantial decreases in destructive practices, such as beam and otter trawl fishing, in some areas were seen as key initiatives. The EU was working to reduce excessive discards of fish (Norway had banned discarding for many years).
- **The inputs of hazardous substances and nutrients had been reduced.** While nutrient inputs had fallen, 17% of the North Sea was still experiencing eutrophication. Industrial inputs of mercury and lead to the sea had also decreased.
- **Good MPA coverage.** In the North Sea 5.4% of the waters and seabed were within Marine Protected Areas (MPAs).

While these successes were acknowledged, a number of continuing and new areas of concern were identified. These included;

- **Coastal eutrophication.** Eutrophication has been identified as a problem in the coastal regions of the eastern North Sea from Belgium to Norway and in some estuaries and bays in eastern England and north-west France. Nitrogen inputs, principally from agriculture, are the biggest cause of eutrophication (Section 4.3). Only a few countries have managed to control nitrogen inputs sufficiently to approach OSPAR's 50% reduction target for nitrogenous inputs to problem areas.
- **Hazardous substances.** Pollution by hazardous substances remains an issue in some areas with concentrations of metals (cadmium, mercury and lead) and persistent organic pollutants above background levels in some offshore waters of the North Sea and at unacceptable levels in some coastal areas. For example, lead concentrations were *unacceptable* at 40% of locations monitored, while polycyclic aromatic hydrocarbons (PAHs) and polychlorinated biphenyls (PCBs) were at *unacceptable* levels at more than half of the monitoring sites.
- **Marine litter accumulation.** Beach litter in the southern North Sea is around 700 items per 100 m of beach, which is the average level for the

coast of the North East Atlantic, but quantities are higher in the northern North Sea. Monitoring of fulmars (surface feeding sea birds) revealed that over 90% have microscopic plastic particles in their stomachs (Section 4.2).

- **Limited progress had been made towards fully sustainable fisheries.** While some stocks showed signs of recovery, many economically important North Sea fish stocks were still outside sustainable limits. While many ecologically damaging fishing practices had been reduced, many still remained. By-catches of rays, sharks, porpoises and dolphins in fishing nets was of particular concern.
- **Sea bird breeding failures.** In the northern North Sea, some sea bird species have suffered breeding failures in most years during the period 1998–2008. This may be due to the combined effects of climate change and fishing on key prey species. Breeding success was good in 2009, but the long-term picture is of serious concern.
- **Seabed habitat damage.** Damage to shallow sediment habitats and reefs has occurred due to the use of heavy bottom contact fishing gears, especially beam trawling.
- **Climate change impacts.** The rate of warming of the North Sea has been between 1 °C and 2 °C over the past 25 years: faster than the wider North East Atlantic. There is evidence from both the plankton and fish assemblages that they are changing in response to warming. Southern species of fish, such as John Dory, sea bass and red mullet are becoming more common further north, whereas North Sea cod numbers appear to be declining faster than is predicted from the impacts of fishing alone (cod is a northern species).

Against this background the QSR also raised concerns about the potential impact of planned responses to climate change, in particular the increasing spread of wind farms and the need to undertake large-scale coastal defence works along most of the southern North Sea coastline. These sea defences will include hard engineering, resulting in habitat loss and beach recharge with implications for the demand for aggregate dredging. Norway is also proposing the use of old oil and gas rigs as sites for sub-seabed carbon dioxide storage schemes. Thus, after several millennia of human impacts and in a region with complex political structures, the North Sea can perhaps be seen as a region that has been pushed close to breaking point and that is now, through international political action, showing signs of recovery. However, even as it recovers from previous damaging activities, a new range of human pressures are emerging.

6.3 The Black Sea

The Black Sea is an inland sea one-fifth the size of the Mediterranean (Figure 6.2; Section 6.5). The north-western region is occupied by a continental shelf 200 km wide and approximately 160 m depth which extends south to Istanbul. The north-west shelf has previously been an important fishing area. Throughout the central, south and eastern areas the water depth ranges from 500–2200 m. The Black Sea receives fresh water from a large catchment area, of 2×10^6 km^2, draining from six major rivers. The Black Sea is connected to the Sea of Marmara, the Aegean and the Mediterranean through the narrow Bosphorous channel. To the north it is connected to the Sea of Azov a shallow inland sea (less than 50 m in depth) that is one-tenth the size. At times during the last glaciation there is evidence that when sea level was lower and there was no connection to the Mediterranean, the Black Sea was a freshwater lake. The Black Sea is thought to have formed in 5600 BC when the Mediterranean Sea breached through the Bosphorous Straits.

The Black Sea is unique in being the world's largest *meromictic* basin: meaning that the surface and deep waters do not mix. One of the reasons for this lack of mixing is its large freshwater inputs, 350 km^3 y^{-1}, and its unusual hydrography. Warm saline water (salinity 38) from the Mediterranean Sea flows in at depth and the cooler less saline water (salinity 18–22) of the Black Sea flows out at the surface. This salinity stratification prevails in the Black Sea and inhibits the mixing of the upper 50 m of the water column with that below. In some areas, where it is more than 2200 m deep, this represents a considerable proportion of the water mass. This lack of mixing means that most of the deeper

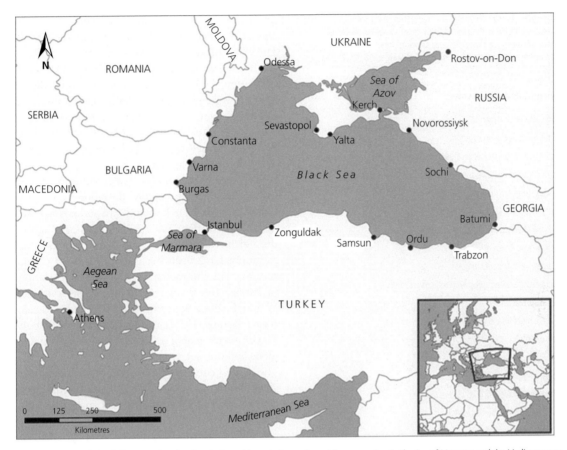

Figure 6.2 Map of the Black Sea region showing the Sea of Azov to the north and the connection to the Sea of Marmara and the Mediterranean to the south. The largest cities on the coast of the Black Sea are indicated. Inset box shows context within Europe. Source: © Norman Einstein 2007. Licensed under the Creative Commons Attribution-Share Alike 3.0 unported license https://creativecommons.org/licenses/by-sa/3.0/legalcode.

parts of Black Sea are *anoxic* (no oxygen), about 90% of its total volume, or *euxinic* (lacking oxygen with hydrogen sulphide present). The Black Sea therefore contains the largest mass of toxic hydrogen sulphide gas on the planet (produced from the anoxic degradation of organic matter). The Black Sea was named the *Euxine Sea* by the Greeks and is the origin of the term *euxinia*. There is some evidence of flaring of flammable hydrogen sulphide gas during thunderstorms where the gas has been ignited by lightning.

Despite the low salinities and anoxic bottom waters, the Black Sea is home to 1618 algae, 1983 invertebrate, 168 fish and 4 marine mammal species (the harbour porpoise, the short-beaked common dolphin, the bottlenose dolphin and Mediterranean monk seal). It is three times less diverse than the Mediterranean Sea due to the brackish conditions. The fauna contains very few echinoderms, and no octopi or squid due to their low salinity tolerances. The Black Sea is home to four introduced species of shark, two native dogfish species and three species of ray: the thornback, common stingray and spiny butterfly ray. Of the approximately 199 total fish species, 8.5% are classified as threatened (Table 6.1). This is compared with the Mediterranean Sea, where of the 519-odd native fish species 8.3% are threatened, and the North Sea, where of the 201 fish species about 9% are threatened.

Table 6.1 Threatened (critically endangered, endangered, vulnerable) and near threatened Black Sea fish species. A further fifty-two species are classified 'least concern', and the remainder have not been assessed. Data from: Froese and Pauly 2016, IUCN Red List 2016.

Species conservation status

Critically endangered

Angel shark *Squatina squatina*

Russian sturgeon *Acipneser gueldenstaedtii*

Bastard sturgeon *Acipenser nudiventris*

Persian sturgeon *Acipenser persicus*

Starry sturgeon *Acipesner stellatus*

European sea sturgeon *Acipenser sturio*

Beluga *Huso huso*

European eel *Anguilla anguilla*

Danube delta dwarf goby *Knipowitschia cameliae*

Endangered

Atlantic bluefin tuna *Thunnus thynnus*

Vulnerable

Spiny butterfly ray *Gymnura altavela*

Smooth hammerhead *Sphyrna zygaena*

Common thresher *Alopias vulpinus*

Spiny dogfish *Squalus acanthias*

Sterlet *Acipenser ruthenus*

Pontic shad *Alosa immaculata*

Green wrasse *Labrus viridis*

Near threatened

Thornback ray *Raja clavata*

Common percarina *Percarina demidoffii*

6.3.1 Contamination

The Black Sea has a complex environmental history with high levels of nutrient enrichment and other contaminants and over-exploitation. It is bordered by six countries and receives the riverine discharges and waste from a large catchment draining from an area representing around 20% of the surface area of Europe. The dumping of waste from ships and the discharge of poorly treated sewage were not well regulated in the Black Sea and oil spills were common before 2000. The volume of oil spilt has decreased five-fold to

approximately 10 t y^{-1} after 2000 (except in 1997 and 2003 when more accidental spills occurred). The concentrations of hydrocarbons throughout the Black Sea exceed the maximum allowable concentration, 0.05 mg l^{-1}, and were five times higher along shipping routes in 2002–04. Oil pollution in surface waters peaks near the largest cities in Georgia, Turkey and Romania due to shipping practices (refuelling and spills), municipal wastes and run-off (Figure 6.3a).

For other contaminants, water column concentrations are mostly below detection limits, but hotspots occur near known point-sources for metals and pesticides (Table 6.2). Metals are high on the north-west shelf (Table 6.2) but mercury and cadmium have been decreasing. Elevated dichlorodiphenyltrichloroethane (DDT) concentrations in Turkish rivers suggest its illegal use continues. Within the Black Sea sediments, DDT and hexachlorocyclohexane (HCH) are five times higher than the maximum allowable concentration in all coastal regions and did not decrease between 1994 and 2007. The highest concentrations occur on the north-west shelf (Table 6.2). Toxic PCBs also occur in high concentrations, but their sources and spatial distribution are not known. Sedimentary hydrocarbon concentrations were high between 1996 and 2006 exceeding 190 µg g^{-1} near the major coastal cities of Constantina, Zonguldak, Samsun, Trabzon, Novorossiysk, Kerch, Yatumi and Batumi (Figure 6.3b). The concentrations of HCHs, DDT and PCBs in fish and mammals are also high compared to the other regions in Europe. The radioactivity from the Chernobyl incident is thought to no longer pose a risk.

6.3.2 Eutrophication

The Black Sea has a long history of human use. It has been an important seaway since 800 BC when the Greeks settled the southern coastline and later the north-west coast in 700 BC. The waste from human settlements, for example, sewerage and crop fertilisers, has entered the Black Sea in increasing quantities ever since that time. This anthropogenic nutrient enrichment led to elevated surface-dissolved nitrogen concentrations. Between 1960 and 1980, dissolved nitrogen increased nine-fold on the western shelf, four-fold on the north-east

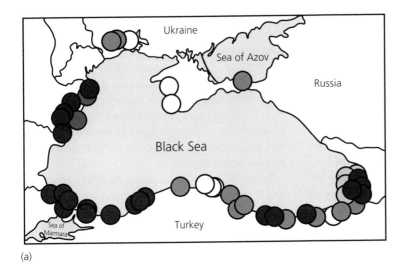

(a)

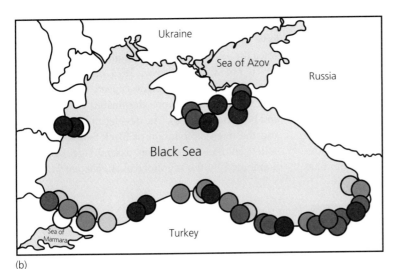

(b)

Figure 6.3 Petroleum hydrocarbons in the Black Sea (a) water column (shading indicates concentrations of petroleum hydrocarbons from pale to dark grey as follows: less than 0.018, 0.018–0.020, 0.021–0.050, 0.050–1.180, and 0.181 mg l⁻¹); and, (b) sediments (shading indicates concentrations of petroleum hydrocarbons proceeding from pale to dark grey as follows: less than 5.5, 5.5–26.0, 26.1–60.7, 60.8–190.0, and more than 190.1 µg g⁻¹). Source: Black Sea Commission (2007). Black Sea Transboundary Diagnostic Analysis. Publication of the Commission on the Protection for the Black Sea Against Pollution, Istanbul, Turkey. (c), Commission on the Protection of the Black Sea Against Pollution.

Table 6.2 Summary of contaminant distribution in Black Sea sediments excluding hydrocarbons that are shown in Figure 6.3. Limited data mean only a very general picture may of the pollution be acquired. Data from: Black Sea Commission 2008.

Region	Sediment contaminants of concern
Romania	HCB, HCH, DDT, Cu, Cr, Cd and Hg
Bulgaria	HCB, DDT, HCH, Cu, Cr, Cd and Hg in NE
Ukraine	HCB, DDT, HCH near Odessa, Cu, Cr, Cd and Hg in SW
Turkey	HCH near Istanbul
Russia	HCH near Sochi

coast, and more than five-fold on the Romanian coast. During the 1990s, the concentrations of dissolved nitrogen at the mouth of the Danube near Sulina were very high, peaking at about 350 µM. Since the 1990s the concentrations of nitrogen and phosphorous delivered by the River Danube have reduced from 7.0×10^5 to 2.5×10^5 t NO$_3$ y⁻¹ and 6.0×10^5 to 1.0×10^5 t PO$_4$ y⁻¹. This is due to the closure of livestock farms, reduced industrial fertiliser production and improved waste treatment. High concentrations of ammonia indicate significant point-source industrial discharges on the Ukrainian and Romanian coasts. The changes in industry were partly driven by economic recession in the early 1990s. In addition to the anthropogenic sewage and fertiliser discharges, other sources of pollution with a chemical oxygen demand would also have contributed to the oxygen consumption.

However, despite the large decreases in industrial discharges, the amounts of dissolved nitrates remained high at 40 µM in the north-west between 2000 and 2010, and were 3–4 times higher than on the north, north-east and southern coasts. The organic nutrient inputs from the Danube are also substantial varying seasonally between 20 and 120 µM dissolved organic nitrogen and 1–5 µM for dissolved organic phosphate at the mouth.

The large increases in nutrient supply resulted in increased phytoplankton blooms on the north-west shelf that halved the euphotic zone depth. Mean annual chlorophyll concentrations in the north-west Black Sea from 1980 to 1995 were more than three times higher than in the western or interior Black Sea. Since 1995 chlorophyll concentrations have decreased, although they remain high during spring in some areas; for example, at Zmeiny, 17 mg m^{-3} in 2007. Phytoplankton biomass in the Ukrainian, Romanian and Bulgarian Black Sea regions also decreased from a mean annual biomass of 10 g m^{-3} in 1980–90 to 4 g m^{-3} in the 2000s, with occasional peaks during coastal blooms. Coincident with these decreases the phytoplankton assemblage showed a doubling of species diversity, decreasing phytoplankton: zooplankton ratio, a reduction in bloom frequency and shifts in bloom timing (from summer to spring and autumn blooms). These changes support those for biomass and chlorophyll content, showing that eutrophication is decreasing.

From 1973 to 1995, most of the Ukrainian shelf and the entire Azov Sea experienced severe seasonal hypoxia linked with the eutrophication. Hypoxia on the north-west shelf was first recorded between the Danube and Dnestr deltas with a 3500 km^2 hypoxic water mass. After this, every few years hypoxia affected water masses of 14 000 km^2 to 20 000 km^2 (Figure 6.4) between 8 and 40 m depth. Hypoxic events have caused mass mortalities of benthic animals since the 1970s, and these mortalities have amassed 6 × 10^6 t of animals between 1973 and 1990 of which 90% were fish. The hypoxia caused Black Sea ecosystems to switch to one dominated by gelatinous and bacterial biomass. The development of hypoxia, and the over-exploitation of Black Sea fisheries resulted in the zooplankton assemblage becoming dominated by the scyphozoan *Aurelia aurita*. These

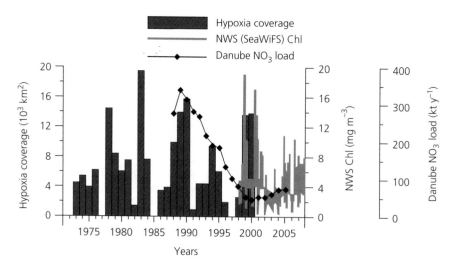

Figure 6.4 Hypoxic (less than 0.2 mg l^{-1}) water mass extent on the north-western shelf (NWS) of the Black Sea from 1970 to 2001, satellite derived surface chlorophyll concentrations from 1998 to 2008, and nitrate discharges from the Danube between 1998 and 2005. Source: Black Sea Commission. (2008). State of the Environment of the Black Sea (2001–2006/7). Publication of the Commission on the Protection for the Black Sea Against Pollution 2008–03, Istanbul, Turkey. Black Sea Commission. © 2008, Commission on the Protection of the Black Sea Against Pollution.

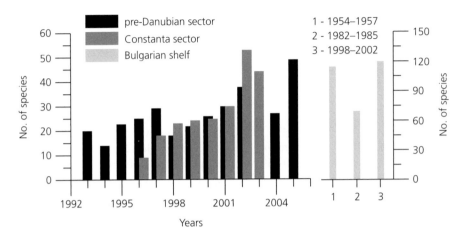

Figure 6.5 Benthic diversity in the Black Sea near the Constanta, pre-Danubian and Bulgarian coasts. Right hand side shows benthic diversity on the Bulgarian shelf during three distinct intervals. Source: Black Sea Commission. (2008). State of the Environment of the Black Sea (2001–2006/7). Publication of the Commission on the Protection for the Black Sea Against Pollution 2008–03, Istanbul, Turkey. Black Sea Commission. © 2008, Commission on the Protection of the Black Sea Against Pollution.

changes were later added to by non-native invasive species introductions (Section 6.3.3).

Hypoxia continued to affect large areas of the north-west shelf in the 2000s although the total volume of hypoxic water declined. An area of 13×10^3 km^2 was hypoxic on the north-west shelf in 2001, representing one-third of the total shelf area (Figure 6.4). Although there are insufficient data after 2001, hypoxia had reduced with small local events occurring near the Danube delta in 2005. Below the euphotic zone in the interior basin, dissolved oxygen concentrations decreased by about 20% between 1995 and 2007 which seems to be linked to a longer-term climate trend.

In the 1960s, extensive fields of the red alga *Phyllophora* occurred on the north-west shelf. These fields were important habitats for invertebrates and up to 40 species of fish. *Phyllophora* biomass declined from 9×10^6 t in the 1960s to 8.0×10^3 t in 2000 (Table 6.3) because of eutrophication-limited light penetration through the water column. These algae were harvested commercially to produce agar up until 1996 when harvesting ceased as biomass dropped below 120 000 t. The biomass of the brown macroalgae *Cystoseira barbata* also declined three-fold between the 1970s and 1998. In 2005, some recovery of macroalgal beds was observed in shallow waters (less than 10 m depth), but little recovery

had occurred at greater depth; the previously extensive beds had extended down to 20 m.

Eutrophication resulted in seafloor communities of reduced diversity (Figure 6.5), with high abundances and biomass of opportunists and invasive species that are hypoxia tolerant. Thus, these communities fit the classic organic enrichment pattern (Section 3.2). Although there are many gaps in the data, it seems that some improvements in benthic diversity occurred between 1993 and 2005 (Figure 6.5). Species that had been extirpated because they were intolerant of hypoxia began to reappear. Indices of benthic community health, AMBI (Section 2.4.5), suggest that the north-west coast benthos is of moderate and degraded state along the coast between Odessa and Constanta. The most degraded communities were concentrated in the Danube delta, and north and south of this communities were in good ecological condition. However, the benthos remained dominated by the snail *Rapana* and the bivalves *Anadara inequivalvis* and *Mya*

Table 6.3 Standing biomass of benthic red macroalgae *Phyllophora* spp. in the Black Sea. Data from: Black Sea Commission 2008.

Phyllophora populations	1960s	1985	1993	2000
Biomass (t)	9×10^6	3.0×10^5	1.2×10^5	8×10^3
Coverage (km^2)	12×10^3	4×10^3	–	1.5×10^3

arenaria that have high tolerance of hypoxia. Crustacean populations had increased somewhat by 2005 but had not fully recovered. The dominance of tolerant opportunists in these communities suggests that these ecosystems remain fragile.

6.3.3 Non-native invasive species

The sea walnut, *Mnemiopsis leidyi*, was introduced to the Black Sea in the 1980s through ships ballast water. Very large populations of the ctenophore followed with a total biomass of up to 1×10^9 t and densities of around 400 m^{-3}, and *M. leidyi* outcompeted the native *A. aurita*. These large populations of *M. leidyi* negatively impacted important commercial fisheries species by consuming their eggs and larvae. The ctenophores competed with the fish larvae and zooplankton for food, and had a detrimental impact on the recruitment of benthic taxa with planktonic larval stages, such as some bivalve mollusc species. At times the gelatinous biomass (dominated by *M. leidyi* and *A. aurita*) comprised 90% of the total zooplankton biomass in the north-west Black Sea.

Populations of the non-native invasive species (NIS) began to decline after the deliberate introduction of another ctenophore, *Beroe ovata*, in 1998 in an atempt to control *M. leidyi*. *B. ovata* predates upon *M. leidyi* and in the Black Sea this successfully reduced the longevity of the *M. leidyi* blooms which had a positive result on the critical life-history stages of commercial fish species. Blooms of the *M. leidyi* continue to occasionally occur in both the north-west and western areas. The Black Sea mesozooplankton assemblage has shown improvements: on both the western and eastern coasts, the edible zooplankton biomass has increased from its 1993–95 to reach 150 mg m^{-3} in the west by 2005 and 20 mg m^{-3} in the east by 2007. Additionally, the abundance of cladocerans and copepods, that were absent during the 1980–90s, increased, and the copepod, *Pontella mediterranea*, that was extirpated because of its sensitivity to eutrophication, has now returned.

The arrival of *M. leidyi* in the 1980s also affected the benthos by increasing benthic organic matter flux and by consuming their planktonic larval stages. The Black Sea was impacted by other NIS including the predatory snail, *Rapana venosa*, introduced in 1947, *Mya arenaria* in the 1960s and the bivalve *Anadara inequivalvis* in 1986. The snail predated on

and caused the loss of Black Sea oyster *Ostrea edulis* reefs previously a commercial fishery. The soft-shell clam, *Mya arenaria*, outcompeted the native bivalve, *Corbula mediterranea*, and *Anadara inequivalis* outcompeted the native fishery species *Chamelea gallina*. When *M. leidyi* populations began to reduce in 1998–99 after the successful biocontrol measures, populations of bivalves increased because their planktonic larvae were released from predation. Populations of the predatory gastropod *Rapana venosa*, however, responded by feeding on the now abundant juvenile and adult bivalves. Consequently, both the bivalves and the snail populations crashed due to predation of the former and resource limitation of the latter. By 2008, the north-east Black Sea benthos became dominated by deposit-feeding opportunistic polychaetes (such as *Prionospio multibranchiata* and *Capitella capitata*). In the north-west there was a similarly high benthic biomass dominated by polychaetes, *Mya arenaria*, *Anadara* and *Rapana*, indicating continuing organic enrichment.

6.3.4 Over-exploitation

From 1960 to 1970, 26 commercial fish species were landed in the Black Sea in quantities of more than 100 000 t y^{-1}; however, after 1980 only five species were commercially viable. The total fishery landings in the Black Sea declined from 900 000 to 100 000 t y^{-1} between 1986 and 1992. Dolphin and porpoise populations were impacted by over-exploitation peaking from the 1930s to the 1950s. Commercial hunting was banned in 1966 in Bulgaria and Romania and in 1983 in Russia.

Over-fishing in the Black Sea resulted in two major ecological *regime shifts*. Declines in dolphin populations, from one million to 400 000 between 1950 and 1987, and declines in mackerel and bluefin tuna occurred due to over-fishing. After the top predators were removed, the first regime shift occurred in the 1970s and only small planktivorous species were landed (Table 6.4). The second regime shift occurred in 1989 when the anchovy fishery collapsed from 322 758 t in 1984 to less than 15 000 t in 1989. Since the pelagic fisheries collapse, landings have been low and the catch has been dominated, 31–75%, by the European Anchovy. Demersal fish catches have been low and generally decreasing since 2002 (Table 6.4), due to over-fishing and benthic hypoxia. Anadromous fish species reached a minimum of 500 t in 1999–2000, and appeared to recover afterwards but this was mostly attributable to landings of Pontic shad. Sturgeon nursery habitats have not recovered. Spratt and anchovy were fished at high levels and whiting and turbot were fished intensively in 2006. The horse mackerel stock remains small and has not recovered.

Commercial shellfish species include *Chamelea gallina*, *Mytilus galloprovincialis* and the snail *Rapana* spp. (Table 6.4). In 2005, mussel and clam stocks were one-third of their size in 1994, but *Rapana* fisheries took off in 1994 and had grown 40% by 2005. Macrophyte species that used to be commercially harvested dropped substantially between the 1960s and 1996 when commercial harvesting ceased.

Table 6.4 Important commercial fisheries in the Black Sea and their landings from 1990s–2000s. Data from: Black Sea Commission 2008 and Shylakohv and Daskalov (2008). *Regular harvests only occur in Turkey.

Fishery	1994	2000	2006
Pelagic fish European anchovy *Engrulis encrasciolus*, spratt *Sprattus sprattus*, horse mackerel *Trachurus mediterraneus*, Atlantic bonito *Sarda sarda* and bluefish *Pomatomus saltatrix*.	3.0×10^5 t	3.8×10^5 t	3.0×10^5 t
Demersal fish Whiting *Merlangius merlangius*, picked dogfish *Squalus acanthias*, turbot *Psetta maxima*, striped mullet *Mullus barbatus*, red mullet *Mullus surmuletus*, and 4 sp. Mugilidae	3.0×10^4 t	3.0×10^4 t	1.5×10^4 t
Anadromous fish Pontic shad *Alosa pontica*, Sturgeon *Acipenser gueldenstaedtii*, *Acipenser stellatus* and *Huso Huso*.	5.0×10^3 t	5.0×10^2 t	2.0×10^3 t

Shellfish	1994	2000	2005
Rapana spp.	5.6×10^3 t	7.2×10^3 t	1.3×10^4 t
Mytilus galloprovincialis	9.3×10^3 t	2.9×10^2 t	3.0×10^3 t
Chamealea gallina*	3.3×10^4 t	1.0×10^4 t	1.0×10^4 t

6.3.5 State and future of the Black Sea

The Commission for Protection of the Black Sea Against Pollution was established by the Bucharest Convention in 1992 and its protocol is to prevent, control and as far as possible eliminate land-based pollution, dumping, oil pollution and other harmful substances, and to achieve good ecological status for the Black Sea. The Commission's State of the Environment of the Black Sea report (2008) concluded the following:

• Anthropogenic nutrient enrichment combined with pressure from over-fishing, pollution, habitat degradation and the introduction of NIS have combined to have major effects upon the Black Sea ecosystems. A number of species are now endangered in the Black Sea including: *Phyllophora* spp., the brown alga, *Cystoseira*, mussels, cockles, the oyster, *Ostrea edulis*, the scallop, *Flexopecten ponticus*, 14 species of crab, more than 20 species of shrimp, three dolphin and five bird species. Mediterranean monk seals, *Monachus monachus*, were extirpated from the Black Sea in 1997. The conservation status of the dolphins has not been assessed but the harbour porpoise is listed as endangered.

• The Black Sea ecosystems are very different now compared with their condition in the 1960s, having undergone large losses of fish and mammals, particularly those at the higher trophic levels, and large losses of key benthic taxa such as molluscs and crustaceans. The ecosystems remain impacted although the plant and animal communities are showing signs of improvement, and anthropogenic nutrients and consequently hypoxia are also decreasing. The amount of oil spilt has declined but remains high along major shipping routes and within sediments near large cities. Other pollutants of concern include DDT, hexachlorobenzenes (HCBs), HCHs and PCBs which occur at toxic concentrations in some sediments.

• Benthic ecosystems in particular continue to be at risk because although improving, they have undergone major changes in the last few decades with large decreases in diversity and major losses of important habitats (e.g. extensive seagrass and macroalgal fields and oyster reefs). Although benthic diversity has increased, these

communities remain dominated by hypoxia-tolerant opportunists, suggesting some organic enrichment still occurs and the NIS remain a threat to recovery.
- Overall, fisheries are also improving compared with the situation after the 1989–92 collapse, but they are diminished compared with the 1979–88 baseline. Stocks are very variable and a lack of regional coordinated fisheries management, illegal fishing and use of destructive gear, plus an unstable food web due to eutrophication, presents risks in the future.

The recommendations of the 2008 Black Sea Commission report were that in order to continue to improve and prevent further declines the following actions were required:

- Further improvements in waste treatment and agricultural activities to reduce terrestrial nutrient loads.
- Improved coordination of fisheries regulation, specifically dredging and trawling.
- Monitoring of benthic communities as indicators of change (organic enrichment and hypoxia) by sampling indicator species seasonally.
- Annual monitoring of pollutants in sediments.
- Monthly measurements of nitrogen, phosphorous, dissolved oxygen, chlorophyll, phytoplankton and zooplankton biomass, and diversity at key sites in the water column, below the seasonal thermocline and near the sea floor.

6.4 Chesapeake Bay

Chesapeake Bay is the largest estuary in the United States. It covers 11 600 km^2 and has an average depth of 6.4 m and is up to 9 m near the Susquehanna tributary and is as shallow as 2 m in other areas (25% of the bay area). Formed by glaciers 10 000 years ago, it is bordered by Maryland to the north and Virginia to the south and opens into the Atlantic Ocean. One-hundred-and-fifty major waterways drain into Chesapeake Bay from a catchment of 166 000 km^2 from six different states. Eighty percent of the freshwater discharge comes from the Susquehanna, Potomac and the James rivers. The bay was named Chesepiooc by the native American Algonquian tribe meaning village at a big river.

Chesapeake Bay has a diverse fauna and flora of 3600 species including more than 348 species of fish, 173 species of shellfish, 2700 species of plant and 16 species of aquatic macrophyte. Twenty-nine species of water bird inhabit the bay, with a further 87 species of migratory birds visiting. One million overwintering birds visit each year because it is an important layover site for Atlantic migratory birds. Five main species of whale visit the bay plus bottlenose dolphins, four species of turtle and occasionally Florida manatees. The bay also contains a number of migratory fish species such as the American eel, *Anguilla rostrata*, shad and the shortnose and Atlantic sturgeons (*Acipenser brevirostrum* and *A. oxyrinchus*). *Acipenser brevirostrum* is listed as vulnerable by the International Union for the Conservation of Nature (IUCN). Larger fish include seven species of shark: basking sharks, Atlantic angel shark, Atlantic sharpnose, bull shark, dusky sharks, sandbar sharks, sand tiger sharks and four species of ray. The bay is thought to be an important nursery area for sharks.

The area was settled by Europeans in the sixteenth century and since then it has experienced significant modification from its natural state. In the eighteenth and nineteenth centuries, forests were cleared for agricultural use. A population of 18×10^6 people inhabit the catchment of Chesapeake Bay today and the local populations increased three-fold in the last century and with this, urban land use has expanded substantially. Forest clearance continues with 70 acres d^{-1} being cleared in 2007. The changes in land use have led to declines in water quality and combined with over-fishing and, to a lesser extent, NIS and pollution have caused deteriorations in the health of the local marine ecosystems.

6.4.1 Eutrophication and hypoxia

In the 1930s, the declining water quality was recognised from the development of hypoxia in parts of the bay. Since the 1950s, hypoxia has increased in degree and extent (Figure 6.6a). Phytoplankton biomass has increased up to ten-fold in some parts of the bay. Hypoxia is seasonal, initiating in the northern region in the spring driven by freshwater influx and changes in temperature, promoting stratification. As summer progresses, the hypoxia extends southwards, driven by eutrophication. In the autumn as temperature declines and mixing increases, normoxia returns. Study of sediment cores from Chesapeake Bay show that sedimentation has increased substantially since European settlement, between 1760 and 1985, when major land clearances occurred. Diatom communities show that eutrophication and turbidity increased alongside these clearances. Hypoxia and anoxia increased throughout this period, peaking in the 1940s.

Chesapeake Bay has an unusual geomorphology that means bay waters have a long *residence time* that, combined with the large freshwater inputs, make it susceptible to hypoxia (data from sediment cores show that hypoxia occurred prior to human settlement). The increases in anthropogenic hypoxia were driven by deforestation, the run-off of fertilisers from agricultural lands and waste-water discharges.

Today the bay has 324 km^2 of submerged macrophytes and 1149 km^2 of wetlands which are thought to have historically covered around 250 000 km^2 and 2873 km^2, respectively. The increased turbidity caused by sedimentation and eutrophication has caused declines in submerged plants of more than 90% in some areas (Figure 6.7a). Fifteen species of salt-tolerant macrophyte were lost from the upper estuary, and *Zostera marina* and *Ruppia maritima* were lost throughout the estuary. Since the 1980s there has been a 30% increase in macrophyte cover but this is less than a third of its standing stock in the 1920s. The loss of these once extensive macrophyte beds has had consequences. It has removed important habitats for invertebrates and fish including the function they provided as a nursery area. In addition, the macrophytes that have been lost had provided a natural buffer against eutrophication and hypoxia through their use of the enriched nutrients. Aquatic plants can also accumulate contaminants such as heavy metals and so can provide some water treatment. They also stabilise the sediments, oxygenate the water and provide natural sea defences.

Chesapeake Bay hypoxia has caused many mass fish kills in the Maryland part of the bay: on average 113 incidents occurred each year going back to the 1984 which killed around 1.4 million fish per year. This is equivalent to approximately 178 t y^{-1} (assuming most were Atlantic menhaden weighing

125 g each). Most of the fish kills were caused by deoxygenation although a few have been attributed to disease. Mass mortalities of benthic invertebrates also occur and are estimated to kill 75 × 10^4 t y^{-1} of clams and worms. Mass walkouts of crabs occur, referred to as jubilees, while attempting to avoid anoxic water masses. The eutrophication may also cause harmful algal blooms (Section 4.3) of the dinoflagellates, *Procentrum minimum* and *Kalrodinium veneficum*, and blue-green algae since 1990s (Table 6.5). Prior to this, *Pfiesteria piscicida* blooms occurred in the 1990s in Chesapeake Bay and during one event, 50 000 fish were killed. These phytoplankton contain toxins that are poisonous to fish, terrestrial animals and potentially also humans.

In some areas of the bay, the bioturbating infauna has been absent for about 100 years. Switches in life habit were observed in 1971 with changes from long-lived species to opportunists. In the middle of the bay, hypoxia causes mass mortality of recruited animals in the spring, and when conditions improve in the autumn, limited species recolonise. The loss of macrobenthos has led to considerable changes in productivity which is around 90% lower under hypoxic conditions compared with during normoxia (Figure 6.6b).

6.4.2 Pollution and non-native invasive species

Pollution concerns in the bay include high concentrations of the legacy contaminants mercury and PCBs, that are widespread throughout the bay sediments. In 2010, other metals present included aluminium, chromium and iron, which are more patchy, and PAHs near the large cities of Baltimore

Table 6.5 Number of harmful algal blooms per year in Chesapeake Bay during the late 1900s and early 2000s. Blooms of *K. veneficum* only began in 2003. Data from: Li et al. 2015.

Species	Year	
	1990s	**2000s**
Procentrum minimum	13 y^{-1}	23 y^{-1}
Cyanobacteria	89 y^{-1}	91 y^{-1}
	2003	**2008**
Kalrodinium veneficum	< 5 y^{-1}	> 30 y^{-1}

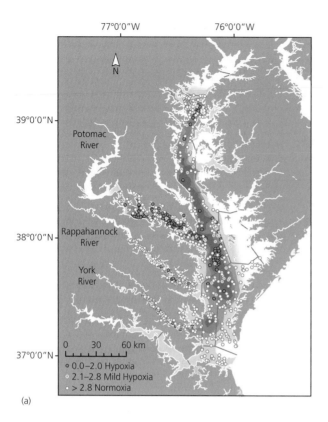

(a)

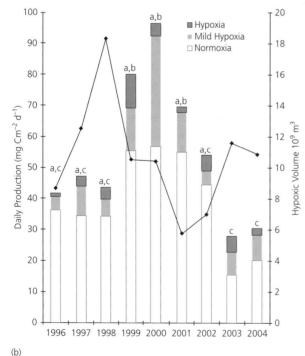

(b)

Figure 6.6 (a) Map of Chesapeake Bay showing main rivers and extent of summer hypoxia between 1994 and 2004. Shaded points: dark shading indicates dissolved oxygen concentrations of 0–2.0 mg l⁻¹ (hypoxic), medium grey indicates mildly hypoxic areas 2.1–2.8 mg l⁻¹ and pale grey normoxic waters more than 2.8 mg l⁻¹. Shaded areas correspond to the same dissolved oxygen concentrations as the shaded points and are interpreted between points. The three main rivers are named. (b) Amount of macrobenthic productivity (bars) each day under different levels of deoxygenation at 250 sites in Chesapeake Bay (indicated by points on (a); letters indicate those categories with significantly different mean productivity where a = normoxia, b = mild hypoxia and c = hypoxia) from 1996 to 2004. Right-hand axis shows hypoxic water volume (black points connected by grey line). Source: Estuaries and Coasts, Relationship between Hypoxia and Macrobenthic Production in Chesapeake Bay, 37, 2014, 1219–232, Sturdivant, K.S., Diaz, R.J., Llanso, R. and Dauer, D.M., © Coastal and Estuarine Research Federation 2014. With permission of Springer.

and Washington, DC and the smaller city of Norfolk. Pesticides are also present in bay sediments and have been problematic in the past. For example, local raptor populations (e.g. osprey and bald eagles) were impacted by DDT in the 1950s (Section 3.4). Pharmaceuticals and personal care products are thought to be an emerging problem in the bay (Section 5.2).

A number of non-native invasive species occur but most are freshwater species including: the Brazilian waterweed, *Egeria densa*, released from aquaria, which has taken hold outcompeting native submerged plants; the brackish water flathead, *Pylodictis olivaris*, and blue catfish, *Ictalurus furcatus*, which were deliberately introduced from the Mississippi between the 1960s and 1980s for recreational fishing. These long-lived species have a broad diet of crustaceans, worms and fish and have spread throughout the bay and could impact fishery species such as menhaden and blue crab in some areas.

6.4.3 Fisheries

Important past fisheries in Chesapeake Bay have included the blue crab, *Callinectes sapidus*, the eastern oyster, *Crassostrea virginica*, Atlantic sturgeon, Atlantic menhaden, the striped bass, American shad and various species of clam (Figure 6.8). Today's fisheries are focused on Atlantic menhaden, striped bass, blue crab and oysters from aquaculture. Many of the Chesapeake Bay fisheries collapsed in the twentieth century due to over-fishing, loss of their macrophyte habitat and the habitat contraction due to hypoxia, although these pressures are difficult to separate from one another. However, some stocks are now recovering due to targeted management practices.

The blue crab was historically a very important commercial fishery in Chesapeake Bay during which time it was the most valuable species landed in the bay. One-third of the United States blue crab catch comes from Chesapeake Bay. Peak harvests—for example, in 2010—were worth US $106 $\times 10^6$. The blue crab stock has been declining since 1990 due to over-fishing and has fluctuated by 50% over the last 60 years or so (Figure 6.8). Increased catches in 2010 were followed by a minima of 21 000 t in 2013. Blue crab populations are assessed annually at 1500 sites throughout the bay. Fishing regulations are set according to the results of these assessment, and fishing is managed at the species level. Blue crabs are a keystone species in the bay because they are prey for fish and bivalves during their larval stages, and for fish (sharks, striped bass, red drum and catfish) and birds when adult.

The *Crassostrea virginica* fishery dates back to at least 1880, and during this time the landings have declined more than seven-fold since 1950 (Figure 6.8) and twenty-fold since 1890 (Figure 6.7b). Large oyster reefs were present in the bay in the seventeenth century, and oysters were initially hand-collected followed by destructive dredging. Over-fishing, disease and habitat loss (destruction of reefs, increased sedimentation and expansion of hypoxic water masses) have all contributed to oyster decline. In 1949 there was an outbreak of *dermo*, a disease that affects oysters caused by the protist *Perkinsus marinus*: the intensity of outbreaks is thought to be linked with the *La Niña* phase of the El Niño–Southern Oscillation (ENSO). Another pathogen, *Haplosporidium nelsoni*, also known as multi-nucleated unknown disease (MSX) appeared in 1959 and by 1963 affected 90% of oysters in the bay, causing the mortality of four-fifths of the population in some areas. The risks from MSX remain high in low-salinity areas. The loss of oyster reefs has contributed to the hypoxia problem: before the 1870s, the Chesapeake Bay oysters filtered all the water in the Maryland portion of the bay in 3.6 days (Figure 6.7b). With the oyster populations present in the bay in 2003, this would have taken two years.

The American shad, *Alosa sapidissima*, harvested by native Americans and early colonialists, was a very valuable fish in the past but in the 1800s there were large declines in stock (Table 6.7; Figure 6.8), and by 1994 all states had closed the fisheries. Similarly, the Atlantic sturgeon *Acipenser oxyrinchus oxyrinchus* another important fishery is now at less than 1% of its 1900s stock (Table 6.6) and the fishery was closed in 1998. Over-fishing, deteriorating water quality and extensive dam construction (being anadromous damming impacted their migration and spawning) led to the collapse of both the shad and sturgeon. The sturgeon population is now listed as endangered in the United States.

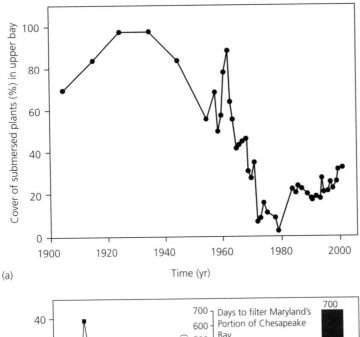

(a)

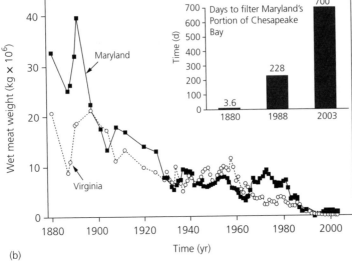

(b)

Figure 6.7 Eighteenth- and twenty-first-century changes in Chesapeake Bay ecosystems. (a) Submerged macrophyte coverage at a site in upper Chesapeake Bay (Susquehanna flats), and (b) changes in harvested eastern oyster biomass in Maryland and Virginia; inset box shows number of days per year required for oysters to filter the portion of Chesapeake Bay in Maryland during three different time periods. Credit: Kemp et al. 2005. Eutrophication of Chesapeake Bay: historical trends and ecological indicators. Marine Ecology Progress Series, 303, 1–20. © Inter-research 2005.

Chesapeake Bay is the primary spawning area for 70–90% of the Atlantic striped bass or rockfish, *Morone saxatilis*, once one of the most important commercial fish species in the bay (Figure 6.8). The declines in striped bass led to a moratorium and since the fishery re-opened catches have remained stable but low (Figure 6.8). Its decline is attributable to over-fishing and habitat contraction that have resulted from the spreading hypoxia. However, there also seems to be a high incidence of mycobacteriosis, a disease that manifests as surface lesions, perhaps due to degraded water quality (e.g. Section 2.3).

One of the key changes in Chesapeake Bay fisheries includes a shift from demersal species such as sturgeon and striped bass towards pelagic species such as the menhaden (Figure 6.8). Populations of Atlantic menhaden *Brevoortia tyrannus* have actually seen an increase over the last 60 years. The menhaden is harvested commercially as bait (77%) for commercial and recreational fishing, to produce fishmeal (23%) for animal feeds and fertilisers and

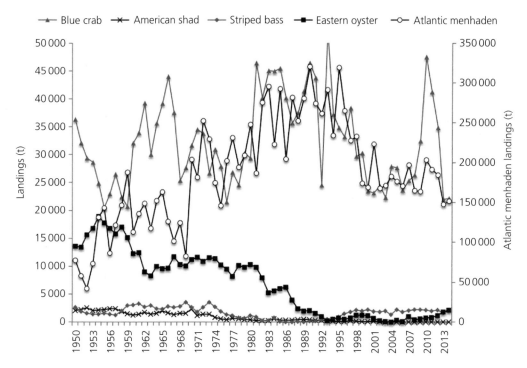

Figure 6.8 Fishery landings for commercially important species in Chesapeake Bay between 1950 and 2014 (Atlantic menhaden are on the right axis and the other four species are plotted on the left hand axis). All American shad fisheries had closed by 1994, bass fisheries underwent a moratorium between 1984 and 1989. Oyster catches are the combined wild caught and farmed. Catches of Atlantic sturgeon are not shown. Data from: NOAA (2016).

Table 6.6 Historic fishery landings of sturgeon and shad in Chesapeake Bay. For historic oyster landings see Figure 6.7b. Data from: NOAA 2016.

Stock	1890	1920s	1990s
Atlantic sturgeon	318 t y^{-1}	10 t y^{-1}	1 t y^{-1}
	1831–50	**1890s**	**1970s**
American shad	41 000 t y^{-1}	8000 t y^{-1}	1000 t y^{-1}

for fish oil. Landings grew eight-fold between 1952 and 1995. However, since 1998 stocks have been lower which is thought to be due to over-fishing, heavy predation or pollution. Menhaden is a keystone species as prey for larger fish and as effective filter feeders (filtering around 15 l min^{-1}). Thus, menhaden can, like the oysters, help to improve water quality by removing excess suspended sediments, nutrients and algal blooms. Their feeding mode and pelagic habit may explain their success in Chesapeake Bay's deteriorated waters.

6.4.4 Management

Activities across states are coordinated by the Chesapeake Bay Programme, created in 1983 to help coordinate the management and restoration of the bay. As part of the 2014 watershed agreement, the programme committed to ensuring that the effects of contaminants in the bay were minimised. Assessments of ecosystem health (based on dissolved oxygen, nutrient concentrations and aquatic macrophytes) in 2015 concluded that only the lower bay could be considered in good health. Of the other 14 regions monitored, 5 were of moderate but unchanging health, 4 were moderate and improving, 3 were poor but improving (Patapsco and Back Rivers, York River, lower western shore and Elizabeth River), and Patuxent River was in poor health and showed no changes. Water clarity and chlorophyll have improved, aquatic grasses are only improving up-river, and nutrient inputs remain a problem.

The Maximum Total Daily Load Project coordinated across six states is tasked with reducing anthropogenic nutrients and sedimentation. Between 2009 and 2014, dissolved nitrogen reduced 6%, phosphorous 18% and sediment load 4%. Thus, the management of nutrient inputs has shown improvements but not in all areas and reductions are slow. A 2015 study by the US Geological Society (Ator and Denver 2015) showed that the nitrates may remain in the ground waters that discharge into Chesapeake Bay for a considerable period, and that the changes instituted by the current management may require decades before they manifest as reductions in the bay.

In 2004, the Chesapeake Bay Oyster Management Plan was introduced which outlined a number of actions: to manage harvests, establish oyster sanctuaries, tackle disease outbreaks and restore oyster reefs using hatchery-raised seed. Aquaculture production of *Crassostrea virginica* is on the rise: using seed oysters raised in hatcheries around 2160 t of oysters were farmed in Maryland and Virginia in 2014 (Figure 6.9). The established oyster sanctuaries include subtidal reefs that are closed to fishing in areas identified as ideal for oyster settlement, and the idea is that these sanctuaries will ultimately seed the fishery areas. Research by the Virginia Institute of Marine Science shows

that oysters are developing disease resistance, and attempts are now being made to breed resistant stock. Projects to restore oyster reefs began in 2010 and to-date, 350 acres of reef have been restored in Harris Creek, in east Maryland. The program aims to establish reefs in ten tributaries by 2025. It is hoped that these reefs will help to treat anthropogenic nitrate pollution, as they did in the past, in addition to providing fishery resources.

6.5 The Mediterranean Sea

The Mediterranean is an almost totally enclosed maritime region lying between three continents; Europe to the north, Africa to the south, with the Levant (a part of Asia) to the east. The Mediterranean connects with the Black Sea in the north-east (Section 6.5), and with the Atlantic Ocean, via the Straits of Gibraltar, to the west. A submarine ridge, running from Sicily to Tunisia, separates the Mediterranean into two basins, the western and the eastern. The Mediterranean is, in fact, the remains of the Tethys Ocean, the water body that originally separated the super continents of Gondwana land and Laurasia. As the African continent moved closer to Europe, the Tethys Ocean became constricted into the narrow sea we observe today, and the

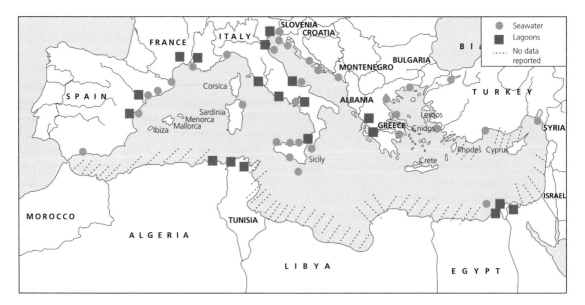

Figure 6.9 Regions of the Mediterranean where eutrophication phenomena have been observed. Data from: European Environment Agency. Environmental Assessment Series No. 5.

continental collision created the Alps, the Pyrenees (Spain) and the Zagros Mountains (Iran), and continues to drive volcanic and seismic activity around the Mediterranean.

This oceanic legacy means that, while appearing as a coastal sea today, the Mediterranean is actually a deep basin, with an average depth of 1500 m and a maximum depth of 5267 m. The Mediterranean has a surface area of around 2 500 000 km^2, a volume of 3 750 000 km^3 and a residence time of 8–100 years. The principal water exchange is with the Atlantic via the Straits of Gibraltar, where a strong inward flow of Atlantic water overlies the outward flow of warm saline water, at 100–300 m depth, formed by evaporation and warming primarily in the eastern basin. This warm saline water with its burden of contaminants spreads out into the Atlantic as a distinct water mass at around 600–1300 m depth. Given the various basins, islands and isthmuses in the Mediterranean, the flow regime is complex and strong seasonal winds blow across the surface from North Africa introducing a seasonal element to the hydrography that drives seasonal surface currents.

The Mediterranean is naturally one of the most nutrient-poor regions of the global ocean, and the system can be characterised as being of high diversity but low productivity. This natural low productivity means that fisheries resources are easily over-exploited and reductions in populations of one species often cause ecological changes in other species through ecological cascades. The system is extremely sensitive to additional, anthropogenic, inputs of nutrients from sewage, agricultural runoff and aquaculture. The most productive ecosystems in the Mediterranean are the seagrass beds that originally covered most of the shallow areas, sometimes down to depths of 40 m, across the entire basin. Damage from vessels and fishing gear, pollution, eutrophication and competition with non-native algae have together driven massive reductions in the extent and quality of seagrass habitats throughout the region.

While significantly lower than in northern Europe, the concentrations of nitrogen and phosphorous in those rivers entering the northern Mediterranean and the Nile are all around four-times background. A further concern is that there is a continuing upward trend. Coastal regions and bays are at the greatest risk of eutrophication (Figure 6.9). Nuisance algal blooms and incidents of toxic shellfish poisoning have been reported from Spain, France, Italy and Turkey in recent years.

The Mediterranean is often characterised as the cradle of civilisation, with first the Egyptian, then the Greek and Roman empires occupying and civilising the lands adjacent to the Mediterranean coast. Subsequent advances and, in particular, the Industrial Revolution and the intensification of agriculture occurred in northern Europe. Southern Europe and much of North Africa and the Levant lagged further behind. This has affected the pattern of human pressures on the region with fishing and shipping dating back at least five millennia. Pressures from urbanisation and industrialisation have been relatively recent. While most of the states bordering the northern Mediterranean are members of the European Union and so are bound by its environmental regulations, most of the remaining independent states are significantly less developed, with attendant less robust environmental protection regimes, a situation confounded by political instability and war. Over 450 million people already live within catchments that drain into the Mediterranean while significant population growth continues in North Africa. These numbers increase each year with the mass arrival of tourists: the Mediterranean is the primary tourist destination in the world. It is estimated that between 235 million and 355 million tourists will visit the Mediterranean coast in 2025. Both the high visitor numbers and the seasonality of the pressures place significant demands on ecosystem services such as sewage treatment and waste assimilation. Tourism also drives urbanisation, land reclaim and coastal defence projects.

One unusual human impact within the Mediterranean is the invasion of non-native species following the linking of two previously isolated oceanic biogeographic provinces (Section 4.8), the eastern Mediterranean with the Red Sea via the opening of the Suez Canal in 1869. The Red Sea is topographically higher than the Mediterranean and so water flows from the Red Sea into the Mediterranean. Water in the Red Sea is warmer, more saline and more nutrient-poor than the eastern Mediterranean. Initially, the hypersaline Bitter Lakes that form part of the canal acted as a barrier to species migrations, but over time they became flushed with Red Sea water and are no longer hypersaline. The physiological barrier was thus removed and the Red Sea species began invading the eastern Mediterranean via the canal: these are known as Lessepsian Migrations. Red Sea species being adapted to warm, high-saline and low-nutrient conditions were able to thrive in the eastern basin, outcompeting the native Mediterranean species. This situation was further exacerbated in the 1960s when construction of the Aswan High Dam on the River Nile reduced freshwater and nutrient flows into the eastern Mediterranean. This made the conditions even more favourable to the invaders. Red Sea species are now an ecologically significant part of the flora and fauna of the Levant Basin, and over 500 Indo-Pacific species have been recorded in the eastern Mediterranean. The Panama Canal is credited with having similar effects in the Caribbean Sea, but the nature of the canal, which has significant sections of essentially fresh water, makes direct migrations less likely and the Pacific species recorded in the Caribbean have probably been transferred within ships ballast.

In recent decades in the Mediterranean, there has been increasing evidence for the occurrence of tropical Atlantic species in the western Mediterranean (over 55 species at the time of writing). These non-natives are the result of range expansions driven by global warming and/or increased transport by shipping. This means that the natural flora and fauna of the Mediterranean Sea are being squeezed from both sides by the invasion of non-native species.

Twenty-three countries and two British Overseas Territories have coastlines that border the Mediterranean (Albania, Algeria, Bosnia-Herzegovina, Croatia, Cyprus, Egypt, France, Greece, Israel, Italy, Lebanon, Libya, Malta, Morocco, Monaco, Montenegro, Northern Cyprus (only recognised by Turkey), Palestine, Slovenia, Spain, Syria, Turkey and Tunisia, and Gibraltar and 'Akrotiri and Dhekelia', which are British Overseas Territories). Given the diversity of nations, traditions and the antipathy between some of the states, achieving international agreements and coordinated actions concerning issues of environmental management has been difficult.

In 1976, 14 of the countries bordering the Mediterranean and the European Union (at that time the European Economic Community) signed the Convention for Protection of the Mediterranean Sea against Pollution also known as the Barcelona Convention. Modelled on the Oslo (1972) and Paris (1974) Conventions (consolidated in 1992 as the OSPAR Convention, see Section 6.2) covering the North Atlantic and the Helsinki Convention (1974) that covers the Baltic Sea. The Barcelona Convention seeks to prevent and abate pollution from ships, aircraft and land-based sources in the Mediterranean Sea. The contracting parties met again in 1995, but all signatories have not yet ratified the amendments. The Barcelona Convention has not seen the same level of engagement and progress as the OSPAR or Baltic Marine Environment Protection Commission—Helsinki Commission (HELCOM) agreements, perhaps due to the economic and political status of many of the signatories. Both the latter have permanent secretariats and coordinated programmes of scientific monitoring, research and reporting. The United Nations Environment Programme (UNEP) has developed a Mediterranean Action Plan to support the Barcelona Convention.

The main human pressures on the Mediterranean Sea are: the poor state of fisheries resources following many decades of heavy and indiscriminate exploitation; the urbanisation of the northern coast with the associated problems of waste-water, eutrophication, industrial discharges and litter; and inputs of oil from shipping, offshore oil and gas exploration and production and oil refining. Coastal development and coastal protection with the associated habitat loss are also amongst the main threats to the Mediterranean environment.

The UNEP Mediterranean Action Plan has identified seven key priorities for the next decade. They are:

- To bring about large reductions in pollution from land-based sources;
- To protect marine and coastal habitats and threatened species;
- To make maritime activities safer and more conscious of the Mediterranean marine environment;
- To intensify integrated planning in the development of coastal areas;

- To monitor the spread of invasive species;
- To limit and intervene promptly in oil pollution incidents.
- To further promote sustainable development of the Mediterranean region

While these are all clearly important goals, they lack a strong political process to drive change in all the nations that border the Mediterranean, and will make achieving progress even at this high level and with these aspirational priorities difficult. Much of the northern coastline now operates under the auspices of the European Union and will be bound by their rules to make progress in delivering marine ecosystems that achieve good environmental status. Thus, there is some hope for improvements in the state of the Mediterranean Sea in the near future.

6.6 The Canadian LOMAs

Canada is the second largest nation on Earth by land area and it has an extensive coastline with many complex estuaries, fjords, bays and offshore islands. Canada also borders three oceans: the North Pacific, Arctic and North Atlantic oceans. Canada's maritime area covers approximately 7.1×10^6 km^2, equivalent to about 70% of the land area. A comprehensive economic analysis of the importance of the marine environment to Canada's economy showed that marine sector activities generated an estimated US \$17.7 billion in direct gross domestic product (GDP) for Canada in 2006, creating over 171 000 jobs. If the indirect impact on GDP and employment were included, then the marine environment supported around 2% of GDP and national employment.

Canada was among the first nations to recognise the need for integrated marine environmental management and to establish a legal framework for planning and management in the marine environment. In 1996, Canada passed the Oceans Act that set out to provide a single framework for the development and implementation of a national strategy for managing estuarine, coastal and marine ecosystems in Canadian waters. The approach was inclusive (in terms of society and stakeholders) and integrated in terms of providing an ecosystem perspective for the planning and regulation of multiple sectors.

To make this approach operational the Department of Fisheries and Oceans (DFO) designated five Large Ocean Management Areas (LOMAs). The LOMAs were delineated so that ecosystem health and economic development issues within their boundaries could be addressed and managed using an integrated approach. This is to say that an approach that explicitly addressed the socio-economic needs of humankind while preserving the health of the marine ecosystem was adopted. One of the key tools in delivering ecosystem protection has been the identification of Ecologically and Biologically Significant Areas (EBSAs).

The five LOMAs designated (Figure 6.10) are:

- Pacific North Coast
- Beaufort Sea
- Gulf of St. Lawrence
- Eastern Scotian Shelf
- Placentia Bay/Grand Banks

The designation of the LOMAs took until 2005 and draft management plans for two of the east-coast LOMAs had been developed and published by 2008. However, the Pacific North Coast plan only emerged in 2013. Changes in the political make-up of the Federal government and the differing priorities of the regional governments and diverse stakeholders have caused progress to slow down.

Managing the LOMAs is a four-step process:

i. *Initiating the planning process* which delineates the eco-region and defines the planning area and team.
ii. *Informing and reporting* on the area that begins by conducting an ecosystem overview and assessment report.
iii. *Setting management objectives* for the area including economic, social, cultural and conservation objectives.
iv. *Developing and implementing an integrated-management plan* for the area, which includes management measures, monitoring and reporting.

In preparing the draft management plans, the DFO conducted an extensive consultation with stakeholders and, as a basis for this, prepared an ecosystem overview and assessment (or 'state of

Oceans Management Areas

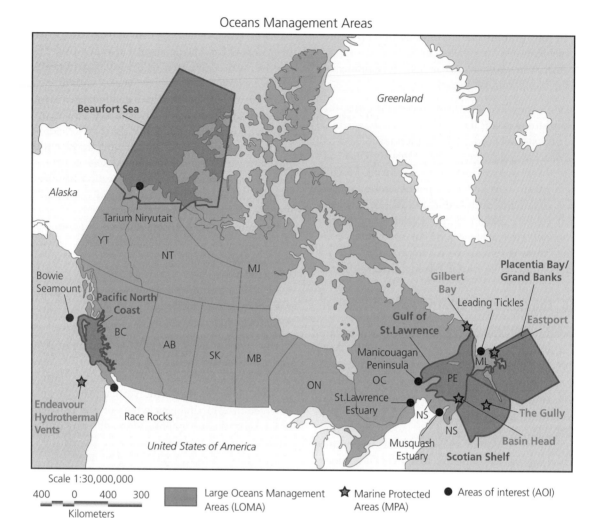

Figure 6.10 Canada is bordered by three oceans and has designated five areas (Beaufort Sea, Pacific North Coast, Gulf of St. Lawrence, Placentia Bay/Grand Banks, and Scotian Shelf) as Large Ocean Management Areas for which integrated management plans are being developed under its Oceans Act. (Redrawn from http://www.oag-bvg.gc.ca/internet/English/pet_354_e_39108.html).

the oceans') report for each LOMA. Similar to the Quality Status Reports for the North East Atlantic produced by OSPAR (Section 6.2.2), this meant that everyone involved in the negotiations started with the same set of facts.

The approach taken has varied between regions but the nature of the reports is broadly comparable. Across all the areas the reports find good evidence of multi-decadal change in the ecosystems. Much of this is driven by changes in the physical forcing of the environment and is potentially attributable to

anthropogenic climate change. The other high-lighted concern is the impact of fisheries, with all regions identifying species whose populations are so depressed, as a result of harvesting, that they are beyond what are considered to be safe biologi-cal limits. While concerns about contaminants and non-native species are also raised in the majority of reports, these tend to be restricted to areas ex-periencing high densities of man-made pressures; that is, around ports, estuaries and major seaways. This probably reflects the scale of reporting: at the

LOMA scale, the widespread impacts of fishing dominate even the most severe impacts of contaminants in a single estuary. The Gulf of St Lawrence LOMA, which is the most industrialised and urbanised of the five LOMAs, considers environmental variation, fisheries, shipping and aquaculture to be more significant than contamination. However, one consequence of the changed nature of the environmental drivers in this area is the altered inputs from the river catchment area and the great lakes.

6.7 The Coral Sea and Great Barrier Reef

The Coral Sea is a marginal sea of the south-west Pacific that is approximately twice the size of the Mediterranean Sea. It is bordered by north-eastern Australia to the west and to the east it is separated from the south-west Pacific by the Solomon Islands, Vanuatu and New Caledonia. At its northern reaches lies Papua New Guinea and the Solomon Sea; to the south is the Tasman Sea that separates Australia from New Zealand. The average depth of the Coral Sea is 2400 m, and the seafloor consists of abyssal plains in the north-west, a large plateau to the west and a series of slopes and deep-water trenches. A 100 km continental shelf runs along the coast of Queensland, around Papua New Guinea, and there is a narrower shelf around New Caledonia. There are three major current systems in the Coral Sea: the Eastern Australia Current running south along the coast from Fraser Island, the South Equatorial Current which flows between the islands from the Pacific where it divides and flows north in the Gulf of Papua as the Hiri Current.

The Coral Sea contains the world's largest modern reef system spanning 14° of latitude, and consisting of more than 2900 separate reefs and 1050 islands. The *Great Barrier Reef* (GBR) has been a UNESCO World Heritage Site since 1981. A diverse range of habitats are represented which support an exceptional biodiversity. The coastal habitats include mangroves 2070 km^2, seagrass of 6000 km^2, saltmarsh 1830 km^2, *Halimeda* banks (calcareous green algae), and 345 000 km^2 of different types of reef ranging from a simple to a complex morphology. The GBR is home to 1625 bony fish species, fringed by 30–40 species of mangrove and 15 species of seagrass, 150 soft corals and sea pens,

411 hard corals and 630 species of algae. Other invertebrates include sponges (2500 species), echinoderms (630), crustacea (1300), molluscs (3000) and worms (500). These groups alone give a total species richness of approximately 7930. There are also many species of bryozoan and ascidian. Many of the species are notable, including the giant triton snail of 0.5 m length and the giant clam which lives for more than 100 years, 30 species of whale, dolphin and porpoise, dugongs, 6 species of turtle, 14 species of sea snake and 136 species of elasmobranch. Many species of birds occur including 20 sea bird and 41 shore bird species, of which 53 are migratory. Between 1.4 million and 1.7 million birds use the islands to breed. Saltwater crocodiles occur at the fringes and there are numerous jellyfish species in the water column.

The Coral Sea region has been populated for around 50 000 years since aboriginal people arrived from the north by crossing the Torres Strait. They lived a hunter–gatherer lifestyle and permanent dwellings seem to have been built for approximately 10 000 years. Queensland is thought to have been the most densely populated region of Australia (with 40% of the total indigenous population). Europeans established their first temporary settlements in *c.* 1824 and following this, a penal colony was constructed. The first immigrant ship arrived in 1848 and rapid development led to the establishment and growth of Queensland's coastal towns and cities over the subsequent 150 years.

There are 13 major ports along the coasts of the western Coral Sea and shipping activities are dominated by exports of coal, sugar, iron ore, oil, chemicals and livestock. A further large port exists in Papua New Guinea, Port Moresby, which exports large quantities of fossil fuels, minerals and timber. The extensive reefs represent a hazard to shipping, and ship groundings and wrecks are relatively common.

The Coral Sea receives freshwater discharges from several large urban areas running off of a catchment of 1.24×10^6 km^2 (26% of the area of the state) and 16 large population centres. Within the GBR region, 74% of the land is used to raise cattle, 11% is used to grow sugar cane and 5% is used for forestry, and land use has not changed significantly since the 1990s. Therefore, the potential nutrient and sediment inputs from livestock farming and agriculture are substantial. Combined with a climatic regime consisting of long, dry periods followed by intense monsoonal rainfall and low vegetation cover means that sediment run-off can be considerable. The Coral Sea generated commercial fisheries landings of 30 000 t in Queensland (capture and aquaculture) in 2009, with approximately 7000 t in the GBR (with the majority from trawl fisheries), plus fishing for the aquarium trade (73 000 fish in 2012) and recreational fisheries.

Queensland has a population of 4.75 million people in an area of 1.73×10^6 km^2 and so population density is fairly low (around 2.8 people km^{-2}) but the bulk of this population is concentrated within the coastal zone (150 km strip inland from the coast) and the effective population density is around 4.5 people km^{-2}. In comparison with India (more than 1000 people km^{-2}), Pakistan (more than 300 people km^{-2}) and parts of Europe, this is still very low and is lower than the states of Victoria, New South Wales and Tasmania. It is thus one of the least populated regions on Earth. However, this resident population increases seasonally with tourist visitors to the GBR region which can be considerable, around 2 million in 2013. Having a fairly low population density and comparatively short industrial and agricultural history, it might be expected that the region would suffer less pollution than the older and more populated regions of the world. In some respects this seems to be true if we compare it to the Mediterranean, the Black Sea, the North Sea and Chesapeake Bay (Sections 6.2–6.4). There is, however, a legacy from a century of historic mining and agriculture that is reflected in some coastal sediments. Some of the coastal habitats of tropical and subtropical systems are particularly vulnerable to pollution and disturbance; for example, saltmarsh, mangrove and reef. The abundance of Australia's natural resources and growing population means that there are many future pollution challenges to come.

Twenty-five species that are resident in the GBR region are listed as nationally vulnerable, threatened or critically endangered (seven reptiles, seven sharks and rays, five mammals, and six sea birds). Another 76 migratory species listed

in Australia's Environmental Protection and Biodiversity Conservation Act of 1999 move through the region. Mammals thought to be at risk include the snubfin and Indo-Pacific humpback whales due to their small and isolated populations. Humpback whales are listed as nationally threatened are in a poor but stable condition (about 14 500 in 2010), and dwarf minke whales are also at risk. Population data for whales and dolphins prohibit a thorough assessment of trends. Seven resident shark species are listed as threatened due to direct over-exploitation or bycatch; those species with slow maturation and growth rate are especially at risk. The status of many species of rays and sharks have not been assessed. Similarly, sea snakes are not monitored, but populations seem to be at risk due to the large numbers (100 000 y^{-1}) caught as fisheries bycatch. All habitat types in the Coral Sea—for example, sea grass, mangrove, sand dune and reef—are in poor condition within some parts of their range.

Six of the total seven marine turtle species visit the Coral Sea and of these five (loggerhead, green, hawksbill, leatherback and flatback) nest on more than 34 islands of the GBR, and the olive ridley occurs in its waters. Of these that inhabit the Coral Sea populations, the IUCN considers three turtle species as vulnerable, one endangered and one critically endangered, while one has not been assessed (the flatback). Dugong populations occur in the north and south of the Coral Sea and were listed as vulnerable by the IUCN at the time of writing.

6.7.1 Nutrients and suspended solids

Nutrient enrichment and sedimentation in the Coral Sea represents one of the main threats to ecosystem health especially in near-shore areas such as reef and seagrass meadows. Anthropogenic nutrient inputs to the Coral Sea are mainly derived from the run-off of fertilisers from crop land, urban run-off and increased sedimentation. Total nitrogen and phosphorous have nearly doubled since the late 1800s to 35,053 t y^{-1} and 5849 t y^{-1} in 2013, respectively. These elevated nutrient inputs cause eutrophication as indicated by chlorophyll a concentrations which have increased in the water column by 10–15% since 2002.

Turbidity in the coastal zone has increased four- to eight-fold since the 1800s, up to 7930 Kt y^{-1}, due to urbanisation, agriculture and changes in the freshwater inputs. These sediments are changing the morphology of estuaries and the distribution of intertidal habitat types and are causing increasing turbidity along the coastal zone. The water-quality guidance for suspended sediments requires more than 2 mg l^{-1} and for chlorophyll *a* more than 0.45 mg l^{-1}. However, the guidance limits for both are frequently exceeded in most regions along the southern two-thirds of the GBR coastline (Figure 6.11).

Increases in the number of algal blooms together with increased turbidity have limited light penetration, causing a 50% decrease in water visibility since 1929. Light is required by seagrasses and corals which contain photosynthetic symbionts. Substantial losses of seagrass have occurred all along the coast with Mourilyan, Cairns, Townsville, Gladstone, Moreton Bay and northern New South Wales being particularly hard hit. Seagrass loss has occurred due to poor water quality (limiting light supply and causing epiphytic overgrowth, that both reduce photosynthesis), sedimentation and habitat loss. Increased freshwater inputs, particularly in the south, have caused stress to seagrass and low-salinity bleaching of corals.

Coral cover has declined more than 20% throughout the GBR since 1985, with the largest losses (of up to 30%) occurring in the south. Baseline data on the amount of coral cover prior to this are limited but it seems likely the decline since the 1800s is also considerable. These coral losses were caused by a combination of declining water quality directly, by restricting light penetration, and indirectly by increasing the stress on corals thus increasing susceptibility to bleaching, disease, competition and predation. A switch in species composition has been observed towards coral taxa that are more tolerant to sedimentation and turbidity.

Reefs usually become established in oligotrophic settings, whereas phytoplankton and macroalgae benefit from nutrient enrichment. Shifts from coral- to algal-dominated systems are attributed to eutrophication both directly, by providing nutrients that support macroalgal growth and thus competition for space and indirectly, by causing coral mortality. Outbreaks of crown-of-thorns starfish,

Acanthaster, a native predator on corals, have been increasing since 1960 with the highest incidence occurring between Cooktown and Bowen. Natural outbreaks are thought to occur every 50 years or so, but poor water quality seems to have increased their frequency. This voracious starfish can consume 478 cm^2 of coral per day. Outbreaks of the snail *Drupella*, another native predator on coral, also occur in the GBR. Elevated incidence of disease (e.g. white syndrome) has been observed for corals which seems to coincide with extreme temperatures. Outbreaks of the cyanobacteria *Trichodesmium* and blooms of the harmful algae *Lyngyba* also occur that can exclude light and smother the corals.

The Coral Sea is an important habitat for dugongs and turtles which were over-exploited between the 1840s and 1970s. Although the northern dugong population is in good health, the southern population has been in decline for several years. Green and loggerhead turtle numbers are improving (Figure 6.12). Declines in dugongs (Figure 6.13) and turtles are due to the loss of seagrass, poor water quality, ghost fishing, inclusion in bycatch, collisions with vessels, ingestion of plastics and occasional poaching. Additionally, turtle reproduction is impacted by habitat loss and light pollution near their nesting sites, and in Moreton Bay reproduction may also have been inhibited by exposure to the *Lyngyba* toxin. Disease affects turtles in some areas, for example, fibropapillomatosis in green turtles, in enclosed bays with poor water quality, and dugongs are also experiencing more necropsies and pneumonia. Turtle and dugong strandings are on the increase; 500 to 1000 turtles and 20 to 160 dugongs per year in 2011–13. A large stranding of dugongs occurred in 2011 when 160 were found along the GBR coast, this was thought to be due to elevated freshwater flow causing salinity stress.

Extensive oyster farming occurs in southern Queensland extending south past Sydney and covering around 43 km^2. The expansion of coastal aquaculture may contribute to the nutrient enrichment problems of the Coral Sea and increase the biosecurity risks. The introduced species *Crassostrea gigas* is spreading and although the ecological impacts are not yet clear, it is likely to compete with native oysters. Also, oyster farms have had a number of disease outbreaks (viral, bacterial, protozoan

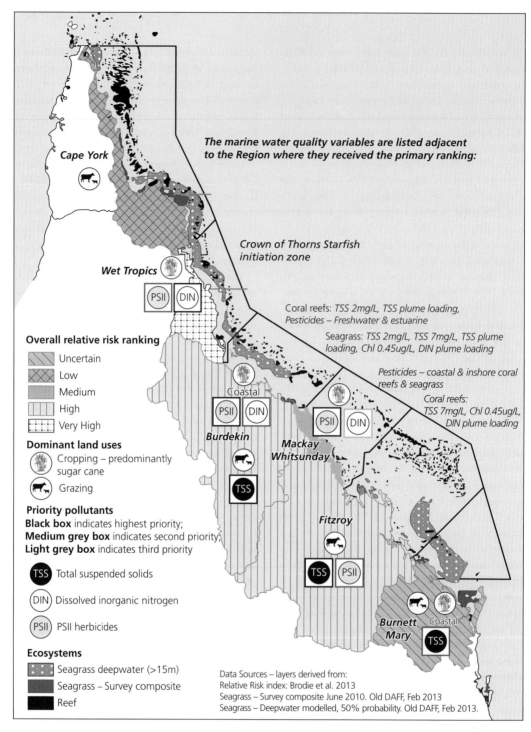

Figure 6.11 Relative risks to the great barrier reef coastline from degraded water quality identified by Brodie et al. (2013) for each of the six natural resource management regions. Showing the extent of coral and seagrass habitats, the main land uses in each catchment, the priority pollutants within each catchment and the risk ranking for adverse ecological impacts. Chl = chlorophyll, DIN = Dissolved Inorganic Nitrogen, TSS = Total Suspended Solids. The crown of thorns starfish initiation zone is where outbreaks begin; these are an important cause of coral loss (their initiation being linked to the runoff of excess DIN). Reproduced with permission from: Brodie, J. et al. 2013. Assessment of the relative risk of degraded water quality to ecosystems of the Great Barrier Reef. A report to the Department of the Environment and Heritage Protection, Queensland Government, Brisbane. TropWATER Report 13/28, James Cook University, Townsville, Australia.

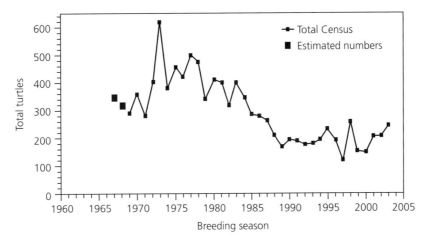

Figure 6.12 Number of loggerhead turtle nesting females from 1967 to 2012 on five beaches of the Woongarra coast (note: 1968 and 1967 are estimates whereas the others are based on tagging census data for the full breeding season). Reproduced from Limpus, C.J. (2008). A biological review of Australian marine turtle species, 1 Loggerhead turtle, *Caretta caretta* (Linnaeus). Queensland Government Environmental Protection Agency, Brisbane. © The State of Queensland (Department of Environment and Heritage Protection).

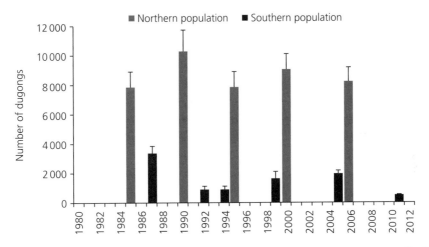

Figure 6.13 Size of the two western Coral Sea dugong populations (mean ± SE) from 1985–2011. © 2014 Commonwealth of Australia (Great Barrier Reef Marine Park Authority 2014) Licensed for use under a Creative Commons By Attribution 3.0 licence (see: https://creativecommons.org/licenses/by/3.0/au/legalcode).

and flatworms), perhaps exacerbated by poor water quality and these could spread to wild populations. Oyster reefs can perform important roles in improving water quality (Section 6.4).

6.7.2 Urbanisation and tourism

Besides nutrients, the other main type of diffuse contaminants of concern in the Coral Sea include pesticides (Figure 6.11) and marine litter. Pesticides are washed off crops or recreational land in rainwater. The amounts of pesticides entering the GBR are estimated at 12,114 kg y^{-1} from agricultural areas. The highest concentrations of pesticides in the GBR occur near Mackay and in the lower Burdekin River, where they frequently exceed the water-quality guidelines by a factor of 10–50. The large pesticide inputs are thought to come from intensive sugar cane cultivation in these areas. In some regions, concentrations exceed the no-observed-adverse-effect level (NOAEL) and have been found to impact corals, algae and perhaps mammals, but the chronic effects have not been established. Being persistent pollutants many of these pesticides have the potential to accumulate in sediments and so will become legacy contaminants (Section 3.5.2).

Growing coastal populations in the western Coral Sea and tourism, in addition to the industries discussed earlier, mean coastal infrastructure will expand and the associated pollution will also increase. In 2014 in most parts of Queensland coastal populations were increasing by 1–2% with some of the fastest being Brisbane, the Gold Coast, Townsville, Cairns, Gladstone and the Cook region in the north. Tourism is big business in the Coral Sea; with about 2 million visitors coming to the GBR in 2013. The industry is forecast to grow at rates of 0.8% for domestic and 4.2% for international tourism from 2014–24. In the GBR tourism is focused in the Whitsundays, Port Douglas and near Cairns. Tourist resorts are also present on 27 of the GBR islands. So, the direct impacts of pollution from urbanisation and tourism will be concentrated in these highly populated/visited areas.

With regard to industrial infrastructure, coastal development has caused habitat loss due to land reclamation and beach modification. Water quality has deteriorated due to sediment, oil, detergents and metal inputs from urban run-off, and these contribute to declining ecosystem health. Increased coastal light pollution affects the behaviour of nesting turtles and sea birds. Point-source disposal of sewage contributes to coastal nutrient enrichment. Improvements in all major sewage treatment plants along the Australian coast of the Coral Sea in recent years have helped to reduce these nutrient inputs (all GBR sewage treatment is at tertiary level, however, sewage discharges from vessels do occur).

As for many coastal regions that attract recreational visitors, the Coral Sea has problems with marine litter. This litter comprises plastic bags, fishing gear, bottles, rubber flip-flops, aerosol and drinks cans, and so is pretty typical of marine litter everywhere (Section 4.1). From 2008 to 2014, 683 000 items of litter, weighing 42 t, were collected from the GBR by the Australian Marine Debris Initiative.

Litter concentrates within the reef lagoons, and in the southern part of the GBR litter composition is of a more ocean-based origin. A considerable proportion of marine litter originates from land being washed into the sea during storms. Other pollution concerns include microplastics and pharmaceutical and personal care products, but these are not currently monitored.

6.7.3 Pollution from hydrocarbon and minerals exploitation

The area of land used for mining in Queensland has approximately doubled since 1999. Traditionally mining focused on the extensive gold, tin, nickel and uranium deposits in the region, but presently interest is concentrated on fossil fuels because of the high global demand. Today, extensive land-based fossil fuel production occurs in southern Queensland, exported through the Coral Sea. Coal production in Queensland has doubled since the 1990s (Figure 6.14a) and is expected to reach $305–447 \times 10^6$ t y^{-1} by 2025. Growth of the industry is expected to focus on three areas: the Surat, Bowen and Galilee basins. All of these are exported through the GBR. Coal-seam gas, of which Queensland has large deposits (Figure 6.14b), is the other expected area of growth and projections suggest it will reach $25–33 \times 10^6$ t y^{-1} by 2020. Furthermore, the bans on shale oil (hydraulic fracturing; Section 5.4.3) and uranium mining have recently been revoked by the state government and so these will probably also become areas of growth.

Oil production in the Gulf of Papua, north-east Coral Sea, generates around 50 000 barrels per day. There is a large oil terminal, the Kumul Export terminal, situated in the gulf. Between 1987 and 2002 there were 282 oil spills there. Extensive oil and gas exploration in the area continues and there are several large construction projects for exploiting the natural gas reserves. Historically drilling was also attempted on a number of the GBR islands but this does not continue today. There have been seven oil spills from large vessels in the Australian part of the Coral Sea, most of which were relatively small (see also Section 4.5.3). The largest was the Oceanic Grandeur which spilt 1100 t of crude oil into the Torres Strait in 1970, followed by the Pacific

Adventurer which spilt 270 t of heavy fuel oil in Cape Moreton in 2009.

The risks of oil spills in the Coral Sea are high because there are many sensitive habitats and species; for example, sea birds, turtles, coral reef, mangrove and seagrass habitats are some of the most intolerant to oil pollution. It is not clear what the impacts of elevated coal dust might be. The Australian coal reserves are relatively low in metals unlike coal from other sources. That said, the impacts could include increasing the turbidity of the water column and the smothering of benthic habitats.

Naturally high concentrations of metals are present in the metal-rich acid sulphate soils (with aluminium and iron) that could be mobilised by erosion. It has been suggested that these acid sulphate soils are linked with blooms of harmful *Lyngyba* algae to which they provide biolimiting trace metals. Also, near-shore sediments in some areas have elevated concentrations of metals from past mining practices. For example, cores from coastal areas near Townsville have high concentrations of mercury, approximately 25 times higher than those prior to the 1800s. This mercury corresponds with a period of intensive gold mining from the 1870s to the 1900s that used mercury to separate out the gold from the surrounding ore.

6.7.4 Shipping and coastal infrastructure

The largest two ports in the eastern Coral Sea are the Port of Gladstone and Hay Point which handle more than 80×10^6 t y^{-1}, followed by Townsville, Abbot Point and Brisbane, which when combined handled $10–14 \times 10^6$ t y^{-1} of cargo in 2011–12 (Figure 6.15). Shipping has grown considerably since 2000 and is predicted to increase more than two-fold by 2032, mostly due to massive growth of the coal and liquid gas industries (currently more than 60% of trade from the largest ports is in fossil fuels). Port developments are planned for the five largest ports and there are plans for greatly increasing capacity to 50×10^6 t y^{-1} at Yamba, New South Wales. In the western Coral Sea the largest port is Port Moresby in Papua New Guinea that does most of its trade with Australia.

Being fringed by a reef system, the potential for ship groundings and collisions in the western Coral Sea, along with the potential for litter, chemical, fuel or oil spills and shipwrecks, is high. However, increased shipping does not seem to have increased the number of collisions over time in the GBR with less than two per year occurring since 1998. Reef-vessel traffic management is good (Figure 6.14) and provides compulsory piloting for ships exceeding 70 m that carry fossil fuels or chemicals to help them navigate the reef complex. One notable exception is the *Shen Neng I* coal carrier that ran aground on the GBR in 2010 at Douglas shoal and damaged up to 400 000 km² of reef and spilt 4 t of fuel oil.

Increased shipping will introduce additional hydrocarbons, litter and sewage through routine operations, and will introduce anti-fouling compounds, some of which are highly polluting (Sections 3.3, 4.2.1, 4.5, 5.7), and will present possible biosecurity risks (Section 4.8). Increased noise pollution from shipping will affect the large resident and migratory cetacean populations (Section 5.7). The pressures from recreational boating are high in some regions: around 250 000 recreational boating licences were issued in 2014 in Queensland. The impacts are the same, albeit of smaller scale, as those created by larger vessels but they are concentrated near the coast and in popular areas. The areas with the highest intensity of recreational boating include Brisbane with 137 925 vessel registrations, Gladstone 47 510, Townsville 23 490, Cairns 21 313 and Mackay with 19 197 registrations.

The development of coastal infrastructure involves dredging to create navigable channels and land reclamation for the construction of ports and their infrastructure. Large amounts of dredging are needed for the construction and maintenances of shipping channels. Since 1981, 8 km² was reclaimed in the GBR region, mostly at Gladstone. As port infrastructure expands, more dredging is required to deepen the shipping channels, and the shift to larger ships with greater draft may require deeper dredging. In addition to damaging seafloor habitats, dredging can resuspend persistent legacy contaminants held within the sediment such as pesticides and metals from intensive mining in the past (Section 3.6). The relocation of the spoil impacts communities at the disposal site by smothering the benthos and introducing legacy contaminants. Increased turbidity negatively influences the resident corals, macrophytes and the communities that depend upon them for food/habitat.

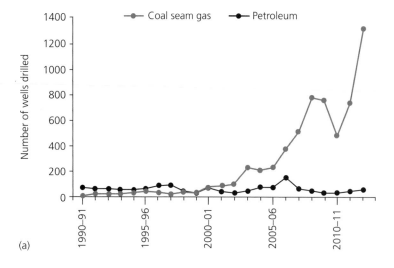

(a)

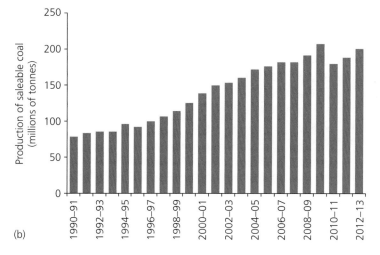

(b)

Figure 6.14 (a) The annual production of marketable coal in Queensland from 1990 to 2013. (b) The number of oil and coal seam-gas wells drilled in Queensland between 1990 and 2013. © 2014 Commonwealth of Australia (Great Barrier Reef Marine Park Authority 2014) Licensed for use under a Creative Commons By Attribution 3.0 licence (see: https://creativecommons.org/licenses/by/3.0/au/legalcode).

Military training exercises in designated areas of the GBR (around 4% of the GBR area) have been conducted for more than a century. Hotspots of defence activity occur in the GBR at Shoal Water Bay, Townsville Star and Amberley. These activities pose a risk of pollution from munitions, for example, perchlorates (Section 5.4.2) and noise (Section 5.7.3). The military has a moratorium on the use of high explosives, except in Shoal Water Bay, in order to reduce the impacts of noise on marine mammals. Pollution threats include the general shipping-related pollution, for example, sewage, fuel, litter, and NIS. In addition, large quantities of unexploded weapons were sunk at the end of the Second World War,

for example, the 8000 t of mustard gas, lewisite, adamsite, teargas, munitions and planes sunk off Cape Moreton, Brisbane.

6.7.5 Management

The GBR marine park protects a large portion of the Coral Sea from anthropogenic impacts by restricting the removal of any fauna or flora without express written permission, regulating tourism, restricting fisheries practice and insisting that ships use designated shipping lanes (Figure 6.15). The reef is managed at several different levels (state, federal and local) with each management level focusing on

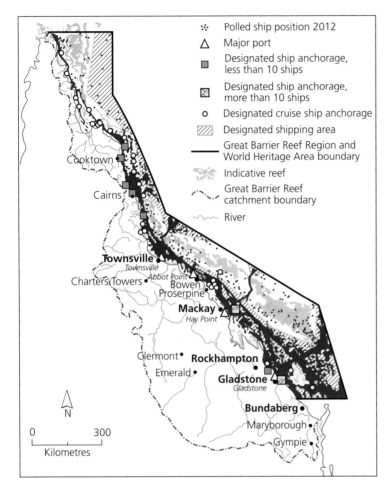

Figure 6.15 Map of shipping anchorages and traffic through the Great Barrier Reef, Coral Sea. Each point represents a vessel position according to polls in 2000, 2006 and 2012. © 2014 Commonwealth of Australia (Great Barrier Reef Marine Park Authority 2014) Licensed for use under a Creative Commons By Attribution 3.0 licence (https://creativecommons.org/licenses/by/3.0/au/legalcode).

different elements and problems (extensive information is available from the Australian government and GBR Park websites; see resources).

The reef water quality protection plan's goal (http://www.reefplan.qld.gov.au/index) is to ensure that 'by 2020 the quality of the water entering the reef from broad-scale land use has no detrimental impact on the health and resilience of the Great Barrier Reef'. The plan has a number of targets relating to land management and three pertaining directly to water quality: to reduce suspended sediment load by 20%, to reduce anthropogenic nutrients by 50%, and to reduce pesticide loads by 60% in priority areas. All targets are to be achieved by 2018.

Since 2009, suspended solid inputs have reduced by 12% and the total amounts of fertilisers used have decreased by 12%, particulate nitrogen and dissolved inorganic nitrogen 17.0% and particulate phosphorous 15%. The provision of financial incentives to farmers to reduce pesticide use has been effective in producing a 31% reduction (Reef Water Quality Protection Plan: Report Card 2014). However, we can see from other systems that have received very high nutrient and sediment inputs (Sections 6.3 and 6.4), that the ecosystem dynamics are complex and achieving the necessary nutrient reductions is not always successful at restoring the system. The recent positive trends in water quality in the western Coral Sea may become compromised by changes in governance. For example, land clearance and land use have been fairly stable in Queensland over the last decade due to restrictions imposed by the Vegetation Management Act of 1999. This stabilisation is believed to have contributed to water quality improvements. However, in 2013, Queensland repealed the Act and has now approved the expansion of high-value intensive agriculture and this will probably result in resurgences in agricultural pollution.

The 2014 State of the GBR report (Great Barrier Reef Marine Park Authority 2014) concluded that there were high risks from climate change, the impacts of urban development and deteriorating water quality, and that the risks from the direct exploitation of the reef were smaller. Climate change, although not pollution *per se*, is affecting the GBR through coral bleaching which has occurred on a large scale in the recent past (e.g. 1998, 2002, 2006, 2016) to extreme effect (up to 50% of the coral in some parts of the GBR were bleached in 2016), and this is increasing in frequency due to elevated seawater temperatures and decreased pH and it is predicted to worsen under most Intergovernmental Panel on Climate Change scenarios (this is discussed further in Section 8.2). The areas of greatest risk from the combined effects of reduced water quality (e.g. nutrients, turbidity, pesticides and disease/predator outbreaks) are in the south, especially near Townsville, Gladstone, Hinchinbrook Island (Figure 6.11) and Moreton Bay.

6.8 The Arctic Seas

The Arctic Ocean extends from the North Pole, 90°N, down to approximately 60°N, except in Hudson Bay, Alaska where it extends to 57°N (Figure 6.16). It is considered the fifth world ocean by the International Hydrographic Organization, and incorporates 11 seas covering an area of 14 × 10⁶ km². In the centre, the Arctic Ocean is divided into two major basins by the Lomonosov Ridge and it attains a maximum depth of 4000–5500 m. It is bordered by Alaska (United States), Canada, Greenland, Norway, Iceland and Russia, and there are 18 ports along its coastline. Extensive shelves are present adjacent to Canada and Russia.

The Arctic is similar to the Mediterranean in being relatively isolated from the other oceans. The Arctic seas are connected to the Pacific Ocean through the Bering Straits, and are connected to the Atlantic through the Fram Strait. The Arctic waters are strongly stratified by the large freshwater inputs (from the Ob, Yenisei, Lena and Mackenzie rivers), and, more importantly, the formation of ice that removes freshwater increasing the salinity. Pacific and Atlantic waters flow into the Arctic and float on top of the cold, very saline and hence dense, deep Arctic water below 900 m. The cold saline Arctic waters are very important because they drive global ocean circulation through the *Global Conveyor Belt*. Dense oxygenated Arctic deep waters flow south through the Atlantic, are replenished in the Antarctic, and then circulate north into the Indian and Pacific Oceans, as Antarctic Bottom Water. Circulation of the warm surface waters proceed in approximately the reverse pattern. The global conveyor transports heat around the world and there are concerns that if the sea ice reduces considerably, this circulation and heat transport will slow down or cease.

The seas of the Arctic experience great environmental variability, with high variability in, for example, salinity, photoperiod and temperature. Large continental shelves cover half its area and there is extensive, permanent sea ice (covering half its surface in the summer). These features are important for structuring the unique Arctic ecosystems. The world's largest colonies of sea birds occur in the Arctic and it is also home to large marine mammal populations. The Arctic Biodiversity Assessment 2013 concluded that there are 64 species of sea bird (6 endemic), 20 species of marine mammal (7 endemic), 243 species of fish, 354 zooplankton taxa, about 4600 benthic animal taxa, 160 species of seaweed (6 endemic) and 50 species of specialised ice fauna including hydroids, nematodes and amphipods. Thirty-nine species of anadromous fish also inhabit the Arctic seas for part of their life cycle. Polar bears, walrus, bowhead whales, narwahls, beluga whales, ringed seals and bearded seals are all endemic to the Arctic. The large Arctic continental shelves are important feeding and breeding areas and they also function as migration corridors for endemics and transients alike. Several whale species visit the Arctic to feed or breed; for example, humpbacks, grey, sei, minke, killer, blue and fin whales, along with harbour porpoise and five species of seal. Notable large fish include the Greenland

Figure 6.16 Map of the Arctic Ocean showing the Arctic seas (dark grey) and those that are the Arctic gateways (light grey). Abbreviations: ACB = Arctic Central Basin, BAF = Baffin Bay, BAR = Barents Sea, BEA = Beaufort Sea, BER = Bering Sea, CAN = Canadian Arctic Archipelago, CEG = Coastal E Greenland, CWG = Coastal W Greenland, CHU = Chukchi Sea, GRS = Greenland Sea, HUD = Hudson Bay complex, KAR = Kara Sea, LAP = Laptev Sea, NOR = Norwegian Sea, SIB = East Siberian Sea, WHI = White Sea. Source: Conservation of Arctic Flora and Fauna/Arctic Biodiversity Assessment (CAFF 2013).

shark, *Somniosus microcephalus*, the Pacific sleeper shark, *Somniosus pacificus*, basking shark, picked dogfish and four species of ray.

Arctic fauna are uniquely adapted to the Arctic conditions and so are very vulnerable to changes in their environment. Many Arctic species also have low reproductive rates—for example, polar bears, Greenland shark, narwahls, walrus and bowhead whales—and so are more vulnerable to anthropogenic pressures. The accessibility of the Arctic Ocean varies according to the extent of the ice cover which ranges from fully covered in winter to half covered in the summer. Thus, the *polynyas* (or recurrent

areas of open water within the sea ice) are hotspots of high productivity, attracting large numbers of sea birds, marine mammals and their predators.

Arctic ecosystems are undergoing unprecedented environmental change (Section 6.8.1) including the loss of large areas of sea ice. The inaccessibility of the Arctic has until now meant that it has remained relatively pristine compared with the other world oceans. For example, the first Arctic crossing was by submarine and did not occur until 1958, and the first ship did not cross until 1977. As the Arctic becomes more accessible to humans, the Arctic ecosystems will experience increasing anthropogenic pressures in addition to extreme environmental change. Thus, although the Arctic has few pollution problems compared with the other world oceans, these are going to increase over the next 50 years.

6.8.1 Current state of the Arctic

In 2010–11, the surface air temperatures over most of the Arctic Ocean were 3 °C above the long-term average for 1981–2010, and the IPCC predicts that temperatures will increase between 2.4 °C and 7.8 °C by 2100. The minimum summer sea-ice cover decreased by 13% per decade between 1981 and 2010, and in 2015 was 29% lower than the 1981–2010 average. The duration of ice cover and its thickness have also reduced. Within the next 30 years the Arctic is predicted to become ice free in the summer (Figure 6.16). Changes in precipitation, run-off, permafrost and glacier melt have also been documented. All of these changes will affect Arctic salinity and temperature regimes that could affect local and global circulation patterns and those species which are intolerant of low salinity/high temperatures.

The biological effects of rising temperatures have already begun to manifest with species geographic ranges shifting; for example, in the North Sea cold-adapted fish are moving north, and a range of Sub-Arctic fish (such as capelin, cod, haddock and herring) and sea birds are moving into the Arctic seas. North Atlantic cold-water zooplankton and phytoplankton species distributions show a northerly shift as do some benthic invertebrates (e.g. crabs, chitons and bivalve molluscs). Increasing numbers of orca have been spotted in Hudson Bay, a nine-fold increase since 1965, and this is thought to be linked to the decreasing extent of sea ice. It is speculated that the environmental changes will alter the patterns of Arctic primary productivity (e.g. the timing and biomass of phytoplankton blooms). Changes in primary productivity together with changes in marine ecosystem composition will impact the structuring of Arctic food webs. While temperate species may move northwards as temperatures rise, cold-adapted polar species will have no refuge. Combined with the increased competition for resources with the invasive sub-Arctic species, in those Arctic taxa that require a narrow range of cold temperatures extinction is highly likely. This pattern

of high extinction rates at the poles is seen during climate change in the fossil record and is predicted by models of range shifts and extinction intensity (Section 8.2).

Because of its inaccessibility most of the Arctic has not experienced extensive human use and consequently the associated local pollutant discharges are low. The most active areas in terms of human exploitation have been the Barents Sea, Iceland Sea and Faroe plateau. However, a number of pollution issues are currently of concern. A sixth plastic gyre was found in the Arctic in 2012. Large amounts of microplastic fibres are present in Arctic Sea ice perhaps due to a convergence of the currents transporting microplastics (Section 4.2.2.3). At present, little is known about the extent of microplastic in the Arctic, but there is an ongoing project looking specifically at the amounts of microplastic in the sea ice.

The Arctic Ocean also contains the largest quantities of high-level nuclear waste (3.68×10^{16} Bq approximately 45% of all ocean-dumped waste up to 1993) compared with the other main oceans. Most of this waste was dumped by the Soviet Union between 1959 and 1992. It includes nuclear reactors loaded with radioactive fuel. There are 20 disposal areas in the Kara Sea east of Novaya Zemlya and six sites in the Barents Sea. Several nuclear test sites were also used at Novaya Zemlya. The impacts of this radioactive waste are not known although most inspections have concluded radioactivity is elevated near the disposal sites but has not contaminated the surrounding environment. However, the problem presented by this waste may change as the area becomes more accessible, as species distributions and Arctic food-web structuring changes and as commercial fisheries expand (i.e. there may be a greater threat to humans through their diet).

Several resident Arctic animal species have been hunted to extinction: the Steller's sea cow, the spectacled cormorant and the great auk. Commercial Arctic fisheries are concentrated in the Barents, Bering, Norwegian and Greenland seas and those near Iceland (Table 6.7). Large quantities of fish and invertebrates are harvested in these areas ranging from 720 000 t y^{-1} to more than 1.5×10^6 t y^{-1}. Landings have been highly variable since the 1950s and several shifts in the species exploited have occurred.

Table 6.7 Fisheries landings in the Arctic seas in 2006. Data from: Sherman and Hempel 2008 and the SeaAroundUs Project 2010.

Region	Landings (t^{-1} y^{-1})
Barents Sea	720 000
Iceland Shelf/Sea	900 000
West Greenland Shelf	200 000
East Bering Sea	1 500 000
Norwegian Sea	1 400 000
West Bering Sea	1 100 000

Damage to benthic cold-water coral communities has resulted from the high intensity of commercial fishing in the Norwegian Sea and Baffin Bay. Commercial fishing contributes pollution in the form of litter, including lost gear. It also damages habitats (by benthic trawling, for example), and removes large fish and invertebrate biomass which can reduce the capability for Arctic ecosystems to withstand the impacts of other stressors such as pollution.

There are several large endemic mammals species of conservation concern in the Arctic including the walrus, *Odobenus rosmarus*, and polar bear, *Ursus maritimus*, both listed as vulnerable by the IUCN (2016). Commercial hunting of walrus in the 1800s and 1900s caused large declines in the north-west Atlantic. There were thought to be approximately 225 000 walrus in 2014, 10% are the Atlantic subspecies that occupy the east Canadian Arctic around to the Kara Sea, and Pacific subspecies inhabit the remainder of the Arctic seas. Although commercial-scale hunting is now prohibited or regulated by quota, walrus are also being impacted by climate change. To feed, walrus need haul-out areas of sea ice above or close to their feeding grounds, and the loss of sea ice is decreasing their access to food. Similarly, polar bears are also threatened by the loss of sea ice that is restricting their hunting range confining them to terrestrial environments. In order to support their long-term energy needs, and the long periods of fasting undergone by females while raising cubs, polar bears need high-fat, high-energy diets of seal, beluga whales, walrus, sea birds and fish.

Other species of concern include the beluga whale, listed as threatened, but data on population size are lacking and the narwhal has not yet been assessed. Bowhead whales, fished extensively between the 1500s and 1946 when it was banned, are increasing in the Arctic. The Greenland shark is listed as near threatened because it was historically over-fished in the 1910s. Artisanal fisheries continue today and the sharks continue to be caught as bycatch in commercial fishing operations. This species was recently found to be the longest-lived vertebrate with a lifespan of approximately 500 years.

6.8.2 Future challenges in the Arctic

The extent of Arctic Sea ice is fast decreasing and the sea is predicted to be ice free in summer by 2040 (Figure 6.18). This has major implications for Arctic ecosystems and will cause increased pollution there. At present the Arctic is closed to navigation between October and June, but as the sea ice continues to diminish it will become accessible for more of the year. It is likely that major shipping routes will develop, considerably reducing the distance and economic cost of passage between the Atlantic and Pacific Oceans. Thus, shipping-related pollution in the Arctic Ocean will increase. Arctic fisheries and other natural resources such as hydrocarbons and minerals will also become more accessible. This will probably mean the introduction of pollutants from oil exploration (Section 4.5), undersea mining (Section 3.6), lost fishing gear (Section 4.2), and litter and pollutants from shipping (Sections 4.2 and 4.5). Many of the Arctic animal migration corridors overlap with the best shipping routes and so the potential for collisions with vessels, and the exposure of Arctic plants and animals to pollution is high.

The US Geological Association estimates that 22% of the world's hydrocarbon reserves are beneath the Arctic (Figure 6.17a). To-date, most hydrocarbon exploration in the Arctic has been conducted on land. Most of the known reserves are in Russia and half of the undiscovered reserves are thought to be in Alaska or just offshore, the north Barents Sea, and west and east offshore from Greenland. The gas reserves are thought to be concentrated in the Kara Sea, the east Barents Basin and Alaska (Figure 6.17a). Shell Oil owns multiple leases in the Chuckti Sea and it wants to begin exploratory drilling, something that has met with protests from environmental groups and is currently being disputed in the courts. British Petroleum owns 8000 km^2 of exploration licenses in the Beaufort Sea and are currently conducting oil exploration there. Greenland has awarded exploration licenses for Baffin Bay to 13 companies, but at present only Cairn Energy are conducting offshore exploratory drilling. These exploratory operations and full-scale drilling will increase oil pollution in the Arctic, and the supporting infrastructure and transport needs would involve greater industrial development which will increase Arctic pollution with contaminants other than oil (e.g. sewage and litter). Furthermore, seismic oil exploration activities produce a significant amount of noise pollution that will negatively affect Arctic mammals (Section 5.6). There is also commercial interest in mining accumulations of minerals in the Kara Sea off the coast of Siberia; for example, for manganese nodules that contain valuable metals.

There may be different risks associated with the exploitation of oil and gas reserves in the Arctic compared with other regions. For example, the risk of collisions will be higher due to the large numbers of icebergs. Thus, there will be increased chances of accidental spills from oil rigs, oil tankers and non-tanker vessels. The removal of spilt oil from sea ice may require new methodologies (Section 4.5), and because temperatures are lower and there is less ultraviolet light available at the poles, the potential for natural biodegradation may be reduced. Mammals and sea birds congregate in large numbers in the Arctic on the sea ice or in the polynyas, therefore oil slicks could cause considerable damage to these aggregations of sea birds and mammals.

The Arctic Marine Shipping Assessment reviewed shipping activity in the Arctic in 2009 and found that the main shipping activity included community re-supply, cargo, fishing and tourist vessels. Most shipping accidents and collisions occurred near the coast and so will affect the resident fauna that inhabit and feed on the continental shelves. Changes in the amount of shipping in the Arctic are already being seen; for example, in the north-west passage of the Canadian Arctic vessels have increased by approximately one-third

between 2000 and 2010 and nearly quadrupled since 1975 (Figure 6.17b). In Greenland, the number of tourist vessels doubled between 2003 and 2008 (Figure 6.17c). At present this number of vessels is small compared to those in the other world oceans, but it is expected to increase considerably as the summer sea ice continues to retreat (Figure 6.18). Due to its remoteness, the potential for rapid emergency responses to clean-up accidental spillages from, for example, ship collisions is reduced, and there are no regular beach clean-up operations by local residents, all of which could mean increased pollution risk. There will also be pollution from regular ship operations; for example, loading/unloading, deck run-off, discharge of bilge water and ballast (Section 4.5) and the associated pollution from it.

Species particularly vulnerable to increased oil and gas exploration and increased shipping include the endemic marine mammals that are already experiencing stress from the environmental changes in the Arctic. The degradation of their prey from spilt oil, sedimentation on the shelves from urban development and damage to the benthic communities from trawl fisheries may all affect these species. Marine mammals are particularly susceptible to noise disturbance, for example from increased shipping and acoustics (Section 5.7). Disturbance of animals mating or pupping on the ice from shipping and tourism has serious effects; for example, walrus have high mortality when disturbed if they are overcrowded and become panicked on the ice.

6.8.3 Management of Arctic resources and ecosystems

According to the United Nations Law of the Sea 1972 (UNCLOS; Chapter 7), each country has exclusive economic claim to any resources in the sea or on the seafloor within 200 nautical miles of their coastline: the exclusive economic zone (EEZ). Canada, the United States, Russia, Norway, Denmark (via Greenland) and Iceland all have claims to some Arctic resources, and none of these countries currently has any claim to the central Arctic Ocean. The exploration and exploitation of any mineral resources in the central Arctic are administered by the United Nations International

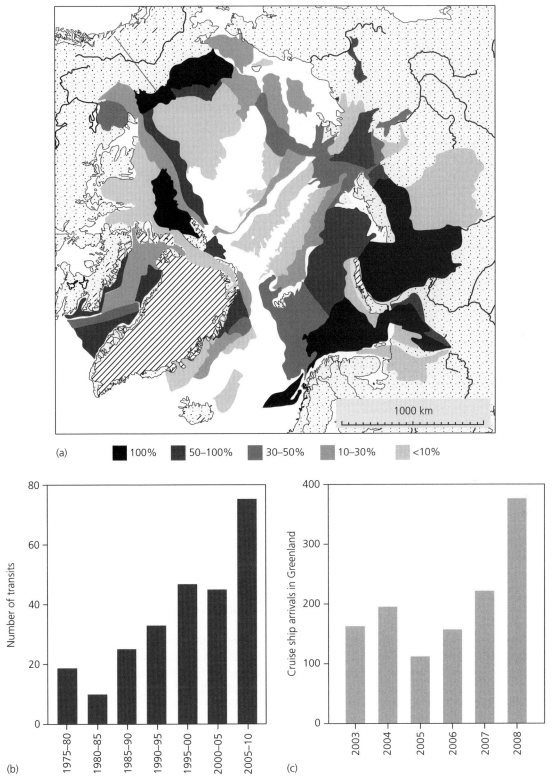

(a)

■ 100%   ■ 50–100%   ■ 30–50%   ■ 10–30%   ■ <10%

Figure 6.17 Hydrocarbon reserves and vessel traffic through the Arctic Ocean. (a) Probability of finding oil reserves in different regions of the Arctic. Credit: the Arctic Portal and U.S. Geological Survey Department of the Interior 2016. (b) Number of vessels transiting through the northwest passage between 1975 and 2010 (five-year averages). (c) Tourist cruiseships docking in Greenland between 2003 and 2008. Source: Conservation of Arctic Flora and Fauna/Arctic Biodiversity Assessment (CAFF 2013).

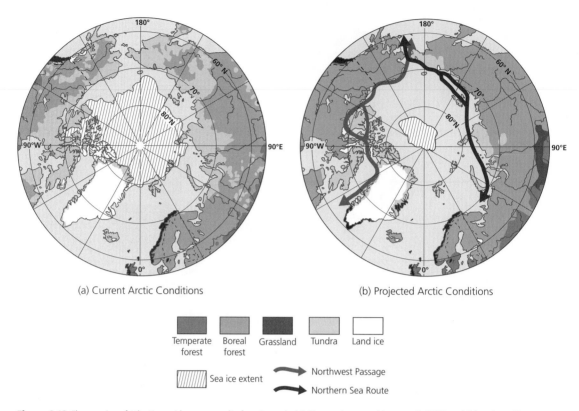

(a) Current Arctic Conditions   (b) Projected Arctic Conditions

Temperate forest | Boreal forest | Grassland | Tundra | Land ice

Sea ice extent

Northwest Passage

Northern Sea Route

Figure 6.18 The opening of Atlantic corridors as a result of sea ice melt. (a) Observed extent of ice cover in 2002, and (b) projected ice cover in 2080–2100. Arrows are the potential/improved sea routes: dark grey = northern sea route, medium-grey = north-west passage. Figure 5.13 from Climate Change 2007: Impacts adaptation and vulnerability. Contribution of Working Group II to the Fourth Assessment Report of the Intergovernmental Panel on Climate Change. (M.L. Parry, O.F. Canziani, J.P. Palutikof, P.J. van der Linden and C.E. Hanson) Cambridge University Press, Cambridge, UK.

Seabed Authority. All Arctic countries wish to have access to the mineral reserves (oil, gas, and manganese nodules) thought to underlie the central Arctic. Under UNCLOS a country may extend its EEZ to 350 nautical miles in areas where it can prove that the shelf is an extension of that country's continental shelf. This proof must include seafloor maps that show this to be the case and the claim must be made within ten years of ratifying UNCLOS. All Arctic countries are now collecting geological data to create such maps. The United States, however, never ratified UNCLOS and so is not bound by this legislation.

With the melting of the summer sea ice the potential for navigable Arctic seaways has increased and is another reason the Arctic countries are all pursuing seafloor Arctic mapping programmes. Disputes have arisen surrounding who has rights of access

over the two potential international sea routes that cross the Arctic (Figure 6.18b). The routes through the north-west passage are claimed by Canada as *internal waters*, being landward of the baseline for the nation's *territorial waters*, meaning that foreign vessels have no right of passage without prior agreement. Other states have insisted these are territorial waters, that is, those which extend 12 nautical miles from the country's coastline, and that under UNCLOS foreign vessels have rights of innocent passage. The north-east passage past Novaya Zemlya, on the other hand, is claimed by Russia as comprising internal waters.

At the 2008 Arctic Conference in Greenland, the member states agreed the Ilulissat Declaration: that demarcation issues would be resolved bilaterally. The environmental impacts of future exploitation and pollution of the Artic seas could depend heavily on the disputed maritime delimitations. The more accessible the Arctic is to other countries, the more shipping traffic and pollution will result. Additionally, shipping routes through international waters have weaker environmental regulations than those under the regional authority of the adjacent country. As for the high seas of the other main oceans, pollution from the dumping of wastes from ships in the Arctic is regulated by the London Convention (1972), the protocol to the London Convention (1996), MARPOL Annex V and its revised 2013 annex (Section 4.2.4).

The key international body for managing and conserving Arctic ecosystems and resources is the Arctic Council, established in 1996 through the Ottawa Declaration (members include all Arctic states plus six representative indigenous organisations). The Arctic Council promotes cooperation and coordination between members concerning sustainable development and environmental issues. The member countries have binding agreements on oil pollution preparedness. The biodiversity working group, for the Conservation of Arctic Flora and Fauna (CAFF), of the Arctic Council conducts monitoring and assessment, convenes expert groups, promotes sustainability and provides data to conserve Arctic resources. The protection of resident marine plants and animals and their habitat is undertaken by the member states individually and through coordinated actions between states because many of the mammal, fish and bird populations migrate between the different regions. For example, for threatened species such as the walrus, protection from commercial exploitation is provided by red-listed status in Norway and Russia, and in the United States this is covered by the Marine Mammal Protection Act. In Canada and Greenland, harvesting is regulated by imposing quotas, and subsistence harvesting continues in all countries. However, many species have no protection, vulnerable life histories and a history of over-exploitation, but little is known about the state of their populations. For this reason, the Arctic Biodiversity Assessment (2013) recommended that permanent monitoring stations be established in the Arctic, and that a better understanding of the effects of individual and multiple stressors is needed.

6.9 Synthesis

- The coastal seas and oceans all show similar patterns of human pressures. The exact timing and patterns of human impacts, however, vary between regions primarily for historical reasons. The Mediterranean and North Sea are heavily affected and significant impacts can be traced back several centuries. By contrast, the Coral Sea and the Arctic have lower levels of impacts but are likely to suffer significant further degradation in the near future unless protective management measures are imposed.
- Nutrients from the sewage of the growing human population and the fertilisers used in agriculture to feed the human population impose a ubiquitous pressure on marine systems.
- Controls of industrial wastes have been successful in halting and even reversing some deleterious trends in some regions (Black Sea, Mediterranean, North Sea and Chesapeake Bay).
- In most sea areas, the impacts from shipping, coastal development and offshore infrastructure (oil/gas, energy) continue to apply serious pressure.
- Most sea areas are now subject to some degree of international agreement for implementing management regimes. The degree of effectiveness varies.
- Recent management approaches seek to develop integrated (i.e. across-sector) and sustainable approaches that recognise the legitimate needs of various user groups, the need for biodiversity conservation and healthy functioning ecosystems to deliver benefits that support human well-being.

Resources

The Arctic

The Arctic Council (2016). http://www.arctic-council.org/index.php/en/.

Conservation of Arctic Flora and Fauna (2016). http://www.caff.is/about-caff.

National Oceanographic and Atmospheric Administration (2016). Arctic Research Programme. http://www.arctic.noaa.gov.

Coral Sea and Great Barrier Reef

Great Barrier Reef Marine Park Authority, Australian Government (2016). http://www.gbrmpa.gov.au.

The Australian Government (2016). The Great Barrier Reef. https://www.environment.gov.au/marine/gbr/protecting-the-reef.

The Australian Marine Debris Initiative (2016). http://www.tangaroablue.org/amdi/amdi-program.html.

eAtlas (2016). Australia's Topical Land and Seas. http://eatlas.org.au/home.

Canadian LOMAs

Fisheries and Oceans Canada (2012). Evaluation of the Integrated Oceans Management Programme. http://www.dfo-mpo.gc.ca/ae-ve/evaluations/11–12/IOM-eng.htm

Office of the Auditor General of Canada (2013). Progress in implementing integrated management of Canada's oceans. http://www.oag-bvg.gc.ca/internet/English/pet_354_e_39108.html.

The North Sea and north-east Atlantic

OSPAR Commission (2010). Quality Status Report 2010. 2 The North-East Atlantic. http://qsr2010.ospar.org/en/ch02_05.html.

The OSPAR Commission (2017). OPSAR Commission. http://www.ospar.org.

Bibliography

Ator, S.W. and Denver J.M. (2015). Understanding nutrients in the Chesapeake Bay watershed and implications for management and restoration—the Eastern Shore (ver. 1.2, June 2015): US Geological Survey Circular 1406. doi: 10.3133/cir1406.

Black Sea Commission (2008). *State of the Environment of the Black Sea* (2001–2006/7). Publication of the Commission on the Protection for the Black Sea Against Pollution 2008–2003, Istanbul: Black Sea Commission.

Black Sea Commission TDA (2008). Black Sea Transboundary Diagnostic Analysis Report. The Black Sea Commission, Istanbul. http://www.blacksea-commission.org/_tda2008.asp

Brodie, J., Waterhouse, J., Maynard, J., Bennett, J., Furnas, M., Devlin, M., Lewis, S., Collier, C., Schaffelke, B., Fabricius, K., Petus, C., da Silva, E., Zeh, D., Randall, L., Brando, V., McKenzie, L., O'Brien, D., Smith, R., Warne, M.St.J., Brinkman, R., Tonin, H., Bainbridge, Z., Bartley, R., Negri, A., Turner, R.D.R., Davis, A., Bentley, C., Mueller, J., Alvarez-Romero, J.G., Henry, N., Waters, D., Yorkston, H. and Tracey, D. (2013). Assessment of the relative risk of water quality to ecosystems of the Great Barrier Reef. A Report to the Department of the Environment and Heritage Protection, Queensland Government, Brisbane. TropWATER Report 13/28. Townsville: James Cook University.

CAFF (2013). Arctic Biodiversity Assessment. Status and Trends in Arctic Biodiversity. Conservation of Arctic Flora and Fauna, Akureyri. http://www.arcticbiodiversity.is/index.php/the-report.

Clark, R.A. and Frid, C.L.J. (2001). Long-term changes in the North Sea ecosystem. *Environmental Reviews*, 9, 131–87.

Frid, C., Hammer, C., Law, R., Loeng, H., Pawlak, J.F., Reid, P.C. and Tasker, M. (2003). *Environmental Status of the European Seas*. Copenhagen: ICES.

Froese, R. and Pauly, D. (2016). Fishbase. http://www.fishbase.org,version 06/2016.

Great Barrier Reef Marine Park Authority (2014). Great Barrier Reef Outlook Report 2014. GBRMPA, Townsville: Australian Government. http://www.gbrmpa.gov.au/managing-the-reef/great-barrier-reef-outlook-report.

International Union for Conservation of Nature and Natural Resources (2016). The IUCN Red List of Threatened Species. http://www.iucnredlist.org.

Kemp, W.M., Boynton, W.R., Adolf, J.E., Boesch, D.F., Boicourt, WC., Brush, G, Cornwell, J.C., Fisher, T.R., Gilbert, P.M., Hagy, J.D., Harding, L.W., Houde, E.D., Kimmel, D.G., Miller, W.D., Newell, R.I.E., Roman, M.R., Smith, E.M. and Stevenson, J.C. (2005). Eutrophication of Chesapeake Bay: historical trends and ecological indicators. *Marine Ecology Progress Series*, 303, 1–20.

Li, J., Gilbert, P. M. and Gao, Y. 2015. Temporal and spatial changes in Chesapeake Bay water quality and relationships to *Procentrum minimum, Karlodinium veneficum* and CyanoHAB events, 1991–2008. *Harmful Algae*, 42, 1–14.

Limpus, C.J. (2008). *A biological review of Australian marine turtle species, 1 Loggerhead turtle,* Caretta caretta *(Linnaeus)*. Brisbane: Queensland Government Environmental Protection Agency.

Arctic Council (2009). Arctic Marine Shipping Assessment Report 2009. Arctic Council. https://oaarchive.arctic-council.org/handle/11374/54

National Oceanographic and Atmospheric Administration (2015). Arctic Report card 2015. http://www.arctic.noaa.gov/reportcard/index.html.

National Oceanographic and Atmospheric Administration (2016). Chesapeake Bay Office. Fisheries. https://chesapeakebay.noaa.gov/fisheries/fisheries.

Parry, M.L., Canziani, O.F., Palutikof, J.P. van der Linden, P.J. and Hanson, C.E. (2007). *Climate Change Impacts 2007: Impacts, Adaptations and Vulnerability. Working*

Group II contribution to the Fourth Assessment Report of the Intergovernmental Panel on Climate Change. Cambridge: Cambridge University Press.

Sea Around Us (2016). The Sea Around Us: Fisheries, Ecosystems and Biodiversity. Tools and data. http://www.seaaroundus.org.

Sherman, K. and Hempel, G. (2008). *The UNEP Large Marine Ecosystems Report: A Perspective on Changing Conditions in LMEs of the World's Regional Seas. UNEP Regional Seas Report and Studies No. 182*. Nairobi: United Nations Environment Programme.

Shylakohv, V.A. and Daskalov, G. (2008). The state of marine living resources. In: T. Oguz (ed.). *State of the En-vironment Report 2001–2006/7*. Istanbul: Commission on the Protection for the Black Sea Against Pollution 2008–2003, 320–60.

State of the Environment 2011 Committee (2011). *Marine Environment 6. Australia State of the Environment 2011*. Independent Report to the Australian Government Minister for Sustainability, Environment, Water, Population and Communities. Canberra, DSEWPaC 2011.

Sturdivant, S.K., Diaz, R.J., Llanso, R. and Dauer, D. (2014). Relationship between hypoxia and macrobenthic production in Chesapeake Bay. *Estuaries and Coasts*, 37, 1219–1232.

CHAPTER 7

Regulation, monitoring and management

7.1 Introduction

Pollution is harmful by definition. Many other substances, and a range of activities, may also cause harm to the environment, human health or our ability to use the marine environment, without individually meeting the strict definition of pollution. For example, activities that cause stress such as noise or lowered oxygen levels may enhance the action of some toxins and contaminants present in low concentrations or which cause harm through accumulation or synergistic interactions with other chemical compounds individually (see Sections 2.2.2 and 2.2.3). In this chapter, we will, for simplicity's sake, use the term *substance* to encompass a wide range of potentially harmful contaminants, other materials (such as litter) and energy. Once the potential for harm—that something is a potential pollutant—has been recognised, there has been a tradition in well-governed societies of introducing regulations to protect the wider society from harm.

Some of the earliest national legislation surrounding the prevention of pollution dates back to the thirteenth century (Chapter 1), but many traditional societies have customs, taboos or laws that seek to restrict the impact of human activities on the environment. This traditional ecological knowledge, however, is limited to traditional activities and materials. It is therefore commonplace to see extremely high levels of plastic litter associated with villages in less economically developed countries in spite of a history of good traditional environmental management. If for thousands of years one's food has been wrapped in leaves and tied with a string made from vines, there was no need for a rule regarding the disposal of the packaging from a meal. The materials biodegraded very quickly. Replace the leaves and vines with plastic bags and polypropylene rope and a problem arises if behaviour does not also shift.

The imposition of laws and regulations to protect society is not straightforward. There are often vested interests on both sides: manufacturers do not want to see the products they develop restricted, and industry does not want to see costs rise (and profits fall) because it must implement additional environmental measures such as waste treatment. Governments often seek to protect society from harm but causing factories to close and people to lose their jobs also causes harm. The most important lesson that needs to be learned is that regulating and managing pollution is a trans-disciplinary and trans-cultural challenge.

Science has two important roles to play in this process: first, the gathering of evidence (data) on the health of the environment (defined in its broadest sense to include human society as well as biodiversity and ecosystem processes) and the threats to its continued health. Second, scientists who understand the complexity of ecological systems must clearly communicate the issues and the need for, and consequences of, trade-offs between costs and benefits to society, regulators and politicians. The delays in developing a response to global climate change shows how vested interests can exploit uncertainty and complexity in environmental systems and our understanding of them to pursue particular individual agendas.

Marine Pollution. Christopher L. J. Frid & Bryony A. Caswell.
© Christopher L. J. Frid & Bryony A. Caswell 2017. Published 2017 by Oxford University Press.
DOI 10.1093/oso/9780198726289.001.0001

In this chapter, we consider some of the challenges of and scientific methods used for monitoring the health of the marine environment before considering how the scientific knowledge is used to underpin measures to prevent marine pollution.

7.2 Monitoring the distribution of pollutants

7.2.1 What to monitor?

Measuring pollutants in the environment is clearly a sensible thing to do. Which of the 5 to 6 million new substances being produced each year should we measure? It is simply not practical to measure everything that human activities release into the environment. Where and how should we monitor pollutants? The answer is ideally everywhere but that is again impractical and so we must decide do we monitor near discharges, near fishing grounds or near recreational areas? Should we take measurements continually or intermittently? If intermittently, how often and when should we take them? There are a variety of methods for determining the biological effects of pollutants on organisms, populations and ecosystems which are discussed in Chapter 2.

Monitoring programmes often incorporate two components as a pragmatic solution to these challenges. First, the discharges are monitored so that the amount and nature of contaminants/energy released is known. This is often a requirement of the discharge licence and failure to abide by the limits set out in the licence can trigger legal action. Increasingly, the monitoring of discharges is done within site boundaries (so that legal responsibility is clearly with the owner) and is carried out by the discharger with periodic audit inspections by the regulatory authorities. The use of automated samplers and real time (see later) technology provides large datasets and instils a degree of confidence (which might be unfounded) in the data.

The monitoring of the discharge concentrations is supplemented by environmental monitoring which often involves background monitoring to establish a baseline, and targeted monitoring addressing critical issues; for example, the accumulation of contaminants within species used for food or the quality of water at bathing beaches. Data loggers recording standard parameters may be used, and these programmes often involve periodic sampling of ecosystem components and subsequent laboratory analyses; for example, to establish the tissue burdens of metals in the flora/fauna or to establish the species composition of the seafloor community through detailed taxonomic analysis.

It is important to collect monitoring data at and near to discharges but the results need to be placed within the context of their toxicity. What does a reading of X mean in terms of deviation from 'normal' or in terms of the risks to humans? This question requires toxicological and epidemiological models to link environmental concentrations (or predicted environmental concentrations) through organismal uptake and transformation pathways to identify the risks of harm to humans. The relationship between observed and 'normal' or background conditions seems, at least initially, to be a much simpler question. We just need to compare our potentially impacted location with a 'clean' site. However, to be a valid comparison, the 'clean' or control site should match the impacted site in every aspect except the presence of the impacting activity. Finding such a site would be difficult in most cases. Thus, we might consider using a number of control sites and establishing a range of values that indicate the 'clean' condition. Of course, the more sites considered, the more robust this understanding will be of the range and variability in the background levels that represent normal conditions. However, each additional site adds to the costs of the monitoring programmes and these are usually recovered from the polluter via their licence fee. The industry will wish to the keep licence fees low and so restrict, for financial reasons, the number of sites used.

7.2.2 Indicators of environmental quality and health

In addition to selecting appropriate locations for monitoring and the frequency of sampling, a decision needs to be made about what to monitor. What parameter will indicate whether our environmental management regime is being effective? We can distinguish two families of indicators: *pressure indicators* that measure the pressure human activities exert on the environment, and *state indicators* that measure the status of aspects of the ecosystem. The former measure the cause and the latter the effect. Pressure indicators would include the volume of a waste stream discharged, the suspended solid load, the concentration/amount of a metal or an organohalide. State indicators include all the measures of the biological health of the ecosystem including, for example, the levels of toxins in the biota, the level of metabolic stress measures (heat shock proteins), the size of populations of species of interest, the disease incidence in wildlife or the taxonomic composition/biodiversity of the system in question (see also Section 2.4).

Within the effluent stream of a factory or treatment works the parameters to be measured are most likely to be pressure indicators, such as the concentrations of the waste substances or other water quality parameters (e.g. suspended solids, biochemical oxygen demand). While these parameters may also be measured in the wider environment, the effects of dilution and mixing may quickly reduce them to levels at which reliable detection is difficult. However, we may still consider them to be ecologically active and so wish to monitor their effects (i.e. cumulative effects over time, changes brought about possibly through interaction with other substances, natural or anthropogenic). This then leads into the field of biological monitoring in contrast to the physical–chemical monitoring.

Biological monitoring provides a direct measure of pollution in that it measures an effect on the ecosystem and not just the presence of a substance. It can integrate temporal variations and impacts arising from protracted exposure. It also provides an indicator of impacts that might arise from multiple stressors: different pollutants and natural stressors that might be present at a location. This is both an advantage and a challenge for management. It is good that it highlights the impacts that are occurring but because it does not link directly to one source of stress it makes a clear management response difficult.

For example, in an estuary a sewage works discharges an effluent with a high BOD, but within its licence conditions. On the opposite bank, a food-processing company discharges the liquor from cooking vegetables; this too has an elevated BOD and this discharge is also within the licence

requirements. At the head of the estuary a fish farm is raising carp in ponds, and the effluent-rich overflow from these ponds enters the river and hence the top of the estuary. A monitoring programme in the estuary shows that growth of mussels is reduced and some of the sediments appear to have fauna indicative of enrichment compared to other nearby (control) estuaries. These estuaries, however, are not identical so perhaps this estuary naturally drains a more productive catchment. In any case, even if the enrichment is anthropogenic, which business should be further restricted as all are currently operating legally?

In this example we see two types of biological monitoring, changes in the growth rate of an indicator species and changes in the biological composition. Biological monitoring can be carried out at the level of the individual, population or community.

Since the adoption of the UN Convention on Biological Diversity there has been a formal requirement on governments to protect functioning ecosystems and this has led to the development of a more systems-level approach (Chapter 1). These approaches use indicators of ecosystem health to identify areas of regulatory concern. Such indicators are by definition *state indicators*. The analogy is often drawn with human health monitoring. On a visit to a doctor we will often have our temperature, blood pressure, and simple aspects of our blood chemistry measured. If any of the parameters are consider to be outside the prescribed limits, even though the cause may be unknown, this is used by the clinician as a trigger for additional tests and investigations. The idea is that the first set of tests is easy to do routinely and quickly and is cheap. If the tests indicate all is well then no other action is needed. However, if they suggest that an issue is emerging they guide the application of more specific investigations that in turn will inform the treatment plan. In the environment we might consider the numbers and types of birds using a site, or the composition of the fish or benthic assemblage as our routine indicators and only if these change over time do we investigate the reason why this change has occurred.

The advantage of this more holistic, two-stage approach is that it tells us about the system. If monitoring was restricted to following a single, or a few, indicator taxa, changes in their number may actually mean very little for the wider eco-system beyond that those taxa had changed in abundance. Integrated holistic approaches are often more cost-effective and have traction with wider society which finds it easy to understand the broad concept of a healthy environment. However, integrated ecosystem monitoring is often a technically difficult approach and while the concept of a healthy ecosystem is intellectu-ally attractive, it is extremely difficult to define operationally. How much change in the numbers of, say common gulls or *Capitella*, makes a system unhealthy? Second, the approach requires two stages of assessment and so is inherently slow to respond to a deteriorating situation; we see some-thing going 'wrong' but need further studies to identify what it is and only then can we begin to remedy it.

Most regulatory regimes are developing moni-toring tools that include both system-level and issue-specific indicators and a range of pressure and state indicators. Such suites of indicators are of-ten referred to as *headline* or *key* indicators and may be reported in terms of pass or fail, as trends (often illustrated by arrows) or by traffic-light-coloured symbols (see examples in Chapter 6).

7.2.3 The automation of monitoring

Ecologically based observations generally require the collection and examination of samples: of the organisms, of biological tissues or collections of the biological assemblages. This means that biological monitoring programmes tend to have intermittent temporal coverage. This is usually not seen as a prob-lem as, after all, one of the advantages of using the biota is that they integrate short-term fluctuations and record a degree of the history of exposure. Until fairly recently, sampling of the physical and chemi-cal aspects of the environment was also based on the collection of samples and subsequent laboratory analysis and so was also intermittent in character.

The last 50 years has seen great advances in the development of analytical devices that can be placed in the environment and left to collect samples auto-matically. These may record the data on storage de-vices for later retrieval or send the data back to base via cables or, increasingly, mobile telemetry and

the mobile telephone network. It is now possible to sample continuously and in near real-time param-eters such as oxygen and suspended solids and to carry out on-site analyses of nutrients, metals and a range of organic compounds. The technology and physical chemistry involved in the detectors is of-ten complex but this represents a major advance in our ability to monitor physical and chemical pres-sures directly.

For example, sensors can record concentra-tions of contaminants to high precision in efflu-ent streams and either store the data or relay the data to a control centre. In many cases, compa-nies are required to provide monitoring of their waste streams before they enter the environment and after treatment and to share these data with regulators. In-line continuous sensing of efflu-ent composition can be used to control the treat-ment regime; for example, by sensing the level of metals in the effluent and ensuring that only the correct amount of chemicals are added to precipitate the metals out. This means that ex-cess treatment chemicals, which can themselves be toxic, are not released, and by reducing the amount of chemicals used the cost is reduced.

Industry uses the data from the automated sen-sors to monitor the performance of its systems and the regulator can use it for enforcement and will of-ten carry out spot checks on the monitoring equip-ment to ensure accuracy. In some situations the monitoring equipment may be made tamper-proof, or be supplied by the regulator to ensure that the data are reliable and consistent for enforcement. However, sensors are technological advanced sys-tems and only available once a sufficient demand for them has been created. They are therefore never going to be available to assist in regulating novel compounds, pharmaceuticals and wastes such as micro-plastics (Chapter 5).

Issues such as cost and reliability are continually being addressed as technology advances, but the availability of so much data creates its own chal-lenges. Environmental data are often extremely 'noisy', that is to say they show a high variance to signal ratio (Figure 7.1a). This makes detecting change difficult, especially if the change is gradual; for example, a slow drift upwards of the concentra-tion of a substance in the waste stream.

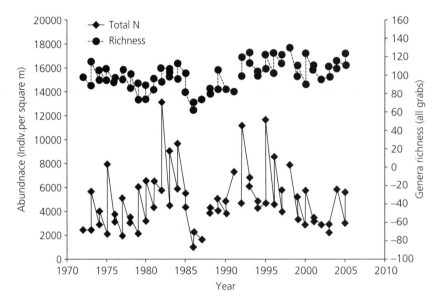

Figure 7.1 Inter-annual variability in the number of individuals per square m and the number of taxa (at the genus level) recorded from five replicate 0.1 m² grabs taken at a permanent monitoring station of the Dove time series in the central western North Sea. Both the number of genera and the number of individuals shows considerable inter-annual variation and decadal trends.

7.2.4 Citizen science

The collection of data on well-established indicators, or proxies, of environmental health can be undertaken by the public as part of *citizen science* initiatives. These programmes produce large datasets very cost-effectively, and have the added bonus of acting as educational exercises that raises societal awareness of environmental issues. Citizen science initiatives are often suited to collecting data on system- or ecosystem-level health indicators.

The largest examples of citizen science are the bird surveys coordinated by various non-governmental organisations (NGOs) and undertaken by amateur bird watchers. For example, the UK Wetland Bird Survey coordinated by the British Trust for Ornithology systematically surveys birds during the annual low-tide surveys of estuaries, non-estuary coastal surveys and counts of upland wader roosts. These surveys cover the entire British coast and estuaries on designated weekends during the autumn and spring migration periods. This survey commenced in 1992 and now provides more than 25 years of data; the estuarine bird counts have been conducted for

over half a century. The data provide a basis for assessing spatial and temporal trends in populations (Figure 7.2). For example, the data were critical for establishing the scale of the collapse, and subsequent recovery, of dunlin following lead pollution in the Mersey estuary (see Section 3.3.6). These data also contribute to wider, international initiatives using citizen science, for example the International Waterbird Census that celebrated 50 years of surveys in 2016.

Charismatic megafauna are often the focus of such initiatives. A number of other species/groups have been targeted for these programmes and these include beach litter surveys (Section 4.1) (a pressure indicator) and the incidence of fish disease as observed by amateur anglers during competitions. Following the collapse of dog whelk populations in Europe in the 1980s, as a result of pollution by tributyltin (TBT) from anti-fouling paints, a series of citizen science surveys were organised by NGOs working with scientists. The objective was to assess the scale of the impacts and the recovery after the 1986 ban on using TBT on small craft (see Section 3.3.8 and Box 7.1).

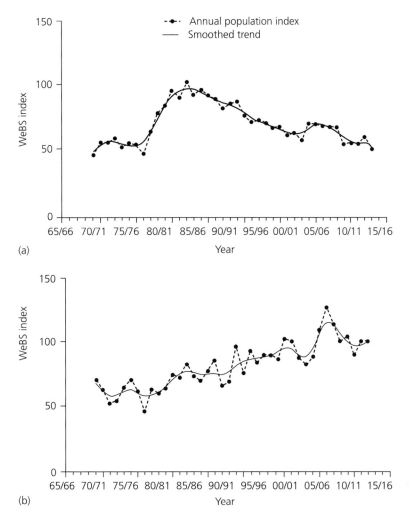

Figure 7.2 In the United Kingdom, amateur bird watchers contribute to a series of annual surveys (WeBS), coordinated by the British Trust for Ornithology, which show the long-term fate (and spatial patterns) in populations of birds species. For example (a) the turnstone (*Arenaria interpres*), which appears to undergoing long-term declines, while (b) the sanderling (*Calidris alba*) has been increasing. (Data from http://www. bto.org/volunteer-surveys/webs/data).

7.2.5 Big data

Data collection is often the most expensive part of any monitoring programme: staff travel to the sites, vessels may be needed, samples need to be worked up in the laboratory, etc. In the past there has been a tendency for each agency or business to commission their own data collection, partly driven by concerns over data ownership and control of the process. Following a commitment by President Obama in the United States in 2013, data collected by US national agencies, paid for, at least in part, with US taxpayers money, should be regarded as publicly owned and placed in the public domain. The growth of the Internet and online repositories has seen this process accelerate, and along with the growth in datasets (driven by increased data automation; see earlier), there is now more data available than ever before. This phenomenon is often referred to as the *Big Data* revolution.

There are strong reasons to use this data in a systematic and collaborative way to further our knowledge. The environment is a complex system (in a mathematical sense; that is, there are multiple drivers and complex non-linear interactions) and so large datasets are required to gain a good understanding of these interactions. The new adage for environmental data is, therefore, *collect it once, use it many times*. Many government agencies are now expected to collect their monitoring data (including that collected by third parties and supplied to them) and to share it openly so that other parties can use it for their own purposes.

This also places an emphasis on selecting parameters to measure that can inform more than one purpose.

7.3 Assessing change

The world is continually changing. This rather trite statement recognises that change is the natural order of the environment and so, in considering pollution and the impacts of human activities on the environment, the pertinent question to ask becomes 'how have we changed this pattern of natural change?' This is a much harder question to answer than simply 'has something changed?' In fact, measuring change in a complex system is difficult. To assess human impacts, we need to measure the change and then understand how much of it is due to natural processes and how much is the result of human actions—an impact (Figure 7.3).

This raises a number of issues; what should we measure to represent the environment? (see Section 7.2). How can we reliably determine change

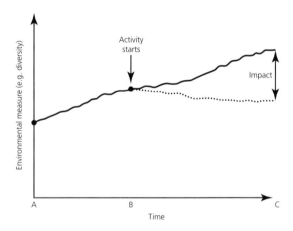

Figure 7.3 An ecological impact is not just a change, as natural systems are generally changing all the time. An impact occurs when trajectory of change of the system is altered by a human activity. Left unaltered by man, the system was trending but the onset at B of an activity has changed the natural pattern. This is only apparent if the trajectory from A to B is known; simply comparing A to C would lead to the conclusion that no impact had occurred.

and how do we partition the changes to pressures (natural and anthropogenic)? How can we easily summarise or communicate this science to the public, policy-makers and regulators (see Section 7.5).

7.3.1 Power: Type 1 and type 2 errors

Most introductory statistics courses introduce the idea of hypothesis testing and the concept of a null hypothesis of no effect/change. In our consideration of the monitoring of the environment for anthropogenic effects, this approach would seem to be adaptable to a 'no evidence of impact' or a 'no observed effect level' (Chapter 2). Such introductory courses also introduce the concept of a statistical test of a hypothesis and the concept of the significance of the test, which in fact is the probability of incorrectly rejecting the null hypothesis, concluding an effect when no effect actually exists. This is known as a type 1 error, a false positive, and the probability, α, conventionally used for screening a type 1 error is 1 in 20 cases ($\alpha = 0.05$).

While a type 1 error sees the incorrect rejection of the null hypothesis, a type 2 error sees the rejection of the null hypothesis when it is correct. This is often referred to as the statistical power and is the probability of obtaining a false negative result, concluding no effect when one exists. The power of a test is influenced by many things including (i) the significance criteria used, (in general by reducing the risk of a type 1 error increases the risk of a type 2 error), (ii) the size of the effect (it is easier to show an effect when the effect is large), and (iii) the sample size.

As with the type 1 error rates, there is no set value for the type 2 error rate (β) but commonly a rate of 4 to 1 is used, that is, $\beta = 0.8$ and $\alpha = 0.05$. However, in medical situations, studies are often designed to accept no false negatives (i.e. $\beta = 0$). This increases the risk of a type 1 error but it is argued that it is better to tell a healthy patient 'we are not sure so we need more tests' (a false negative) than to tell a patient 'you are OK', when in fact they have a problem (false positive). One might run a similar argument for the 'health' of the environment. Better to conclude we need more data than to conclude no impact and continue adding pollution that may then have long-term detrimental impacts.

Fisheries science has a history that can be traced back to the end of the nineteenth century but concerns have recently been raised about the power of many fisheries monitoring surveys. In the case of North Sea fisheries, recent analyses show that the surveys do not have sufficient statistical power. In most cases, they were unable to discriminate changes in abundance of a magnitude that should trigger management action. It is only after the stocks have suffered a major decline that the surveys recognised that decline and triggered action. These surveys have been operating for nearly 100 years and comprise a coordinated international programme. This finding emphasises the challenge of detecting ecologically meaningful change in marine ecosystems that are inherently difficult and expensive to survey and which show high degrees of natural spatial and temporal variation.

7.3.2 Statistical testing

Focused studies of the impacts of particular contaminants are often much easier to understand than the large-scale and more complex holistic ecosystem approaches. Well-controlled laboratory trials as used in toxicity assessments (Chapter 2) are the most straightforward to carry out. Field trials using techniques such as the incubation of mussels in mesh bags at varying distances from an outfall and subsequent monitoring of their growth rate (see Box 7.2) are also relatively straightforward to analyse using simple regression or one-way analyses of variance approaches.

In assessing impacts on the environment using environmental samples, we must recognise the underlying, natural, spatial and temporal variation. The ideal survey design therefore involves samples from multiple (replicate) locations from before and after the potential impacting activity and from sites that are potentially affected and from control locations. These designs are known as Before-After-Control-Impact (BACI) studies and are well matched to a variety of statistical analyses including multiple regressions and analysis of variance approaches that can partition out the variance/effect of the spatial, temporal and anthropogenic factors.

Box 7.2 TBT, dog-whelks and citizen science

Following the identification of TBT as a major pollutant of inshore waters in Europe (see Section 3.3.8), the UK Marine Conservation Society organised a series of surveys to assess the health of UK dog whelk populations. The public were asked to search for egg masses and juvenile and adult dog-whelks on rocky shores and take measurements of individual size. These data were used to determine population age structures. Severely impacted populations were extirpated whereas the less impacted populations became effectively sterile, producing populations comprising old adults with no juveniles. The surveys showed the spatial extent of the effects of TBT around the United Kingdom and, because they were repeated annually, showed that by the early to mid-1990s, some sites were recovering.

The absence of dog whelks or sterile populations (no juveniles) are both indicative of severe impacts. TBT affects dog-whelk reproduction in a dose-specific manner and the chronic effects can be quantified by examination of the internal tissues of the snails. As part of a programme to raise public awareness about human pressures on the North Sea, Stewart Evans and colleagues from the Dove Marine Laboratory (University of Newcastle) instigated a volunteer survey programme of dog-whelk health (Evans et al. 2000). Volunteers were trained to identify and measure dog-whelks in the field, make morphological observations and carry out simple dissections. Evans' team not only showed the recovery of dog-whelks was occurring at many sites around the North Sea, following the ban on TBT use for small vessels. It also found a very strong correlation between the data collected by the community groups and professional scientists working independently. Thereby showing that citizen science not only engages the public and raises awareness but it can also be a valid and robust means of cost-effectively increasing the extent and comprehensiveness of surveys.

Box 7.3 The truth behind 'lies, damned lies and statistics'

This phrase was first used by Lord Courtney in New York in 1895 but made popular by Mark Twain (who incorrectly attributed it to former British Prime Minister, Benjamin Disrali). It appears to derive from an assertion, which also has relevance for the issues considered here, made in *Nature* (26 November 1885, p. 74):

> A well-known lawyer, now a judge, once grouped witnesses into three classes: simple liars, damned liars, and experts. He did not mean that the expert uttered things which he knew to be untrue, but that by the emphasis which he laid on certain statements, and by what has been defined as a highly cultivated faculty of evasion, the effect was actually worse than if he had.

Ecosystems are complex phenomena and environmental data are usually noisy. Statistics are routinely employed to try and isolate a signal (trend, effect, change) from these data. However, both the meaning of the data and the causal drivers are often subject to interpretation. Thus, regulators, industry and pressure groups all have their own perspectives and often seek to use different statistics and even different experts to add weight to their own particular interpretation.

Other powerful approaches are those that use spatial (i.e. increasing distance from an outfall) or temporal (i.e. effects coincide, maybe with a lag, with changes in the amount of an emission) correlations to suggest a relationship. As always, correlation does not prove causation but when combined with laboratory toxicity data or body burden (tissue concentrations) then a logical case can be constructed.

Such well-designed and controlled studies can form the basis of regulatory action including infringement prosecutions and sanctions for the polluters. Poorly designed or controlled studies make regulation ineffective and lead to protracted legal arguments, doing little for environmental protection but enriching the legal profession.

7.3.3 Impacts vs shifting baseline

Given that impacts are defined as changes from what was to be expected if only natural processes were occurring, there is a scientific challenge of determining what the expected should be. For example, when we compare the current state of a parameter to its state 10 years ago we might see no change, but if we were to compare it to 100 years ago there may be a big difference. This difference is the result of both the natural variation/change plus any anthropogenic changes. However, we need reliable data to make comparisons, and there is a tendency for us to compare today with the recent past, say 10–20 years ago rather than three or four decades ago. We tend to think things were OK 20 years ago and we believe the data were reliable.

Large-scale degradation of the environment has been occurring for as long as humans have lived in dense aggregations; see the examples of health impacts of poor air quality from ancient Egypt and widespread metal contamination from Roman times referred to in Chapter 1. Since the Industrial Revolution and the subsequent increase in urbanisation over the last 250 years, the human footprint on the environment has grown in scale; the Great Stink in London, widespread industrialisation, urbanisation (and habitat loss), sewage discharges and the mechanisation of fishing, for example. If we only make our comparisons to a baseline of two decades ago we may seriously underestimate the scale of the change and hence the urgency and size of any management response. This phenomenon has been termed the *shifting baseline*.

The problem becomes even more significant when we consider that there are data that show that even the ancient Egyptians and Romans were affecting their local environments. For the Romans, this included changes across most of the Mediterranean and much of Europe. Stone-age peoples over-exploited fish stocks, and the large-scale forest clearances associated with the hunter-gatherer to agrarian transition, which started in the Middle East 13 000 years ago, will have altered the flux of sediment and nutrients into watercourses and hence the sea. Selecting a baseline for comparison is therefore extremely difficult.

Our baselines are restricted by the limits of the available monitoring data that only extend 100–200 years into the past. However, it is possible to attain useful data by looking retrospectively, as illustrated by the sediment cores from Chesapeake Bay (Section 6.4), the ice-core records of changes in atmospheric lead (Section 3.1), and historic tax records that show over-fishing was exerting pressure on fish stocks back in the twelfth century (Section 6.2). These time periods provide a more convincing

environmental baseline for the degree of ecological change, especially in areas with long periods of human habitation that have caused, for example, nutrient enrichment.

7.3.4 Global environmental status

Given the challenge of making a judgement about the impacts of individual human activities on the health of particular ecosystems, trying to give an assessment of the health of the planet is a massive undertaking. However, in 2005 the UN General Assembly committed to implementing a regular process for reporting on the state of the global marine environment. That process was envisaged as building on the existing regional assessments, but also filling in knowledge gaps, and would include not just the natural environment but also the socio-economic aspects.

The process became known as the UN Assessment of Assessments (AoA) as it sought to summarise and build on existing regional assessments. The first report was released in August 2009. The first AoA drew on 21 regional assessments and concluded that data on living marine resources (i.e. fisheries) were well covered in all regions with good coverage of water quality (i.e. pollution). Monitoring of the effects of human activities on habitats and on protected species was less well developed. The first assessment therefore focused on making recommendations for the ongoing process rather than drawing conclusions about the state of the seas.

Publication of the AoA and the bringing together of the regional datasets stimulated a number of academic studies that sought to utilise the data to assess the state of the global marine environment. These have used trend analyses of fisheries data (Figure 7.4) and more complex algorhythms (Figure 7.5) to assess the changes in ecosystem health. The conclusion is that however it is assessed, the last few decades have seen a decline in the health of the seas, and the main effect at the global scale is from fishing but in some coastal areas pollution, particularly eutrophication and enrichment (with associated deoxygenation) are also considerable (Section 4.3).

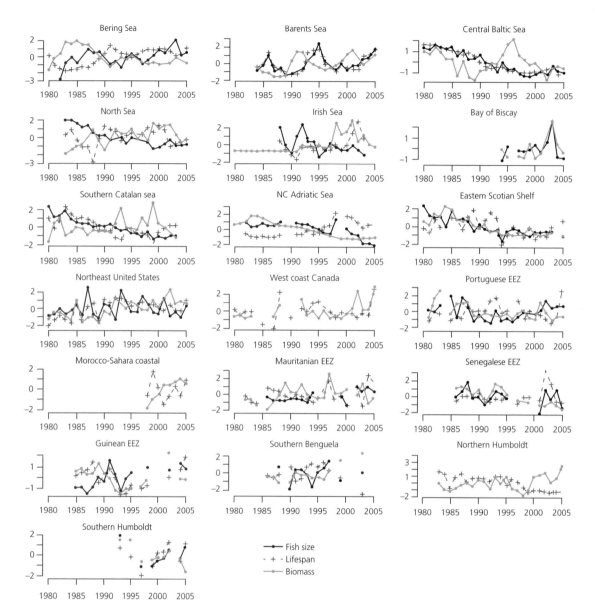

Figure 7.4 Normalised (i.e. mean of 0 and units are standard deviations) time series (1980–2005) for metrics of assessed fish stocks (mean fish size, mean fish life span and total biomass) from several high, mid and low latitude ecosystems and upwelling regions (redrawn from Blanchard et al. 2010).

7.4 Pollution regulation and enforcement

There are approximately 196 countries in the world today (the United States recognises 195, the United Nations 196; the United States did not recognise Taiwan as being independent at the time of writing), and while not all of these countries have an area of direct marine jurisdiction (152 countries claim a marine exclusive economic zone (EEZ)), they all have river systems that eventually drain into the sea. In that sense, therefore, there are 196 (or 195) different

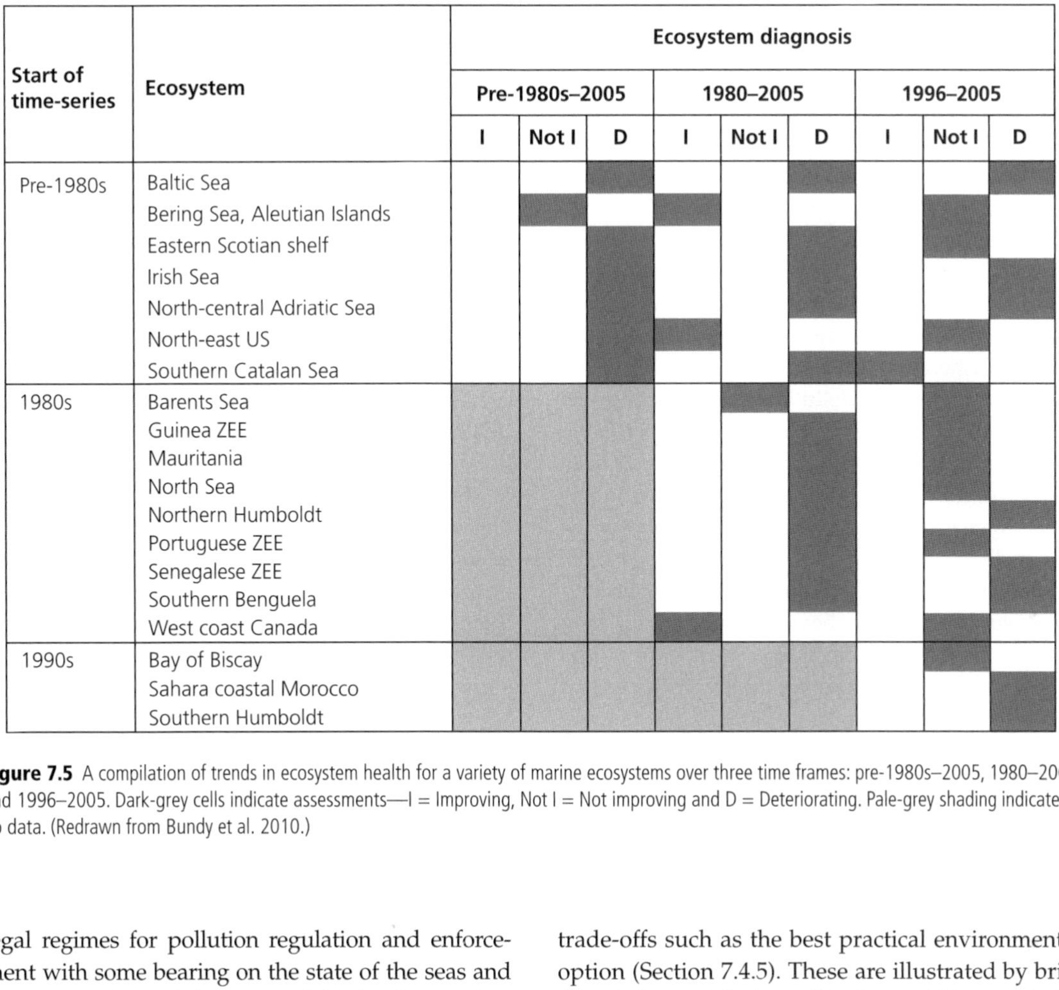

Start of time-series	Ecosystem	Ecosystem diagnosis								
		Pre-1980s–2005			1980–2005			1996–2005		
		I	Not I	D	I	Not I	D	I	Not I	D
Pre-1980s	Baltic Sea									
	Bering Sea, Aleutian Islands									
	Eastern Scotian shelf									
	Irish Sea									
	North-central Adriatic Sea									
	North-east US									
	Southern Catalan Sea									
1980s	Barents Sea									
	Guinea ZEE									
	Mauritania									
	North Sea									
	Northern Humboldt									
	Portuguese ZEE									
	Senegalese ZEE									
	Southern Benguela									
	West coast Canada									
1990s	Bay of Biscay									
	Sahara coastal Morocco									
	Southern Humboldt									

Figure 7.5 A compilation of trends in ecosystem health for a variety of marine ecosystems over three time frames: pre-1980s–2005, 1980–2005 and 1996–2005. Dark-grey cells indicate assessments—I = Improving, Not I = Not improving and D = Deteriorating. Pale-grey shading indicates no data. (Redrawn from Bundy et al. 2010.)

legal regimes for pollution regulation and enforcement with some bearing on the state of the seas and oceans. It is beyond the scope of this book to review all of these. Furthermore, the legislation is continually evolving and some laws that do not initially appear to deal with pollution still might have impacts on the environment. For example, the European Bathing Waters Directive was focused on the safety and quality of places where people traditionally went swimming. It sets standards around the provision of safety equipment and life-guards, about amenities and parking/transport but it also sets standards for the water quality and for the presence (or rather absence) of litter.

In this section, we consider some of the regulatory approaches or frameworks that can be used for pollution management and regulation. These include high-level principles such as the polluter pays (Section 7.4.1) to different regulatory approaches (Sections 7.4.2–7.4.4) to systems designed to deal with trade-offs such as the best practical environmental option (Section 7.4.5). These are illustrated by brief descriptions of their applications within certain jurisdictions, by way of examples.

7.4.1 The polluter-pays principle

A key principle of international pollution management is the polluter-pays principle. In its simplest form it seeks to prevent the deliberate release of pollution as a 'cheap' option for an industry or organisation. Rather, it seeks to ensure that those who produce a waste are financially responsible for the treatment/management of that waste. The challenge, of course, is that treating a waste carries costs such as those incurred by a treatment plant and monitoring facilities, or regulatory (i.e. licence) costs, whereas simply dumping it into the nearest waterbody is very cheap.

If waste is illegally dumped, in addition to the cost of dealing with any act of pollution, a polluter will also be faced with a fine or other punitive measures. This can be illustrated using the example of an oil spill. By definition, an oil spill is an accidental release, and the company responsible will have to fund the clean-up programme using their own workforce, and the state and government agencies that mobilise will also invoice the company for their costs. Additionally, the company will have to pay compensation to businesses adversely impacted, for example, fisheries and tourism operators. These are the direct costs of the pollution. The company will almost certainly also be prosecuted and, depending on the extent to which the accident might have been prevented, may face a significant fine, have to pay further damages, and face the possibility that some staff, including senior executives, could face personal fines and prison terms (although this is unlikely).

As we have noted, the environment can treat many contaminants, breaking them down into harmless forms (i.e. sewage), or it can dilute and disperse them to such an extent that no risk of harm arises (e.g. many metals which are present in sea water naturally). It is accepted that waste producers should be able to utilise the natural capacity of the environment for safe waste disposal, thereby reducing the costs and increasing the capital available for other investments, so stimulating economic activity.

The polluter-pays principle thus involves some form of permitting of waste disposal. An agency, usually a government body, regulates the use of the environment in such a way as to prevent harm to humans, economic activity and the environment. Typically this involves the issuing of a licence or permit to discharge. The permit then sets specific conditions for the discharge including its location, the nature (composition) and quantity of the discharge and any other restrictions deemed necessary. The waste producer then has to design its operation to meet these conditions, and failure to do so becomes a legal offence. Different governments use a variety of different philosophies and administrative approaches to such schemes.

The polluter-pays principle is achievable for point-source contaminant discharges, but many of the ongoing (Chapter 4) and emerging challenges (Chapter 5) are actually diffuse in nature

such as run-off from land carrying nutrients and pesticides, marine litter, discharges of pharmaceuticals down the drain, or run-off of veterinary medicines from farmland, releases of nanomaterials, etc. With these pollutants it is more difficult to determine how much is released and from whence it came. Sometimes there may be several polluters contributing to its release into the environment. These pollutants have to be tackled at the sector level and may require different approaches. Some of the ongoing contaminants have been managed with some success, for example, fertiliser reductions and upgrading of sewage treatment works in Chesapeake Bay and the Great Barrier Reef (although the environmental impacts of the changes in inputs have not yet become entirely clear). The legislation and regulation is in its infancy for many emerging contaminants. European policies have focused on extended producer responsibility for plastic pollution. This extends the producer's responsibility to incorporate almost the complete lifecycle of a product. This approach might also be useful for other diffuse pollutants.

7.4.2 Unified emission standards

Under a unified emission standards (UES) approach, every licenced discharger has to observe the same limit; for example, the discharge must contain no more than x µg l^{-1} of copper and the total volume of the effluent must not exceed y l d^{-1}. The values used for x, the concentration of the contaminant, are usually set by what is technically achievable with the available technology (see Section 7.4.5).

This seems very fair (all dischargers have to meet the same level), transparent and robust (the level is technically possible but ambitious). However, UES has a major drawback when used for discharges to receiving waters that serve many businesses. There is no control over the total input, so along a river like the Rhine or in a large estuary like the Humber, if every discharger discharges within its limits, the total amount of a contaminant discharged may exceed the capacity of the receiving body to assimilate it and negative effects may occur. This is especially problematic for long continental rivers flowing through multiple jurisdictions: the water entering one of the down-stream countries may already be carrying a high contaminant load and the environmental quality of that countries rivers and their ability to utilise the river for waste disposal are compromised. The UES approach is used by most European nations, the United States and Canada.

7.4.3 Environmental quality standards

In order to avoid the problem of additive impacts from the UES, the environmental quality standards (EQS) approach makes a case-by-case assessment. Each applicant for a discharge permit is required to show that its discharge will not compromise the quality of the environment, with the environment having been zoned or designated for particular purposes, each with an accompanying EQS.

For example, a coastal bay may be designated for contact water sports (i.e. swimming and surfing), and the lower estuary for non-contact water sports (i.e. boating and angling), whereas the middle reaches may be designated for the migration of salmonids. A potential discharger will need to show that its discharge, if made to the bay, will not compromise the bay's use for water sports, and if made to the estuary will not compromise salmon migrations. The former will require a low biological oxygen demand (BOD), no pathogens and low turbidity, while the latter will require a low BOD but less stringent controls on pathogens, colour and turbidity.

The EQS approach is more complex to administer than UES (and hence more costly), less transparent to the wider public (as complex modelling is often needed to establish the distribution and fate of the discharged material) and gives an advantage to established businesses, as they may have been able to use a large proportion of the available assimilation capacity. However, it does allow a more strategic planning approach with areas zoned for different activities and with water quality matched to use.

The United Kingdom traditionally used an EQS approach. Most European legislation sets requirements on member states but allows them to achieve those requirements using either a UES or EQS approach, and increasingly a blended approach has been recognised as being the most effective. The Water Framework Directive and the Marine Strategy Framework Directive both make use of environmental quality standards; that is, 'good environmental status' and absolute emission levels set by technology.

7.4.4 Hit lists

In addition to the general regulatory frameworks, substances of high concern have often been identified and subjected to direct policy intervention using a targeted list approach. Examples would include the European Black and Grey lists, the Oslo–Paris Convention for the Protection of the Marine Environment of the North-East Atlantic's Red list (Table 7.1), the UN Environment Programme Persistent Organic Pollutants list, the Stockholm Convention (see

Table 7.1 The OSPAR list of *Chemicals for Priority Action*, the red list, as at 2016.

Group of substances/ substance	Function
Cadmium	Metallic compound
Lead and organic lead compounds	Metal/organometallic compounds
Mercury and organic mercury compounds	Metal/organometallic compounds
Organic tin compounds	Organometallic compounds
Neodecanoic acid, ethenyl ester	Organic Ester
Perfluorooctanyl sulphonic acid and its salts (PFOS)	Organohalogens
Tetrabromobisphenol A (TBBP-A)	Organohalogens
1,2,3-trichlorobenzene	Organohalogens
1,2,4-trichlorobenzene	Organohalogens
1,3,5-trichlorobenzene	Organohalogens
Brominated flame retardants	Organohalogens
Polychlorinated biphenyls (PCBs)	Organohalogens
Polychlorinated dibenzodioxins (PCDDs)	Organohalogens
Polychlorinated dibenzofurans (PCDFs)	Organohalogens
Short-chained chlorinated paraffins (SCCP)	Organohalogens

(continued)

Table 7.1 (*Continued*)

Group of substances/ substance	Function
4-(dimethylbutylamino) diphe-nylamine (6PPD)	Organic nitrogen compound
Dicofol	Pesticides/biocides/organo-halogens
Endosulfan	Pesticides/biocides/organo-halogens
Hexachlorocyclohexane iso-mers (HCH)	Pesticides/biocides/organo-halogens
Methoxychlor	Pesticides/biocides/organo-halogens
Pentachlorophenol (PCP)	Pesticides/biocides/organo-halogens
Trifluralin	Pesticides/biocides/organo-halogens
Clotrimazole	Pharmaceutical
2,4,6-tri-tert-butylphenol	Phenol
Nonylphenol/ethoxylates (NP/NPEs) and related substances	Phenol
Octylphenol	Phenol
Certain phthalates: dibutylphtha-late (DBP), diethylhexylphtha-late (DEHP)	Phthalate esters
Polyaromatic hydrocarbons (PAHs)	Polycyclic aromatic com-pounds
Musk xylene	Synthetic musk

Section 3.4) and various prioritisation lists produced by the Group of Experts on the Scientific Aspects of Marine Environmental Protection (GESAMP).

In general, the lists identify substances or groups of substances of particularly high concern, usually due to toxicity and persistence and set these as priorities for action. In some cases, for example the EU lists and the Stockholm Convention, the requirements for action are legally binding, but in most cases they are advisory.

For persistent substances identified on the EU Black list and for most classes of POP on the Stockholm Convention list, the requirement is for a phasing out of emissions, essentially moving to a UES of zero.

7.4.5 Best practical environmental option (BPEO)

In many cases a waste material might be disposed of in a number of ways with each option having different associated financial and environmental costs. For example, sewage treatment generates sewage sludge. The higher the degree of traditional treatment given, the more sludge generated at the sewage treatment works. A case can be made that depending on the receiving waters, discharging a low BOD waste with some pathogens and particulates allowing final breakdown in the sea is a better option than providing tertiary treatment (and generating more sludge) and a high-quality effluent that then gets discharged into an estuary already contaminated by industry.

Given that some treatment has occurred, a sludge will have been produced. This waste now also needs disposal. Sludge from secondary and tertiary sewage treatment can be used as a soil conditioner in forestry, in parkland and on some crops but this is only an option if these users are nearby (otherwise there is a transport cost, with associated emissions). For primary sludge this is not an option and the main alternatives are to incinerate it, digest it in order to generate methane, bury it or dump it at sea.

To bury sludge uses limited landfill capacity and so is expensive and not favourable. Dumping at sea has the lowest carbon dioxide emissions of any option and it enriches the seafloor (but may over-enrich it), but this is no longer possible in Europe due to policy restrictions. Digestion requires capital investment but the methane can be used to power the plant (although this emits carbon dioxide), while simple incineration requires a fuel (sludge is often co-combusted with flammable industrial waste or oil refinery residues) and emits carbon dioxide. So which option is best? That depends on priorities: at-sea disposal has the lowest carbon dioxide emissions but is now socially unacceptable, and simple co-incineration with other wastes requires the least capital investment but emits the most carbon dioxide and possibly harmful reaction products.

In 1988, the UK Royal Commission on Environmental Pollution recognised these trade-offs and recommended that regulators take a holistic view and seek to apply the *best practical environmental option.* At that time in the United Kingdom, access to landfill sites was controlled by local government, air pollution was regulated by both local government and a national agency, a different national agency regulated offshore disposals, and regional bodies administered discharges to rivers, estuaries and coastal waters. This created a situation where each body tried to protect its bit of the environment and there was no recognition of the trade-offs between the different receiving environments. After all, a large proportion of air pollution ended up being washed out into the seas and oceans, without specific consideration by any organisation.

While recognised as being a sensible way forward and in many ways a driver of the subsequent unification of environmental regulators into cross-sector environment or environmental protection agencies, the application of the BPEO was difficult. Where does one draw a line around activities? Are carbon dioxide emissions from transporting the waste included? What about the emissions from making the lorry or mining the iron ore and producing the steel that make up the lorry?

The operational approach that emerged was the adoption of what became known as the *Best Available Technology Not Entailing Excessive Cost* (BATNEEC). Like the UES approach, this is guided by the technical realities of dealing with the waste, but it also includes economic criteria. The latter is, of course, a controversial subject because an industry views any additional costs as excessive while for an NGO the issue revolves around what price a clean and healthy environment?

7.4.6 Enforcement and prosecution

Unlicenced releases of potentially polluting materials/energy are generally an offence under national environmental protection legislation. We might distinguish two types of release: accidental spills and illegal operational waste discharges. Accidental spills can occur at any time and could be of raw materials on their way to a factory, a manufactured product (or intermediary) or of a waste material en route to treatment or disposal. By their nature, they are accidental and operating procedures will be designed to minimise the risk. Common sources

of spills are tanks splitting or being overfilled when the bund is not closed, spillage of drummed material on deck or a quayside, dropped material during loading/unloading, etc. Generally, operators should have emergency procedures in place to deal with a spill and these will include contacting and working with the authorities.

The environmental consequences of a spill are as diverse as the materials that might be spilt and can be felt at both a significant temporal and physical distance from the original accident. Oil spills are perhaps the most obvious example of this kind of accident, but following a fire at a chemical plant in Switzerland, a slug of highly toxic water moved down the entire length of the Rhine before emerging into the North Sea several weeks later. Similarly, the collapse of a dam at a mine in Brazil released highly turbid water containing suspended solids, metals and metal colloids that travelled the length of the Amazon before emerging into the South Atlantic.

The polluter, of course, is expected to pay but will also be liable for prosecution under environmental and/or health and safety legislation. In general, fines are inversely proportional to the degree of diligence exercised by the organisation. Spills resulting from poorly maintained bunds, corroded tanks, poor training of staff will lead to larger fines. Fines amounting to millions, or even billions, of pounds for single offences are not uncommon, particularly for oil spills. For example, the release of just 20 tonnes of oil by SHELL into the Mersey Estuary in 1989 resulted in a fine of £1million as the operator made a conconous decision to empty the fractured pipeline rather than let the tar like oil solidify in the pipe. The 2010 BP Deepwater Horizon spill in the Gulf of Mexico resulted in a fine of $5.5billion and a further $7.1billion in environmental damages after the company was found to have been grossly negligent. Rapid contact with the authorities and a positive engagement often mitigates the size of the fines.

Discharges of waste that breach the terms of the discharge licence are illegal. Breaches may be due to the volume discharged per unit time, the concentration of the material in the waste stream or the inclusion of additional unlicensed material. The inspection/automated monitoring of discharges seeks to ensure that industries do not routinely do this, but unexpected fluctuations in material quality, volumes processed, failures in treatment plants can all lead to licence terms being exceeded. The consequences of this are generally dependent on the nature and polluting impact of the material and the previous conduct of the business. Often less serious breaches will be given a warning and companies will have their monitoring regime tightened, but any failure to comply with licence terms is an offence and could lead to prosecution. In most regimes, the courts would levy a fine but they also have the power to revoke a licence or alter its terms.

7.5 Synthesis

- Pollution is by definition detrimental, but the effects of pollution can arise from the direct effects from one contaminant or activity or from a combination. Generally, we are concerned with the impacts on humans health and activities, biodiversity and ecosystem functioning. Measuring pollution is therefore not a simple exercise, although it sounds like it should be.

- We can measure the levels of activities that we know can cause pollution (pressure monitoring), we can measure the levels of contaminants or potentially damaging material (including energy) in the environment, and/or we can measure aspects of the environment that concern us—health of the biota, biodiversity, etc. In most cases environmental monitoring includes all three aspects.

- Environmental monitoring programmes are costly and there is increasing use of technology (data loggers, automated analysers) and citizen science (e.g. volunteer wild-bird surveys) to generate data to expand these programmes.

- Management of the activities that can cause marine pollution is usually achieved by laws and regulations. Both the drafting of such laws and regulations and their enforcement requires scientific input. However, they are also responsive to societal needs and priorities—the need for the products, the manufacture of which might cause pollution and the need for employment often being key drivers.

- The marine environment is a complex system and environmental data are often inherently variable ('noisy'). These means that their interpretation relies on statistical analysis and interference. The systematic collection of data is a relatively new phenomena (data have been collected for just over 100 years) and advances in analytical technologies also confound temporal comparisons. This can create an issue of shifting baselines where we are not comparing the present levels to a pristine state but merely one that had been altered in a different way/different extent.

- A major internationally accepted principle of pollution management is that the *polluter pays*. In reality, this is often difficult to implement, particularly for diffuse sources of pollution. Licence fees and the increasing use of environmental taxes seek to implement this by altering the economic framework to include the true environmental cost of activities such as manufacturing and the application of fertilisers.

- Recent decades have seen shifts in environmental policy to a more holistic approach and the development of approaches that recognise that controlling one type of waste may simply generate more of a different type and that the management needs to consider both the whole ecosystem and the trade-offs between activities. The economic valuation of ecosystems and their services may provide new tools for making these judgements.

Resources

Burden Frank, R., Foerstner, U., McKelvie, I.D. and Guenther, A. (eds). (2002). *Environmental Monitoring Handbook*. New York, NY: McGraw-Hill.

BTO Waterbird Surveys. http://www.bto.org/volunteer-surveys/webs.

Marine Conservation Society Beach Clean-up Schemes. http://www.mcsuk.org/beachwatch/.

UN Assessment of Assessments. First Report. http://www.unga-regular-process.org/images/Documents/aoa%20sdm%20(english).pdf.

Waste Water Monitoring. Automation Case Study. http://www.ecotech.com/case-studies/real-time-water-monitoring-solution-for-qld.

Bibliography

Blanchard, J.L., Coll, M., Trenkel, V.M., Vergnon, R., Yemane, D. Jouffre, D., Link, J.S. and Shin, Y.-J. (2010). Trend analysis of indicators: a comparison of recent changes in the status of marine ecosystems around the world. *ICES Journal of Marine Science*, **67**, 732–44.

Blanchard, J.L., Maxwell, D.L., Jennings, S. (2008). Power of monitoring surveys to detect abundance trends in depleted populations: the effects of density-dependent habitat use, patchiness, and climate change. *ICES Journal of Marine Science*, **65**, 111–20.

Bundy, A., Shannon, L.J., Rochet, M.-J., Neira, S., Shin, Y.-J., Hill, L. and Aydin, L. (2010). The good(ish), the bad, and the ugly: a tripartite classification of ecosystem trends. *ICES Journal of Marine Science*, **67**, 745–68.

Evans, S.M., Birchenough, A.C. and Fletcher, H. (2000). The value and validity of community-based research: TBT contamination of the North Sea. *Marine Pollution Bulletin*, 40: 220–5.

ICES (2001). *Report of the Working Group on the Ecosystem Effects of Fishing Activities 2001*. Copenhagen: ICES,Copenhagen.

CHAPTER 8

The future ocean

8.1 Introduction

What will the future ocean look like? As the Nobel Prize-winning Danish physicist, Nils Bohr, noted: 'prediction is very difficult, especially if it's about the future'. We see this phenomenon played out every day in the form of weather forecasts. Using the world's largest and most powerful supercomputers, including data from a network of ground stations and satellite systems, it is still only possible to predict the weather, with any degree of confidence, a few days in advance. This is because the weather (or at least the atmospheric circulation) is a complex system, in the mathematical sense; it contains a number of non-linear relationships, and strong feedbacks that mean small differences can diverge, producing a wide range of possible outcomes. The physicochemical ocean system of currents and mixing and all the biological systems from ecosystems down to physiological and cellular processes are also mathematically complex and highly interconnected. Thus, predicting the future state of the ocean is equally difficult.

While detailed predictions are impossible, we can make some educated guesses; we can model it and produce forecasts (Section 8.3). The oceans will be warmer than they are now, they will be more acidic, and there will be less water locked up as sea ice (all because of global warming). Ocean circulation will change as a consequence, and there will be more areas of low dissolved oxygen. Changes in the hydrological cycle are expected to cause changes in land runoff and inputs of particulates and nutrients (due to greater storminess/flash floods) in many regions. As human populations continue to increase, many pollution problems will also continue to grow. More sewage and litter will be generated (Chapters 3 and 4) as population size increases, and a greater proportion of the population will live in the coastal zone. These populations will need more food that will in turn produce more agricultural run-off and so increase the delivery of nutrients to the oceans (Section 4.3). There will be fewer large fish (due to continued heavy fishing pressure) so biomagnification is less likely. The demands for energy (Sections 4.5, 4.7, 5.6, 5.7), raw materials (Section 3.6) and industrial products used by society (Sections 3.7–3.8, 4.3, 5.8) will generate more point-source discharges. The demand for medicines and non-degradable materials in our day-to-day lives will increase. There will be inputs of novel compounds as the chemical and pharmaceutical industries continue to develop increasingly complex new products and micro- and nano-particulate size materials (Chapter 5).

The future ocean will continue to be used by humanity for waste disposal and treatment, for transport, food supply and recreation while still offering a cultural and spiritual heritage. In concluding this work we wish to consider some of these issues while recognising that commenting on the future ocean is a highly risky activity for anyone without a crystal ball.

8.2 The 'Grand Challenges'

Pollution arises from human activities and, in particular, the disposal of waste. The number of people on Earth has been increasing essentially exponentially since the Middle Ages; around 1 billion individuals in 1800, passing the 6 billion mark in 1999 and 7 billion in 2012. The United Nations estimates that the global human population will be approximately 10 billion by 2050 (Figure 8.1). The

Marine Pollution. Christopher L. J. Frid & Bryony A. Caswell.
© Christopher L. J. Frid & Bryony A. Caswell 2017. Published 2017 by Oxford University Press.
DOI 10.1093/oso/9780198726289.001.0001

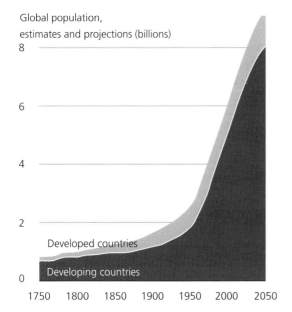

Global population,
estimates and projections (billions)

Developed countries

Developing countries

Figure 8.1 Human global population growth since 1750 and projections up until 2050. Credit: Hugo Ahlenius and GRID-Arendal http://www.grida.no/graphicslib/detail/trends-in-population-developed-and-developing-countries-1750–2050-estimates-and-projections_1616.

amount and nature of waste generated has grown with global populations. The shifts in the standard of living and lifestyle changes mean that the rate of waste production has far exceeded the rate of population increase. Approximately 44% of the global population lives in the coastal zone (within 150 km of the coast), a figure that is increasing as this zone becomes more urbanised. Therefore, much of the pressure exerted on the environment by humans occurs in the marine environment, either directly from activities in the coastal seas or indirectly via atmospheric or river transport of materials to the oceans.

The first wave of environmental awareness occurred in the middle of the twentieth century and by the end of that century human activities were having global effects on the planet, potentially impacting key components of the systems that support human life. Examples include the destruction of the ozone layer, climate change and the loss of habitats. To mark the turn of the Millennium, the United Nations held a Summit at which the 189 UN member states and 22 international organisations agreed to work towards achieving 8 goals by 2015

(Table 8.1). These became known as the Millennium Development Goals (MDGs).

The MDGs focused on poverty alleviation and improved education and healthcare, and environmental sustainability was incorporated as goal number 7. The issues identified by the MDGs were often referred to as the 'grand challenges' facing society in the twenty-first century (Table 8.1). One of the great strengths of the approach was that for each goal there were one or more specific numerical targets. While there was much debate about the suitability of some of these metrics (and the reliability of the data for measuring them), they provided a major impetus for political action to back the MDGs.

The general consensus in 2015 was that a great deal of progress had been made and several of the MDGs had been achieved. In order to continue this momentum, a new set of goals was agreed. These goals link back to the concept of sustainable development, a core aspect of the 1992 UN conference in Rio de Janiero, and so placed the environment at the core of the development agenda. It recognised the links between human activities, the environment and human well-being. The 17 objectives adopted in 2015 guide UN actions from 2016 to 2030 and are known as the UN Sustainable Development Goals (SDGs) (Table 8.1).

Like the MDGs, the SDGs also have associated quantitative and qualitative targets. For SDG 14, *Conserve and sustainably use the oceans, seas and marine resources for sustainable development*, there are 7 quantitative targets and 3 qualitative (Table 8.2). The first target is a significant reduction in marine pollution. Unfortunately, the document does not define what a significant reduction is or against which baselines it should be measured. The second target includes a commitment to reducing all adverse effects on marine and coastal ecosystems and to undertake restoration.

Thus, while addressing marine pollution is a key target, under SDG 14, marine pollution also affects the potential to achieve other SDGs. For example, fisheries contribute about 6% of the total protein consumed by the human population (and some consumed by agricultural animals). The accumulation of POPs, particularly mercury and some organohalides, restricts the use of some fish stocks for human consumption even now (see Sections 3.3

Table 8.1 The 8 Millennium Development Goals set in 2000 with a target for achievement date of 2015 and the 17 Sustainable Development Goals, adopted in 2015 to guide UN policy and funding to 2030.

UN Millennium Development Goals (MDGs)
1 To eradicate extreme poverty and hunger
2 To achieve universal primary education
3 To promote gender equality and empower women
4 To reduce child mortality
5 To improve maternal health
6 To combat HIV/AIDS, malaria and other diseases
7 To ensure environmental sustainability
8 To develop a global partnership for development

UN Sustainable Development Goals (SDGs)
1 End poverty in all its forms everywhere
2 End hunger, achieve food security and improved nutrition and promote sustainable agriculture
3 Ensure healthy lives and promote well-being for all at all ages
4 Ensure inclusive and equitable quality education and promote lifelong learning opportunities for all
5 Achieve gender equality and empower all women and girls
6 Ensure availability and sustainable management of water and sanitation for all
7 Ensure access to affordable, reliable, sustainable and modern energy for all
8 Promote sustained, inclusive and sustainable economic growth, full and productive employment and decent work for all
9 Build resilient infrastructure, promote inclusive and sustainable industrialisation and foster innovation
10 Reduce inequality within and among countries
11 Make cities and human settlements inclusive, safe, resilient and sustainable
12 Ensure sustainable consumption and production patterns
13 Take urgent action to combat climate change and its impacts
14 Conserve and sustainably use the oceans, seas and marine resources for sustainable development
15 Protect, restore and promote sustainable use of terrestrial ecosystems, sustainably manage forests, combat desertification, and halt and reverse land degradation and halt biodiversity loss
16 Promote peaceful and inclusive societies for sustainable development, provide access to justice for all and build effective, accountable and inclusive institutions at all levels
17 Strengthen the means of implementation and revitalise the Global Partnership for Sustainable Development

and 3.4), thus impacting the achievement of SDG 2. Concerns about noise pollution from offshore energy schemes could constrain developments that contribute to SDG 7 (Table 8.2).

So, the grand challenge we face now is how to accommodate a growing human population, making increasing per-capita demands (including waste disposal/assimilation capabilities) on the finite resource that comprise planet Earth. The word that most often features in discussions of these issues and in the UN policy statements is *sustainable*. Since the UN Convention on Sustainable Development was held in Rio in 1992, the term has been widely used and misused, has come to mean different things to different people and has been split into the three pillars of sustainability (environmental, economic and social). As environmental scientists, it seems obvious to us that sustainability has to be based on the natural limits set by the environment: for example, the sustainable fish catch is set by ocean productivity; the sustainable level of urban waste-water disposal to the sea is determined by the mixing regime, the dilution capabilities, the organisms able to break it down and the oxygen supply. While these limits are important, a counterargument is that if these levels fail to provide an economic return, then there will be an incentive to cheat; for example, to exceed fishing quotas or dispose of extra waste. If the imposition of these limits undermines social structures—for example, by not adequately controlling disease risks from waterborne pathogens—then they will not be obeyed. Thus, while population growth is the grand challenge and sustainability is a critical aspect of the solution, it has to respect the ecological, economic and social constraints.

8.3 Future scenarios

8.3.1 How has the global ocean changed over the last century?

Global ecosystems are undergoing unprecedented changes and have done so over the last 1000 years in response to anthropogenic stressors such as climate change, the over-exploitation of resources, changing land use and environmental pollution.

Table 8.2 The UN targets associated with Sustainable Development Goal 14: Conserve and sustainably use the oceans, seas and marine resources for sustainable development. Data from the UN General Assembly 2015, Transforming Our World: The 2030 Agenda for Sustainable Development.

UN Numbering	Target
	Quantitative Targets
14.1	By 2025, prevent and significantly reduce marine pollution of all kinds, in particular from land-based activities, including marine debris and nutrient pollution
14.2	By 2020, sustainably manage and protect marine and coastal ecosystems to avoid significant adverse impacts, including by strengthening their resilience, and take action for their restoration in order to achieve healthy and productive oceans
14.3	Minimise and address the impacts of ocean acidification, including through enhanced scientific cooperation at all levels. (Acknowledging that the United Nations Framework Convention on Climate Change is the primary international, intergovernmental forum for negotiating the global response to climate change.)
14.4	By 2020, effectively regulate harvesting and end over-fishing, illegal, unreported and unregulated fishing and destructive fishing practices and implement science-based management plans, in order to restore fish stocks in the shortest time feasible, at least to levels that can produce maximum sustainable yield as determined by their biological characteristics
14.5	By 2020, conserve at least 10% of coastal and marine areas, consistent with national and international law and based on the best available scientific information
14.6	By 2020, prohibit certain forms of fisheries subsidies which contribute to overcapacity and overfishing, eliminate subsidies that contribute to illegal, unreported and unregulated fishing and refrain from introducing new such subsidies, recognising that appropriate and effective special and differential treatment for developing and least developed countries should be an integral part of the World Trade Organization fisheries subsidies negotiation
14.7	By 2030, increase the economic benefits to small island developing States and least developed countries from the sustainable use of marine resources, including through sustainable management of fisheries, aquaculture and tourism
	Qualitative Targets
14.a	Increase scientific knowledge, develop research capacity and transfer marine technology, taking into account the Intergovernmental Oceanographic Commission Criteria and Guidelines on the Transfer of Marine Technology, in order to improve ocean health and to enhance the contribution of marine biodiversity to the development of developing countries, in particular small island developing States and least developed countries
14.b	Provide access for small-scale artisanal fishers to marine resources and markets
14.c	Enhance the conservation and sustainable use of oceans and their resources by implementing international law as reflected in the United Nations Convention on the Law of the Sea, which provides the legal framework for the conservation and sustainable use of oceans and their resources, as recalled in paragraph 158 of 'The Future We Want'

These pressures act in concert to decrease the health of marine ecosystems and their inhabitants. In this section we will briefly consider the environmental changes marine ecosystems have undergone, the changes that are predicted for the next 80 years, the impacts on marine ecosystems and the resilience of marine flora and fauna to pollution in this rapidly changing world.

Since the 1900s sea surface temperatures have increased more than 1°C, the sea level has risen around 23 cm and pH has declined by approximately 0.1 pH units. Over the last 60 years the amount of oxygen in the oceans has decreased at a mean rate of 0.06–0.43% per year, and in some areas, for example, the Pacific Ocean at 50°N, up to 22%. These changes are concurrent with an increase

in the number of areas, now totalling 500, afflicted by low oxygen (hypoxia), this number has been doubling every decade since 1960 (Section 4.3).

Superficially these changes might seem small but they have major implications for ecological systems that have already begun to manifest. The time required for the oceans to adjust to a new climatic state is substantially longer (decades to millennia) than for the lower atmosphere (days to months) due to its greater heat capacity, and so we may expect changes in the ocean to become more pronounced in the coming years. These problems, of course, are another manifestation of pollution: the release of large amounts of carbon previously locked up in the fossil fuels buried beneath continental and ocean sediments or carbon that was locked up in the trees and plants lost to deforestation and other habitat destruction.

The most noticeable current ecological manifestations of climatic change include the changing biogeographic ranges of species. Species that are intolerant of the new climatic conditions in their existing ranges are beginning to shift, mostly, towards polar latitudes where the temperatures more closely reflect those to which they are physiologically adapted. There is strong evidence that these biogeographic shifts (involving more than 129 marine species in 2010) are leading to the redistribution of species and are changing ecological interactions. Meta-analyses show that these changes have so far amounted to geographic shifts of on average 19 ± 3.8 km y^{-1} across all trophic levels. The shifts in marine species are occurring 20 times faster than on land. The majority of these 70%, are shifting towards the poles. For example, data from the Continuous Plankton Recorder have shown that North Atlantic zooplankton are moving northwards at a rate of around 23 km y^{-1}. Also, fish species are shifting to deeper, cooler waters. For example, in the North Sea, fish species in the cool water assemblage are shifting northwards and also into deeper water: 28 species shifted on average 3.6 m deeper per decade. Other changes include shifts in the timing of spring events such as migration and larval recruitment, which have moved two to three days earlier each decade. Shifts in species can have major implications for the species which previously occupied the 'new' habitats and these changes are similar to those for non-native invasive species (Section 4.8) and include competition, altered predator–prey dynamics, and the introduction of disease/parasites (Section 6.4). These changes in species geographic ranges can lead to local and global extinctions.

8.3.2 What will change in the future?

The Intergovernmental Panel on Climate Change (IPCC) has made a number of medium and long-term climate projections in the 28 years since its establishment. Five assessment reports have been published that include climate projections based on a number of different scenarios of greenhouse gas emissions that are linked to scenarios of potential socio-economic growth (Figure 8.2). Many scenarios have been proposed and we will focus on the intermediate scenarios A1B and an earlier derivation IS92A (Figure 8.2). Scenario A1B includes rapid economic development with the energy needs delivered from a mixture of sources. The climate projections for these scenarios show a warming of up to 2.5°C by 2029 and 7°C by 2099 under the intermediate emissions scenario (Figures 8.3 and 8.4).

In addition to impacting the biota directly, these changes in temperature affect physical processes such as ocean circulation and mixing, which have implications for spatial and vertical heat and oxygen transport and the solubility of essential gases such as oxygen. Weather patterns are expected to change with an overall increase in global rainfall and a strengthening of the existing regional patterns. Changes in rainfall will produce changes in the spatial patterns of surface run-off with up to a 20% decrease between 20° and 50° North and South and up to a 40% increase above 50°N. The increased freshwater inputs can influence estuarine and coastal salinity, enhancing the stratification of water bodies that inhibits vertical mixing. The greater the run-off, the greater delivery of pollutants, nutrients and sediments that cause turbidity and eutrophication problems in coastal waters. Increased freshwater inputs combined with changes in circulation, decreased gas solubility and greater nutrient inputs can lead to hypoxia. The high-emissions scenario, A2, predicts a continued decline of more than 7% of the amount of dissolved oxygen in the oceans by 2100 in all areas except the Arctic. Eutrophic and deoxygenated conditions are therefore highly likely to spread, seasonal eutrophication and/or hypoxia may become more frequent and prolonged. Changes in salinity, nutrients and low oxygen conditions can cause stress to marine organisms, inhibit growth and cause mass mortalities (Sections 6.3, 6.4 and 6.7) of intolerant taxa that may result in lost biomass or loss of habitat. Under deoxygenated conditions, the body size of taxa become reduced, and models predict that between 2000 and 2050 this could be as much as 20% for fish under high-emission scenarios (A2; Figure 8.2). Also, ocean acidification can cause a thickening of, for example, mollusc shells. These changes in body size and shifts in the proportions of shell and soft-tissue biomass will affect the viability of some

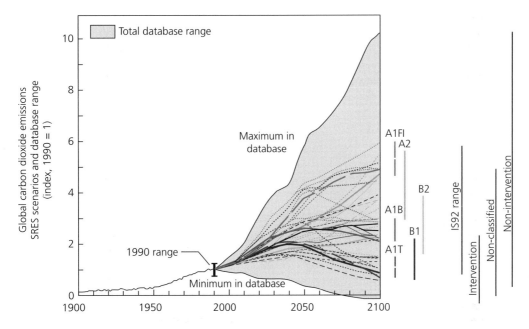

Figure 8.2 Observed and projected emissions (as an index relative to 1990) for IPCC scenarios from the first two IPCC assessments (right: the 1995 scenarios; left: the scenarios used in the third and fourth assessment reports). Scenario A1: rapid global economic growth with 1.4–6.4°C temperature rise; scenario B1: global environmental sustainability 1.1–2.9°C increase; scenario A2: regional economic development 2.0–5.4°C warming; and, scenario B2: regional environmental sustainability. IS92A: business as usual with a warming of 2.4°C. None of the scenarios include any present or future measures to reduce emissions. Figure 2a from Emissions Scenarios. A Special Report of IPCC Working Group III of the Intergovernmental Panel on Climate Change (2000). (Nakicenovic, N. and Swart, R.) IPCC, Geneva, Switzerland.

species for food production. Predictions for sea-level change suggest that it will increase a further 0.21–0.28 m under emissions scenario A1B. Rising sea levels will also contribute to habitat loss.

Atmospheric carbon dioxide is absorbed by the ocean and causes decreases in ocean pH and it is predicted to decline 0.3–0.4 units by 2100 under intermediate emissions scenarios (Figures 8.2 and 8.5; IS92A). The changes in pH are more dramatic at the poles due to lower temperatures, and significant changes could occur within decades. pH at depth will also be affected and there will be a shallowing of the aragonite saturation horizon (a regularly used indicator of acidity) of up to 2000 m in the Atlantic at 80°N by 2100, and so will affect deep-dwelling species. Decreases in pH can cause the dissolution of biogenic calcium carbonate and/or inhibit its formation, making the production of skeletal material more difficult for marine organisms. At present the effects of ocean acidification are not well understood

and seem to vary greatly between species, but it is an active area of scientific investigation. However, the effects for some taxa are severe; for example, inhibition of egg fertilisation and larval development for sea urchins and bivalve molluscs. Coral reefs are also expected to be under considerable threat from ocean acidification due to their $CaCO_3$ composition and when pCO_2 exceeds 500 ppm 30% of warm water, coral reefs will be lost (Figure 8.6b), along with many of the species that inhabit them. The other main group affected by ocean acidification will be the molluscs with approximately 50% being lost once pCO_2 exceeds 500 ppm (Figure 8.6b).

Bioclimatic ecological envelope models (those based on known statistical relationships between species distributions and physiological tolerances) have been produced to help understand and forecast the species shifts that might arise from the IPCC climate scenarios. Under scenario A1B, shifts of at least 1000 species are forecast before 2050 with

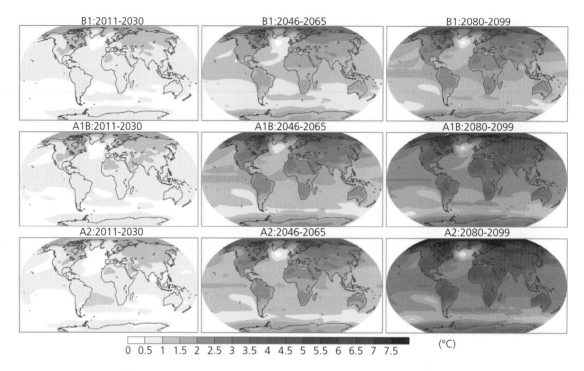

Figure 8.3 IPCC-predicted average global surface warming for scenarios B1, A1B and A2 (Fig. 8.2) for 2011–30, 2046–65 and 2080–99 (left to right). Data are the mean temperature anomalies (air temperature °C) across several climate models relative to the 1980–99 baseline. Figure 10.8 from Climate Change 2007: The Physical Science Basis. Contribution of Working Group I to the Fourth Assessment Report of the Intergovernmental Panel on Climate Change. (Solomon, S., Qin, D., Manning, M., Marquis, M., Averyt, K., Tignor, M.M.B., Miller Jr., H.L. and Chen, Z.) Cambridge University Press, Cambridge, UK.

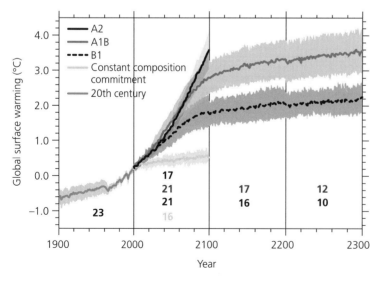

Figure 8.4 Predicted changes in global surface temperatures for the different IPCC scenarios (Fig. 8.1) up to 2100; those after 2100 represent stabilisation scenarios (twentieth century data are also simulated). Lines represent the mean (and shaded regions ± 1 standard deviation) from several different model outputs (indicated by numbers below). Figure 10.4 from Climate Change 2007: The Physical Science Basis. Contribution of Working Group I to the Fourth Assessment Report of the Intergovernmental Panel on Climate Change. (Solomon, S., Qin, D., Manning, M., Marquis, M., Averyt, K., Tignor, M.M.B., Miller Jr., H.L. and Chen, Z.) Cambridge University Press, Cambridge, UK.

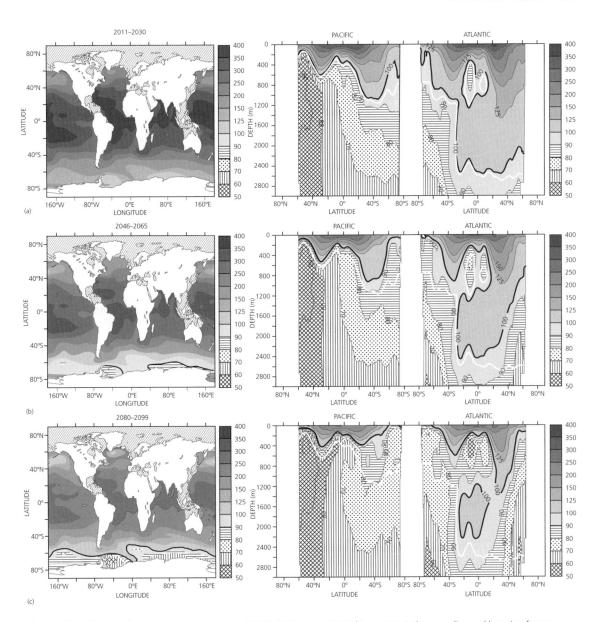

Figure 8.5 Predicted median aragonite saturation states (%) for IPCC scenario IS92A (Figures 8.2–8.4) over medium and long time frames (compared to a 1994 baseline) globally (left), and with depth and latitude (right). Where the aragonite saturation state is less than 100, CaCO$_3$ begins to dissolve (white line = 1994 depth, and black line = predicted depths). (a) From 2011–30 pCO$_2$ is estimated to be 440 ppm, during which time the aragonite saturation horizon is predicted to shallow by 100 m at 80°S and 10–100m at 80°N. (b) From 2046–65, pCO$_2$ is expected to be 570 ppm, and the aragonite saturation horizon is expected to shallow to 700 m at 80°S; and 100 m to 2000 m at 80°N in the Pacific and Atlantic, respectively. (c) From 2080–99 pCO$_2$ expected to be 730 ppm, with no further shallowing of the aragonite saturation horizon at 80°S. No further changes are predicted in the Pacific at 80°N but a further 600 m shallowing is predicted in the Atlantic Ocean during this interval. Diagonal line pattern = geographic areas with insufficient data. Figure 10.23 from Climate Change 2007: The Physical Science Basis. Contribution of Working Group I to the Fourth Assessment Report of the Intergovernmental Panel on Climate Change. (Solomon, S., Qin, D., Manning, M., Marquis, M., Averyt, K., Tignor, M.M.B., Miller Jr., H.L. and Chen, Z.). Cambridge University Press, Cambridge, UK.

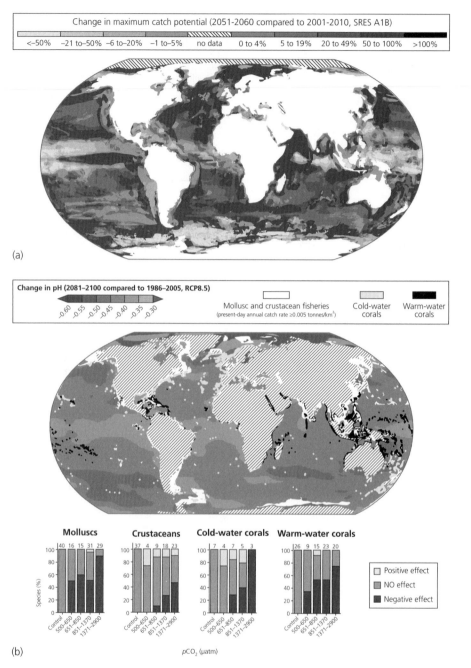

Figure 8.6 Threats to fisheries from climate change. (a) Projected redistribution of around 1000 species of exploited fishes and invertebrates as maximum catch potential between 2001–10 and 2051–60 (decadal averages) in response to the elevated temperatures projected for climate scenario A1B (see also Figs 8.2–8.4). (b) Global distribution of potentially vulnerable present-day mollusc and crustacean fisheries, warm- and cold-water coral reefs, and projected changes in surface ocean pH by 2081–00, compared with 1986–2005, under scenario RCP8.5 (similar to IPCC scenario A2, Fig. 8.2). Stacked histograms compare the proportion and nature of species sensitivity to ocean acidification under five different scenarios projected for 2100 (number of species analysed shown at the top of each). The five pCO_2 categories correspond to the IPCC scenarios as follows: no changes (380 µatm), 500–650 µatm (RCP4.5), 651–850 µatm (RCP 6.0), 851–1370 µatm (RCP 8.5); and, 1371–2900 µatm represents scenario RCP8.5 by 2150. (µatm are approximately equivalent to ppmv). Figure 2.6 from Climate Change 2014: Synthesis Report. Contribution of Working Groups I, II and III to the Fifth Assessment Report of the Intergovernmental Panel on Climate Change (Core Writing Team, Pachauri, R.K. and Meyer, L. (eds)). IPCC, Geneva, Switzerland.

poleward shifts of up to 6 km y^{-1}. The largest shifts are predicted at the Tropics and the Equator. This redistribution of species will affect fisheries production, and after 2050 the catch potential will be up to 50% lower at the Equator and the South Pole as species shift towards the Tropics (under Scenario A1B, Figure 8.6a). Many polar species show smaller shifts because they have no refuge: there is no habitat remaining with the required water temperatures. The forecasted shifts in demersal species are slower than for pelagic taxa in most areas. Demersal taxa inhabit environments that are more buffered at least with respect to temperature changes (because it takes time for the heat to transfer to depth in the oceans). So, deeper waters will clearly represent one refuge up to a point. As anticipated, the shifts are larger under higher emissions scenarios.

Our experiences of the responses of ecosystems to non-native invasive species (NIS) can help us to understand how the receiving systems will respond to these new arrivals, because the ecological effects of the two are very similar. However, species range shifts may also be compounded by the presence of NIS (Section 4.8) that may have the advantage of arriving in systems that are already stressed/perturbed, for example, nearshore polluted coasts, estuaries and ports. The speed of NIS spread has so far been shown to be approximately twice that of the climate-driven shifting species as they occupy their new niches, perhaps because NIS is facilitated by human vectors. The melting of sea ice and opening of Arctic corridors (Section 6.7) may facilitate the spread and mixing of Atlantic and Pacific marine taxa. This may result in greater homogeneity between the Pacific and Atlantic and can impact ecosystem dynamics in unpredictable ways.

Studies of the changing global patterns of biodiversity under the IPCC scenarios from 2005–2050 have shown that local species extinctions will be high in subpolar regions, at the Equator and in enclosed water masses. This is probably because warming will be fastest at the poles (Figure 8.3), contractions in their viable habitat (unlike non-polar species that can shift) and the ecological pressures from the immigration of many shifting species. These models predict a 60% species turnover by 2050. If we look at the fossil record we see similar patterns of biogeographic range shifts and elevated extinction at polar latitudes that are associated with past climatic change.

8.3.3 What will these changes mean for marine pollution?

As we have seen throughout the preceding chapters, the health of an organism and its nutritional status can contribute considerably to its ability to cope with exposure to toxicants. The poorer an organism's condition, the less able it is to metabolise, detoxify or bind contaminants. So, stress caused by temperatures or salinities outside those that are optimal, or decreased oxygen and pH may add to the stress of toxicant exposure. Similarly, chronic toxicity from other pollutants or hazardous algal blooms (HABs) may contribute to the poor health of organisms and thus compromise their ability to withstand pollution. Additionally, environmental conditions—for example, temperature, pH and oxygen—can alter the toxicity of contaminants by affecting their environmental availability or their biological uptake (Section 2.2.1).

The bioaccumulation of persistent pollutants means that the more pollution we introduce into the environment, the greater the chance of exposure and the greater the body burden. This affects the health of the animal and the health of predators that are not directly exposed to the pollutant and this presents an increased risk to human health through biomagnification in the food chain (Section 3.3).

In terms of population and community health, any anthropogenic activities that cause habitat loss, reduce food supply or have reduced the biomass of taxa considerably (i.e. from fishing) can reduce the viability of populations. Many marine organisms are now exposed to endocrine-disrupting chemicals (EDCs) that can impair species fertility and reproduction (Section 5.4). Both ocean acidification and deoxygenation can also impair larval development and so add to the pressures on reproductive processes. Similarly, the large losses of important nursery habitat—for example, seagrass, macroalgal and mangrove habitat—are occurring for many species.

Less biodiverse ecosystems are less resilient to change and so the synergy between the many different anthropogenic pressures, whether they be climate, pollution, urbanisation or fishing, combine to reduce the capability of the system to tolerate and adapt to further change. In addition to changing the species composition of communities and ecosystems, environmental changes can influence the behaviour of marine organisms. Some of these changes can have serious implications for ecosystem processes or functions. For instance, deoxygenation can cause a shift from an infaunal community that mixes the sediment, in turn cycling nutrients through the ecosystem to one where there are no infauna present (e.g. Section 6.4) and so nutrient cycling is reduced. These changes have the potential to affect the productivity of whole ecosystems. One study calculated that US$350 billion of ecosystem services are already lost each year to hypoxia.

As we have seen from the various case studies (Chapter 6), sufficient pressure can result in a system that undergoes major changes or *regime shifts* whereby one stable ecosystem is replaced by another. For example, the Black Sea has undergone at least two major shifts in ecosystem composition

over just a few decades due to NIS, over-fishing and anthropogenic eutrophication. These major reorganisations of ecosystems can have far-reaching effects that we could not have anticipated. In Chesapeake Bay, the loss of extensive oyster reefs, wetland and seagrass habitat compounded the bay's eutrophication problems and combined with over-fishing have led to the collapse of several important commercial fisheries (Section 6.3). The chronic stress from such combined pressures can promote continued degradation or can slow recovery (Chapter 6). In the Arctic seas, over-exploitation has similarly remapped the trophic pathways between predators and prey. In the Arctic, over-fishing of walrus and bowhead whales in Svalbard (Figure 8.7a) has caused a reorganisation and increased the production of pelagic fish and seabird populations (Figure 8.7a-b) at the expense of whale biomass. Over-fishing can result in *trophic cascades* as top predators are removed: lower trophic levels are released from predation pressure and the pressure on the basal trophic level, often the

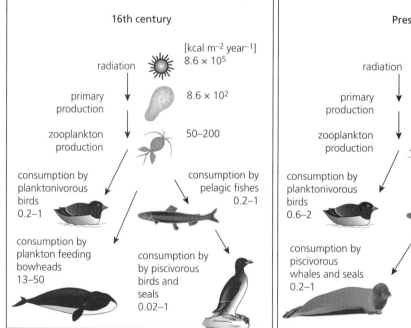
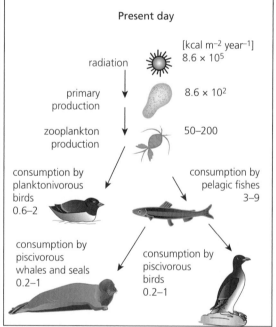

Figure 8.7 Changes in Arctic pelagic food-web structuring at Svalbard before (a) and after commercial exploitation (b) of walrus and whales during the sixteenth century (if primary and secondary production remained constant). Production of each component is in Kcal m^{-2} y^{-1}. Reproduced from Weslawski et al. (2000), © 2000 by Institute of Oceanology PAS.

ecosystem engineers—for example, kelp, coral, sea-grass—can intensify shifting the system out of its natural balance. These ecosystem shifts can lead to a redistribution of biomass (e.g. the Black Sea and Chesapeake Bay are now dominated by microbes) and/or reduced productivity in some ecosystem components; for example, lowered benthic production in Chesapeake Bay (Section 6.4, Figure 6.8b).

8.4 Interactions, conflicts and compromises

The marine environment is vast, covering around 70% of the Earth's surface with a mean depth of 3.72 km and a total volume of 1335 million km^3. All of this water is connected and in constant movement. It requires approximately 1000 years for the oceans to turn over, and so there is no part of the ocean in which pollutants will be isolated. Ultimately, all the materials we have added to the oceans will be mixed (and diluted) throughout the world ocean. This vast volume of water and its capacity for dilution was the key to the traditional approaches for the management of most wastes, and it has worked for many of the substances released during the pre-industrial and early industrial phases of human civilisation. Biodegradable materials become degraded as they are mixed, and the metals and other contaminants are diluted below the levels of biological activity. However, for the last 200 years, or one-fifth of the residence time of the Ocean, we have been producing novel compounds, some of which are not degraded, or the time required for degradation is so long that we consider them essentially non-biodegradable (e.g. the POPs, radioactivity and most plastics). Although they are diluted, they may remain active at very low concentrations. We have seen in Section 5.3 that some organic molecules act as hormone mimics and disrupt the endocrine systems that control reproductive and developmental processes at concentrations many orders of magnitude below that which they are considered toxic.

The vastness of the ocean provides a massive dilution capacity and it also effectively prevents the post-disposal recovery of pollution, plastic litter and oil being limited exceptions to this generalisation. The dilution and mixing may also give rise to further challenges. When chemicals are brought together, they often react to form new compounds. In complex biological and chemical systems, they will often interact to produce other unexpected effects. In Chapter 2, we examined how some compounds reduced the toxicity of one another (antagonistic interactions) and others enhanced their action, such that a mix was sometimes more potent and had a greater toxicity than that of the individual constituents (synergistic effects). Such interactions occur not just between individual chemical contaminants but also between classes of pollutants (low oxygen caused by an oxygen-demanding waster increases the toxicity of metal contamination) and between pollution and other human activities (a fished ecosystem is more vulnerable to invasion by non-native invasive species: biological pollution).

The toxicity of many substances is greater when organisms are stressed, for example by higher temperatures, altered pH, lowered oxygen or if they are underfed. Therefore, the global-scale changes associated with climate change—for example, warming, lowered ocean pH and a greater extent of ocean hypoxia—could all, individually and collectively, enhance the toxicity of other pollutants (Section 8.3).

The assessment of toxicity has traditionally focused on death as an endpoint, because it is easy to measure unequivocally, but most substances cause sub-lethal effects that require some time to manifest (see Chapter 2). These effects can, in fact, often be severe, such as those which impact reproduction and development, or they may be genotoxic or neurotoxic. Even if no obvious changes are induced, metabolic energy is usually required to detoxify the contaminant, excrete it and/or biochemically isolate it (e.g. metallothioneins). These processes divert energy away from other activities (e.g. the basis of the scope for growth approach to pollution assessment; Section 2.4.4) and may compromise an organism's the ability to deal with other pressures. Thus, even the sub-lethal effects can increase sensitivity to other pollutants. At the population level, lowered fecundity due to resources being used for detoxification can, for example, reduce a population's resilience to both natural phenomena such as predator outbreaks or disease and can slow the recovery following hurricanes/cyclones or the cessation of a polluting activity.

Estuaries are naturally species-poor ecosystems, due to the biological challenges of coping with the naturally fluctuating environmental conditions and the strong physiological, particularly salinity, gradients. Within ports, poor water quality (e.g. low oxygen, raised temperatures from cooling water discharges, contaminants and eutrophication) may further reduce the pool of species present. Thus, pollution has created additional vacant niches that non-native invasive species (NIS) might colonise (Section 4.8). Therefore, the global spread of NIS by ship ballast water, hull fouling and floating debris (Sections 4.8 and 4.2.3) may become enhanced by increased vulnerability to invasion caused by pollution effects: another example of a synergy at the ecosystem level.

The UN's second Sustainable Development Goal (Table 8.1) is to alleviate hunger. Marine fisheries currently supply about 6% of the protein consumed globally each year. Given that most fish stocks are now either fully or over-exploited, there is little scope for capture fisheries to add to the food supply. However, Persistent Organic Pollutants (POPs) and metals are actually restricting the use of some of the wild fish stocks due to health concerns (see Chapter 3). It is also possible that other fish stocks may be less fecund than they would be naturally due to the sub-lethal, in particular endocrine-disrupting, effects of pollutants. Thus, marine pollution may be restricting both the food quantity (reduced fecundity and body size or shell thickening in shellfish) and quality (contamination) of global fisheries at a time when the growing global population is in need of large amounts of high-quality food.

This example nicely illustrates some of the difficult societal choices that marine pollution raises. Society has a choice about how money is spent: we could pay for more waste treatment and have more fish fit for human consumption from the sea, or we can accept that we will reduce the food supply from the sea as a result of using cheap waste-treatment options (and hence cheaper end-products). To compensate for the lost productivity, we also need to spend more on intensifying agriculture to make up the food gap. However, increasing agricultural production will also increase nutrient and silt inputs to coastal seas making them less attractive to recreational users and uninhabitable for some species.

In democratic political systems, this clash between economic, cultural/societal and environmental science priorities should be resolved through the democratic process.

Recent years have seen considerable campaigning by environmental lobby groups for marine protected areas (MPAs); marine reserves, marine conservation zones, etc. Governments have responded by establishing an increasing the number of MPAs. Many international agreements, including the European Marine Strategy, the OSPAR Convention and the UN Environment Programme all have policies calling for the expansion of MPAs. Greenpeace has declared a target of 40% of the ocean to be within MPAs, and the International Union for the Conservation of Nature (IUCN) has called for 20–30% of each habitat type to be protected by MPAs.

This government-sponsored growth of MPAs could lead one to believe that the widespread adoption of MPAs will solve all the ocean's problems by preventing over-fishing and marine pollution. However, although there is some evidence of a direct fishery benefit of MPAs in areas with relatively static fish populations (e.g. reefs), the evidence for direct fisheries benefits in more typical areas of the ocean that have highly mobile stocks is limited. MPAs certainly can protect seabed habitats from direct damage but pollutants move through the water column and so regularly cross the boundaries of MPAs without regard for the legislation. However, this is not to say that MPAs do not have a powerful indirect role in providing marine environmental protection. Although an MPA boundary does not stop pollutants or fish moving across it, the existence of an MPA does, in general, lead to changes in the marine management of the surrounding area. For example, dredge disposal may not be permitted adjacent to an MPA or licenced discharges may be subject to greater control if occurring near to an MPA.

Marine Protected Areas are examples of ocean zoning and there is an increasing use of these zoning approaches and marine spatial planning as a means of bringing the regulation of the impacts of the different sectors into a single, unified, holistic management regime. The Canadian LOMAs and the Great Barrier Reef Marine Park discussed in Chapter 6 provide examples of these spatial management approaches.

8.5 Can marine pollution ever be 'solved'?

So, back to the question posed at the start of this text: can marine pollution ever be solved? The answer to this apparently simple question is both yes and no. First, we should note that stopping pollution does not mean stopping the addition of substances or energy to the environment. It means restricting the levels and locations where contaminants are discharged so as to prevent harm to living resources, hazards to human health or the impacts on human activities (the ecosystem services). In structuring the main body of this book we considered those pollutants where the nature of the challenge and the technical/management measures are well understood: our solved problems. However, we also note that there are groups of pollutants where this knowledge is lacking and these were either already established as ongoing problems or were still emerging problems, as they are only just being recognised now.

We can stop future pollution using a combination of setting zero emission levels for substances of high toxicity or biological activity (e.g. the EDCs; see Section 5.3) and mandating comprehensive treatment and strict controls on the remaining emissions. This approach would impact the economy in numerous ways. Thus, while such changes are entirely possible, there are however a number of technical and societal challenges to these solutions so that in reality we are unlikely to ever solve marine pollution in this way. The societal challenges include the economic benefits accrued from the activities that generate pollution (Section 8.2). A discussion is therefore needed surrounding the benefits to society versus the benefits to the marine environment and the additional costs of pollution to society. This debate is further complicated by the often-promoted idea that while what we do today may be setting up a future potential problem, by (or before) the time it manifests technological advances will have solved it and prevented significantly damaging consequences. This argument underpins the current nuclear energy programmes where waste is processed and concentrated but then placed in storage until a long-term solution is found. However, out-of-sight can mean out-of-mind: driving

technological advancements forwards requires a sustained stimulus and a steady commitment.

New substances are continually being produced and causing recognisable changes in the biosphere. There are risks to humans from even extremely low levels of some substances and it is difficult to see how a regulatory regime could ever keep pace with the advances in chemical/material sciences. The adoption of the precautionary principle in the 1980s sought to address this challenge. In essence the precautionary principle advocated that no substance be released into the environment unless it has been shown to be 'safe'. Proponents argue that traditional approaches allowed contaminants to be released and regulation was only applied after the negative effects have been observed. This means that damage has already occurred and it often makes finding a solution more complicated; for example, we need to find a replacement product or an alternative treatment method. As it is rarely the case that substances can be recovered from the marine environment, we are faced with either waiting for natural dispersion/dilution or biogeochemical cycling to remove material from the active biosphere and then we can consider the need for any restoration.

The precautionary principle is, therefore, a risk-management strategy that places the burden of proof on the discharger. Following the adoption of the precautionary principle by the UN General Assembly in 1982 it has featured in many international policy accords including those on climate change and biodiversity protection. It is also part of the legal framework for pollution regulation in the European Union and parts of the United States and Canada. In other areas, including Australia, Japan and The Philippines, aspects of the approach have been established in case law. Thus, although marine pollution can be solved, it is unlikely that it will ever be completely solved due to technical and socio-economic challenges. The next section considers some of the approaches that might facilitate this process.

8.6 Will marine pollution ever be 'solved'?

The challenges to solving marine pollution are twofold: technical-regulatory and socio-economic. The former can be addressed via approaches such as the precautionary principle, meaning that although new substances are continually developed, they can only be brought into use when they are shown to be safe. Such high-level principles and the enabling regulations reduce the time lags associated with the development of new legislation for every new substance that needs controlling (e.g. see Chapter 5). The counterargument is that the development of regulations slows down the adoption of beneficial new technology and/or materials. The adoption of precautionary principles, or at least their operational applications, is confounded by consideration of the economic benefits each new substance, product or process will bring. In recent decades, two approaches have attempted to align environmental protection with socio-economic considerations. First, through the adoption of direct economic instruments to drive environmental behaviour, and second, using valuations of environmental processes as economic resources.

The idea that the polluter pays is a guiding principle of international pollution policy (Chapter 7). Of course, in reality the manufacturer passes the cost of pollution prevention and waste treatment on through the pricing of their products or services, and so ultimately the consumer pays. This is consistent with the principle that the final consumer of the product can be seen as the ultimate cause of the pollution and so he or she is the polluter who has to pay. Using the simple economics of supply and

demand it is possible to generate an economic incentive for cleaner production processes.

In a free market the consumer has choice: a choice between products and the choice not to buy a given product at all. The lower the price, the more product is sold; the higher the price, the less product is sold (Figure 8.8a–b). Therefore, if we add a tax to polluting goods but not to the non-polluting goods, this will raise the price of the polluting goods and thus will reduce their sales. Consumers either do not buy the more expensive products or they switch to alternative, less polluting (and untaxed) alternatives. This pollution tax model can be extended further and the monies raised from the taxes can be used for environmental projects and/or to subsidise the development, or the price of, new less polluting technologies. This is the *environmental tax* model. The level of tax is critical: at one level it might simply be a market intervention to try to alter consumer behaviour. While applied at a higher level and with hypothecation (the tax revenue does not become part of the general tax income but is reserved for particular types of activities, for example, ecological restoration), it both provides an incentive to change behaviour and can directly fund pollution prevention, research, restoration and other ways to compensate for the damage done. In this

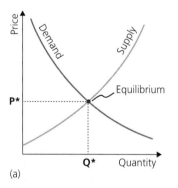
(a)

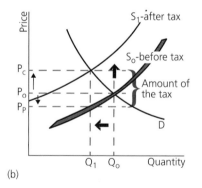
(b)

Figure 8.8 Supply–demand curves that are used by economists to understand the effects of the market on pricing. (a) When there is a high demand for something the price will rise (the demand curve slopes up to the left). If there is an excess of something the price declines (the supply curve slopes down to the left). Where the two curves cross, the free market is in equilibrium regarding the price per unit of the goods/service. If a manufacturer makes more of the product, the quantity supplied increases and the price falls. The costs of manufacturing usually go down with scaling up, so the lower price is initially compensated for by greater sales, but beyond the equilibrium price, further economies of scale in supply do not compensate for lost revenue. (b) Applying an environmental tax to each unit of production moves the supply curve up (S_0 moves to S_1). The new equilibrium consumer price (P_c) suggests lowered demand (Q_0–Q_1): this is the environmental saving, but also note that as the tax is a flat rate (P_c–P_p) of the new price, the supplier only receives P_p, the difference P_0–P_p is the contribution to the tax paid by the supplier that cannot be recovered from the final consumer. This is the price incentive to switch to a 'greener' product and so avoid having to pay the tax.

way environmental taxes could alter the market and fund the clean-up or offset the investments that address the consequences, of pollution. Environmental taxes could provide the funds for investments in the development of alternative non-polluting products, either by funding research or offering incentives to bring new technology/products to market.

Due to the nature of the supply–demand curves the manufacturers/producers cannot actually pass all the taxes on to the consumers (Figure 8.8b). Thus, irrespective of the level of tax levy, there will always be an element paid for by the final consumer and an element paid for by the business affecting its profitability. This provides an incentive for cleaner production.

The end of the twentieth century produced a number of major international environmental policy initiatives. Alongside the adoption of the UN Millennium Development Goals, the United Nations also supported a Millennium Ecosystem Assessment. The Millennium Assessment (MA), as it became known, was a joint initiative of the World Resources Institute, the UN Environment Program, the World Bank and the UN Development Program. The MA involved 1360 experts from across the world and was finally completed in 2005.

In developing the methodologies for completing the assessments, and in reporting them, the researchers focused on the links between ecosystems and human well-being. This involved new methods for quantifying the links between ecosystems and human health, ecosystem processes and the food supply. Much of the quantification of these impacts relied upon placing monetary values on ecosystem functions as well as natural products. This approach was developing in the preceding decades as the Valuation of Ecosystem Goods and Services (Section 1.1). By using monetary values, the MA was able to illustrate how dependent human well-being is on ecosystems to provide goods (e.g. products such as fisheries, clean water) and services (e.g. nutrient cycling, waste assimilation/breakdown). Like the International Panel on Climate Change (Section 8.3), the MA reviewed a large body of existing scientific data from all around the world and provided expert consensus, reviews and conclusions drawn from this body of data.

The message from the MA was that the last 50 years has seen substantial and increasing use of ecosystems to support human well-being, that the means and extent to which these benefits had been obtained had damaged ecosystems, but that the recovery of ecosystem health was possible. We had not yet reached the point of no return in most ecosystems. However, the MA also cautioned that policy and regulatory changes needed to be made to reverse the damage, and while the responses/changes that needed to occur were known, they were not being implemented. The MA therefore illustrates that at the scale of the global impacts on the biosphere, just as we have shown for marine pollution, it is a case of 'problems solvable but solutions not implemented'.

8.7 Concluding thoughts

At the time of writing it is expected that the global population will nearly double in the next 50 years and that the changes in the distribution of people mean that the population of the coastal zone will more than double. The degree of urbanisation in the coastal zone will therefore also double. There is little doubt that supporting the food, energy, transport and waste-disposal needs of this larger global population will place huge demands on the Earth's ecosystems. Covering 70% of the planet, the marine system will come under increasingly heavy pressure from human activities and the inherent natural resilience of marine ecosystems will be tested to what may be their limits.

It is easy to see that these pressures will cause changes in marine ecosystems and that much of that change will be down to marine pollution. It would also be easy, albeit rather trite, to say that technological advances will provide solutions to these problems: we learnt how to treat sewage to prevent anoxia and outbreaks of typhoid and to clean-up oil spills (Sections 3.2 and 4.5). The reality is that changes in ecosystems will occur and we must have a scientific understanding of the nature and scale of these changes, their consequences on complex systems and the available alternatives so that prudent strategies may be implemented. Politicians talk of evidence-based policy decisions but most evidence is not unequivocal and contains some degree of uncertainty. The information used to manage human activities and to control marine pollution needs to be the best available, but the challenges of anticipating the future means that nothing can never be known with certainty. Society, politicians and their advisors in particular must learn this and stop expecting definitive answers when asking questions about complex systems. Managing society's expectations would also reduce the scope for both industry and environmental lobby groups to promote particular points of view by exploiting uncertainty. Nobody has the crystal ball that will allow accurate testing of alternative scenarios.

Science is stimulating because it seeks to understand the unknown; it is therefore an ongoing process and the science that underpins marine pollution management is no different. As new materials are developed, as the environment responds to natural and anthropogenic drivers of change, and as society seeks different goods and services from the oceans, novel questions and problems will arise. Marine pollution science, economics and social science all need to continue playing a role in making marine pollution a solvable problem.

Resources

Intergovernmental Panel on Climate Change (IPCC). (2016). https://www.ipcc.ch.

The Millennium Ecosystem Assessment (2005). http://www.millenniumassessment.org/en/Index-2.html.

Bibliography

Cheung, W.W.L., Lam, V.W.Y., Sarmiento, J.L., Kearney, K., Watson, R. and Pauly, D. (2009). Projecting global marine biodiversity impacts under climate change scenarios. *Fish & Fisheries*, 10, 235–51.

Dulvy, N.K., Rogers, S. I., Jennings, S., Stelzenmuller, V., Dye, S. R. and Sjkolal, H. R. (2008). Climate change and deepening of the North Sea fish assemblage: a biotic indicator of warming seas. *Journal of Applied Ecology*, 45, 1029–39.

Nakicenovic, N. and Swart, R. (2000). Emissions Scenarios. A Special Report of IPCC Working Group III of the Intergovernmental Panel on Climate Change. Geneva: Intergovernmental Panel on Climate Change.

Pachauri, R.K. and Meyer, L. (2014). Climate Change 2014: Synthesis Report. Contribution of Working Groups I, II and III to the Fifth Assessment Report of the Intergovernmental Panel on Climate Change. Geneva: Intergovernmental Panel on Climate Change.

Pereira, H.M. et al. (2010). Scenarios for Global Biodiversity in the 21st Century. *Science*, 330, 1496–501.

Solomon, S., Qin, D., Manning, M., Marquis, M., Averyt, K., Tignor, M.M.B., Miller Jr., H.L. and Chen, Z. (2007). Climate Change 2007: The Physical Science Basis. Contribution of Working Group I to the Fourth Assessment Report of the Intergovernmental Panel on Climate Change. Geneva: Intergovernmental Panel on Climate Change.

UN General Assembly (2015). Transforming Our World: The 2030 Agenda for Sustainable Development. UN reference A/RES/70/1: http://www.un.org/ga/search/view_doc.asp?symbol=A/RES/70/1&Lang=E.

Weslawski, J.M., Hacquebord, L., Stempniewicz, L. and Malinga, M. (2000). Greenland whales and walruses in the Svalbard food web before and after exploitation. *Oceanologia*, 42, 37–56.

Glossary

Abiotic referring to the non-living component of the environment; for example, climate, temperature, pH (*cf.* **Biotic**).

Acidification increase in the acidity of water. Normally refers to anthropogenic activity.

Acidity the concentration of alkalinity-buffering ions in water. Normally refers to the concentration of hydrogen ions, expressed as pH.

Active biomonitoring similar to biomonitoring except artificially reared organisms are transplanted to the target and reference sites.

Active Pharmaceutical Ingredient (API) the active ingredients in pharmaceuticals.

Additive toxicology the sum of the toxic effects of several different contaminants.

Advect to move laterally within a water column.

Agenda 21 a comprehensive plan of action to be taken globally, nationally and locally by organisations of the United Nations, governments and major groups in every area in which humanity impacts on the environment. It derived from the United Nations Conference on Environment and Development, held in Rio de Janeiro in Brazil in June 1992.

Alga (plural 'algae') photosynthetic organism in the phylum Protoctista. Includes unicellular forms (e.g. diatoms) and multicellular forms (e.g. seaweeds).

Alkalinity the concentration of acidity buffering ions in water.

AMBI the AZTI tecnalia Marine Biotic Index.

Amphiphillic compounds with both water-attracting and water-repelling parts.

Anadromous migrating from the sea to freshwater to breed.

Anoxic the absence of oxygen.

Antagonistic with respect to pollution: the term describing the process by which one pollutant decreases the adverse effect of another.

Anthropogenic caused by human activity.

Antifouling see fouling.

Antioxidants a compound which stops the oxidation of other compounds.

APHA American Public Health Association.

Aquaculture raising of aquatic organisms in culture. Includes **mariculture.**

Artificial reefs man-made structures that function as habitat for marine organisms; for example, shipwrecks, oil rigs or reefs constructed of modular units.

Assimilation the process of incorporating ingested items (food) into body tissues.

ASTM American Society for Testing Materials.

ATP adenosine triphosphate, the main chemical energy-storage molecule in cells.

Autotroph (primary producer) an organism that obtains energy from inorganic substances through primary production (*cf.* **Heterotroph**).

Autotrophy primary production.

Ballast water water loaded into tanks on unladen or partly laden vessel to lower it in the water and so increase its stability and seaworthiness.

Benthic referring to organisms living on or in the bed of a water body.

Benthos collective term for such organisms.

Bioaccumulation the increasing accumulation of a contaminant in the body tissues over time.

Bioavailability whether a compound is biologically active and can interact with the biota.

Bioassay a test to determine the biological effects of a compound.

Biochemical oxygen demand the amount of oxygen consumed by a sample of water. Oxygen consumption is a product of microbial respiration (biological oxygen demand) and chemical oxidation of contaminants (chemical oxygen demand).

Biodegradable referring to materials that break down in the environment as a result of microbial activity.

Biodiversity see biological diversity.

Bioextraction the use of animals and plants to remove excess nutrients (see eutrophication).

Biogenic of biological origin.

Biogeography the study of past and present geographical distribution of plants and animals at different taxonomic levels (see also **Range shift**).

Bioindicator or sentinel a species (or a group of) that indicate biological change at the organism level and above.

Biological diversity the variability among living organisms from all sources including diversity within species, between species and of ecosystems.

Biological oxygen demand see Biochemical oxygen demand.

Biomagnification process whereby pollutants in the body of organisms increase or are 'magnified' throughout the food chain.

Biomarker a biochemical, cellular, physiological or behavioural change which can be measured in body tissues or fluids or at the level of the whole organism that reveals the exposure at/or the effects of one or more chemical pollutants.

Biomarkers of effect (or damage) hose that indicate exposure to pollutants have caused damage.

Biomarkers of exposure (or defence) those that indicate exposure to pollutants has occurred.

Biomonitoring using a biological response, of a sentinel species, as an indicator of change in the environment.

Biosecurity measures taken to prevent the invasion of non-native species, in particular diseases and parasites, to a region.

Biomass the mass of living organisms.

Biota the living organisms within an area.

Biotic referring to living components of the environment, or to products derived from living components; for example, detritus (*cf.* **Abiotic**).

Biotic Index a water-quality scoring system based upon the presence of living organisms.

Biotransformation the chemical alteration of a compound in the body.

Bioturbation sediment disturbance or mixing by the animals that inhabit it; for example, worm burrows.

Body burden the amount of contaminant contained within an organism.

Bloom a rapid increase in the population of algae in response to a sudden input of nutrients.

Biochemical Oxygen Demand (BOD) the quantity of oxygen required to oxidise the organic matter and any chemicals present in a reduced state (it is sometimes used as an abbreviation for Biological Oxygen Demand).

Bubble lesions a decompression-like condition in marine mammals.

Buffering capacity the degree to which a water body can maintain its pH in the face of the addition of acid or alkali material.

Byssal threads (or byssus) the attachment mechanism of bivalve molluscs, secreted by the byssal gland. Composed of a very strong and flexible collagen-like protein the threads are glued to the substrate via the byssal plates.

Calefaction artificially induced change in water temperature.

Catadromous (catadromy) migration from freshwater to the sea to breed.

Catchment the land surface that drains into a given river.

CFC chlorofluorocarbon.

Channelisation engineering a channel to make it straighter and more homogeneous in structure than the natural channel.

Charismatic species species that attract much human interest and sympathy, normally because of their large size or visual beauty.

Chemical oxygen demand (COD) see biochemical oxygen demand.

Chemosynthesis (chemoautotrophy) primary production in which energy is derived from chemical oxidation of simple inorganic compounds.

Chemoreception the detection of chemical signals from the water column that may be of biotic or abiotic origin.

Community the assemblage of interacting living organisms within a location or habitat.

Contamination the introduction, directly or indirectly, of substances or energy into the environment such that levels are altered from those that would have existed without human activity.

Contaminant the substance added that causes contamination.

Co-tolerance tolerance to a contaminant that is acquired indirectly by existing tolerance to another toxicant.

Critical pathway the series of steps that brings a pollutant from its source to its point of biological impact.

Crude oil the raw or unrefined form of oil.

CSO Combined Sewage Overflows: mechanisms that allow sewage pipes to overflow into rivers, estuaries and the sea in the event of major flooding or heavy rain.

Cyanobacteria 'Blue-green algae': photosynthetic bacteria formerly known as Cyanophyta and classed as algae.

DBPs Disinfection by-products from the treatment of drinking water.

DDE dichlorodiphenylethane; a breakdown product of DDT.

DDT dichlorodiphenyltrichloroethane; a persistent organochlorine compound widely used as an insecticide.

Demersal living adjacent to, and in association with, the seafloor; for example, demersal fish.

Dermo a disease of that affects bivalve molluscs caused by a protest *Perkinsus marinus*. Outbreaks are thought to be caused by large fluctuations in temperature or salinity.

Detritus particulate organic matter derived from formerly living organisms.

Diffuse source referring to input of a pollutant over a large area.

Discharge a) with reference to river flow, the total volume of water moving past a given point.

Discharge b) with reference to pollution and contamination, the input of a contaminant into a water body.

Dissipative or dispersive pollutants those that are dispersed in the environment.

Disturbance a discrete event that removes, damages or impairs the normal function of organisms.

Diversity see biological diversity.

DOC Dissolved Organic Carbon.

DOM Dissolved Organic Matter.

Ecopharmacology study of the fate of drugs, and their metabolites, in the environment and the effects on organisms within an ecosystem (including humans).

Ecosystem the biotic components of the environment and the abiotic components with which they interact.

Ecosystem goods and services the economic benefits taken from natural ecosystems in the form of products such as fish, reeds, etc. (the 'goods') or as 'services' such as waste breakdown and assimilation and nutrient recycling. The term *ecosystem services* is increasingly used to include both the physical goods and the services elements.

Ecotourism tourism whose aim is to observe wildlife or experience 'natural' environments.

Ecotoxicology a sub-discipline of toxicology that considers the effects of toxins on ecosystems.

EC50 the concentration at which a toxic effect is observed for a specified species, toxicant and effect under the stated test conditions.

Endocrine-Disrupting Compounds (EDC) those that disrupt hormone-mediated processes.

Endoplasmic reticulum an organelle that functions in protein synthesis.

Environmental toxicology the impacts of toxins in the environment.

EPA United States Environmental Protection Agency.

Epibenthic attached to or living on the bed of a water body.

Epifauna organisms living or active on the surface of a water body.

Epiphyte an organism growing on a plant.

Epiphytic attached to plants.

Euphotic zone see photic zone.

Euryhaline tolerant of a wide range of salinities.

Eutrophic referring to water containing high concentrations of nutrients, relative to an oligotrophic or mesotrophic water body.

Eutrophication the increase in nutrient concentrations in a water body, normally used to refer to the effects of anthropogenic pollution.

Euxinic the condition whereby there is an absence of dissolved oxygen and hydrogen sulphide (H_2S) is present.

Evaporite the solid residue remaining when water evaporates. It comprises salts formerly in solution.

Exploitation the use by humans of water or its products for their own benefit.

Exposure time the length of time that organisms are exposed to toxicants for (in toxicity testing programmes).

Extended producer responsibility (EPR) a policy that holds the producer of a waste responsible for its ultimate disposal even after it has been sold on to retailers and consumers (e.g. plastic pollution).

FAO United Nations Food and Agriculture Organization.

Fish aggregation device (FAD) an object used to attract fish.

Food chain a linear representation of feeding interactions, incorporating a single species (or biospecies) at each trophic level.

Food web a representation of feeding interactions within a community.

Fouling the attachment of organisms onto solid surfaces, leading to an impairment in function of the structure affected. Treated by physical removal or application of **anti-fouling** agents, pesticides that kill the biofouling agent.

Genotypic derived from the genotype (the sum of all genetic material transferred from parent to child).

Ghost fishing the continued operation of fishing gear lost or discarded by fishers.

Global conveyor belt the global ocean circulation that begins in the Arctic, travels down through the Atlantic and extends through into the Indian and Pacific oceans (completing a circuit). It is driven by differences in density due to salinity and temperature differences at the poles. A major mechanism for global heat transport.

Glycolysis the biochemical breakdown of glucose.

Golgi apparatus (or Golgi Body) an organelle consisting of folded membranes that functions to package proteins in vesicles for transport and secretion outside of the cell.

Growth promotion relates to the use of antibiotics, at sub-therapeutic levels, to improve the health and therefore growth potential of livestock reared through intensive farming practices.

Harmful Algal Bloom (HAB) blooms of phytoplankton that produce toxins, usually dinoflagellates, diatoms or cyanobacteria. The blooms cause harm to animals and humans through the consumption of contaminated seafood, often shellfish, in which toxins may have bioaccumulated.

Heterotroph an organism that obtains its energy from eating other living organisms or their byproducts (*cf.* **Autotroph**).

Histocytology the study of the pathology of cells and tissues.

Holoplankton see **Plankton**.

Hydrophobic water repelling, cf. hydrophilic water attracting.

Hypertrophic extremely eutrophic.

IBM Integrated basin management.

ICES the International Council for the Exploration of the Sea.

ICM a) Integrated Coastal Management.

ICM b) Integrated Catchment Management. To avoid confusion with **Integrated Coastal Management**, IBM is preferred.

ICZM integrated Coastal Zone Management.

IFREMER the *Institut Français de Recherche pour l'Exploitation de la MER*.

IMO International Maritime Organisation. An international governmental organisation responsible primarily for issues relating to shipping, including pollution from ships, ballast water and ship safety.

Imposex the condition whereby females develop male sexual features in addition to their normal female reproductive structures. Most often seen in neogastropods such as dog whelks and whelks.

Infauna benthos living within the bed sediments, rather than on its surface.

Interbasin Transfer (IBT) artificial transfer of water from one river basin to another.

Internal waters in the United Nations law of the sea internal waters are defined as being landward of a nation's territorial waters (see territorial waters).

Intertidal referring to the part of the coastal zone that is periodically exposed and inundated by tidal movements.

IRBM Integrated River Basin Management.

Irrigation application of water onto agricultural land to enhance growth of crops.

ISO International Organisation for Standardisation.

Keystone species a species whose role is critical to the maintenance of a community.

Krebs cycle respiration in aerobes comprising a series of chemical reactions through which oxygen is used to create energy as ATP; the process occurs within intracellular organelles termed mitochondria.

LC50 Lethal Concentration of a specified toxicant that causes the mortality of 50% of the individuals of a particular species tested under the stated test conditions.

Lessepsian migration the migration of species from the Red Sea through the Suez Canal into the Mediterranean.

Lethal effects those that kill.

Liming the addition of calcium carbonate to a water body to raise its pH.

Linkage density the mean number of feeding links per species in a food web.

Lipofuscin an insoluble granular matrix formed by the peroxidation of lipid membranes; can bind metals.

Littoral zone the intertidal zone.

LOEC The lowest observable effect concentration for a specified toxicant and species. The concentration at which an effect is first observed.

LOMA Large Ocean Management Area.

Lymphocystis cauliflower-shaped growths on the skin, fins and gills of fish. Caused by a viral infection. Growths may become so extensive that they inhibit the ability of fish to swim or eat.

Lysosome a membrane-found organelle which functions in the degradation of endogenous compounds and xenobiotics.

Macrofauna (macroinvertebrates) animals retained on a 0.5 mm (500 µm) sieve.

Macrophyte a large multicellular photosynthesising organism; generally applied to vascular plants but also refers to multicellular algae.

Management mechanisms whereby human activities are controlled or regulated.

Marine Protected Areas (MPAs) designated areas of the sea that are protected as reserves/conservation zones where human activities are restricted.

MARPOL Marine Pollution Annex to **UNCLOS**.

Maximum allowable concentration is the highest safe concentration of a toxin that an employee may be exposed to under law over an eight-hour period.

Meiofauna animals retained on a 0.045 mm (45 µm) sieve.

Merocmictic meaning that the surface and deep waters of a water body do not mix; for example, the Black Sea.

Mesotrophic referring to water containing a concentration of nutrients intermediate between that of a eutrophic and an oligotrophic water body.

Metallothioneins a family of cysteine-rich low molecular-weight proteins with various functions including metal homeostasis, metal detoxification and nerve cell growth.

Millennium Development Goals (MDGs) eight goals agreed by the United Nations members at the Millennium summit. With the overall objectives of poverty alleviation, improved education and healthcare and environmental sustainability.

Mixed function oxidases also known as cytochrome P450 enzymes occur within the cells of animals where the function in biotransformation by catalysing oxidation-reduction reactions.

Mixed layer also known as the epilimnion. The upper layer of a watercolumn that is mixed by the surface winds.

Model species a convenient and well-known species that can function as a model in laboratory studies, and is taken to be representative of other taxa.

Monitoring carrying out periodic measurements of parameters.

MSFD the European Union's Marine Strategy Framework Directive.

Multi-nucleated unknown disease (MSX) affects oysters and is caused by the protists *Haplosporidium nelson*.

Mussel watch an international, active biomonitoring programme that uses mussels to characterise contaminant levels in coastal areas (including pesticides, PAHs, PCBs, organochlorines and heavy metals).

Niche the definition of the physical, chemical and biological conditions in which a species exists. The Fundamental Niche comprises the conditions under which the species could exists and the Realised Niche is the actual envelop of conditions in which the species does occur.

Nanometre one millionth of a millimetre.

Nanotechnology the manipulation of matter at the atomic scale between 1 nm and 100 nm in any one dimension.

Nekton organisms living within the water column that are able to swim or otherwise move independently of the current.

New molecular entities the term used in pharmacology to refer to a new drugs that are not derivative, for example complex molecules purified from natural materials or combination products where at least one component is new.

NIS Non-native Invasive Species, also often referred to as alien species; species that have become established at a geographical location following their introduction, deliberate or accidental, by human activities.

NOEC the no observable effect concentration within toxicity testing refers to the concentration at which no adverse sub-lethal effect has been found.

NOAA National Oceanic and Atmospheric Administration.

Nutrient cycle the pathway followed by a nutrient as it changes from an inorganic to an organic form and back.

Nutrient spiral a condition in which the cycle is completed only when the nutrient has moved to a different geographical location.

Oleophilic materials which attract oils, rather than water, such materials are used in oil spill clean-up operations.

Oligotrophic referring to water containing low concentrations of nutrients, relative to a eutrophic or mesotrophic water body.

OECD Organisation for Economic Development.

Oocytes immature egg cells.

Osmoregulation controls the osmotic pressure within a biological organism/organ/ tissue/cell/organelle to maintain the required water and salt content.

OSPAR Oslo–Paris Convention for the Protection of the Marine Environment of the North East Atlantic.

Oxycline a region of rapid change in oxygen concentration.

PAH polyaromatic hydrocarbons.

Papilloma a benign tumour that grows in the skin; the name derives from the nipple-like appearance.

PCB polychlorinated biphenyl; a persistent substance widely used in electrical components.

PEC the Predicted Environmental Concentration of a contaminant.

Phenotypic derived from the phenotype (the expression of a gene in response to the environment).

Pelagic referring to the water column, as opposed to its bed or edges.

Persistent pollutants remaining in the environment in its polluting state for long periods.

PCPs Personal Care Products (see PPCPs).

Pharmaceutical and personal care products (PPCPs) considered to be any products used by individuals for health or cosmetic reasons (e.g. prescription drugs, non-prescription drugs, perfumes or cosmetics) AND any products used by the agricultural industry to promote the health or growth of livestock animals.

PICT Pollution-Induced Community Tolerance.

PFA Pulverised Fuel Ash: the predominantly silty fuel residue that remains following combustion of coal.

pH see **Acidity**.

Photic zone (Euphotic zone) the depth of water to which sunlight penetrates. The bottom of the euphotic zone is defined as the depth at which only 1% of light intensity at the surface remains.

Photosynthesis see **Primary Production**.

Phytoplankton see **Plankton**.

Plankter an individual member of the plankton.

Plankton (planktonic) Organisms inhabiting the pelagic zone which have no independent means of propulsion or are too small and weak to swim in the horizontal plane. **Holoplankton** – organisms which spend their entire lives as members of the plankton. **Meroplankton** – organisms which spend only part of their life cycle as members of the plankton. **Phytoplankton** – the photosynthesising component of the plankton, mainly single-celled algae but including autotrophic bacteria. **Zooplankton** – animal components of the plankton.

PNEC the predicted no-effect concentration. Guidelines for calculation are available from the European Commission's technical guidance; calculations include toxicity values, the number of values used, the species used and their trophic position.

POC Particulate Organic Carbon.

Point source a single location from which a pollutant is derived.

Pollution the introduction of substances or energy into the environment resulting in deleterious effects to humans, human activities or other living components of the environment.

Pollutant a substance that causes pollution.

Polyculture aquaculture of two or more species simultaneously.

Polynya recurrent areas of open water within the sea ice. Usually are hotspots with high productivity.

POPs Persistent organic pollutants (see also **persistent pollutants**).

Precautionary principle a strategy that requires a potential discharger to demonstrate no environmental effect before a discharge is licensed.

Precipitation a) water that falls from the atmosphere to the surface of land or sea; including rain, snow, etc.

Precipitation b) the process by which a substance dissolved in water separates out of solution and becomes a solid.

Primary production (autotrophy) production of organic compounds from inorganic components, using energy fixed from an external source.

Producer responsibility organisations (PRO) bodies established by polluters to meet their recovery and recycling obligations.

Planktivorous animals that feed on plankton.

Photosynthesis primary production in which energy is obtained from sunlight.

Quality Status Reports (QSR) outputs form the North Sea Conferences (1987–2006).

Range shift the expansion and/or contraction of a species' biogeographic range.

Redox an abbreviation for oxidation reduction, a chemical reaction in which an atom or ion loses electrons to another atom or ion.

Redox Discontinuity Layer (RDL) the layer at which a sediment changes from an oxygenated to an anoxic state.

Redox potential the capacity of a substance to be reduced i.e. gain electrons (aka Reduction potential, Oxidation/Reduction Potential (ORP) and Eh)

Refractory not readily assimilated and therefore slow to decompose.

Residence time the amount of water in a region/sea/lake divided by the time taken to replace it.

Resources the components of a water body that are exploited. The term is normally applied to components that can diminish through overuse, such as stocks of fish.

Run-off rainfall that is not absorbed by soil but passes into surface water bodies.

Salinisation increase in concentration of salts in water or soil.

Salinity the concentration of dissolved ions in water.

Scope for growth a method for quantifying the energy an organism has available for growth.

Secondary microplastics those produced by the breakdown of larger plastic particles (more than 5 mm) (*cf.* primary microplastics).

Primary microplastics the small plastic particles (less than 5 mm) produced as feedstock for the plastics industry.

Sewage human and domestic waste dissolved and/or suspended in water.

Sessile taxa that have limited mobility.

Socio-economic pertaining to human society and cultural interactions.

Solute a substance dissolved in another substance.

Stratification development of discrete vertical layers within a water body.

Stress a state in which the normal physiological functioning of an organism is impaired.

Sub-lethal effects those other than lethality; for example, cancers, tumours, behavioural change or reproductive failure.

Succession directional change through time at a site by means of colonisation and local extinction of species (N.B. succession is a controversial process with several, often contradictory, definitions in ecology).

Sustainability the management of resources in a way that does not deplete them and therefore ensures their continuation.

Synergistic toxicity whereby the toxicity of a mixture of compounds exceeds their summed individual toxicity.

Synoptic referring to the examination of multiple influences.

Taxon (plural: taxa) organisms classed together within the same taxonomic group (differs from species in that the taxonomic group may be genus, family, order, etc.).

TBT tributyltin: a toxic organotin compound used as an anti-fouling pesticide.

Test battery a suite of toxicity tests that encompass multiple trophic levels.

Territorial waters in the United Nations Law of the Sea territorial waters are those that are within 12 nautical miles of a country's coastline.

Thermocline zone between the epilimnion (or mixed layer) and hypolimnion (or deep ocean) within which the temperature changes abruptly.

Toxin a substance that has a toxic effect.

Toxic (toxicity) a poisonous quality; and the magnitude of this quality.

Toxicology the scientific discipline focused on the study of toxins (the effects in toxins, their mechanisms and their quantification).

Translocation movement and release of individuals of a species beyond its native range.

Trophic level an identifiable feeding level within a food web.

Tsunami a long wavelength ocean wave generated by a seismic disturbance that has devastating effects when it breaks on the coast, often referred to incorrectly as a 'tidal wave'.

Turbidity the concentration of suspended particles in water.

UNCLOS United Nations Convention on the Law of the Sea.

Upwelling the appearance at the surface of a water mass previously within the depths of the water body.

Vitellogenin the precursor to egg-yolk formation in vertebrates and invertebrates; used as a biomarker of endocrine disruption.

Water mass a body of water characterised by certain values of its conservative properties (i.e. those not altered except by mixing); for example, salinity, temperature.

WFD the European Union's Water Framework Directive.

Xenobiotic a foreign compound within an organism (i.e. those which it does not produce itself).

Zooplankton see **Plankton**.

Zwitterions molecules with both positively and negatively charged parts.

Index